Remote Sensing;
People in Partnership with Technology

Proceedings of the Sixth Forest Service Remote Sensing Applications Conference

Jerry Dean Greer, Editor

Denver, Colorado
April 29–May 3, 1996

Sponsored by:
United States Department of Agriculture,
Forest Service and the
Society of American Foresters,
Remote Sensing Working Group

ISBN-1-57083-040-1

Published by
American Society for Photogrammetry and Remote Sensing
5410 Grosvenor Lane • Suite 210
Bethesda, Maryland 20814

Printed in the United States of America

ON THE COVER

The stereo photo pair on the cover is of an island and lake on Sealaska Corporation land at Cabin Creek, Prince of Wales Island, Alaska. Two metric Rollei cameras mounted forty feet apart on a helicopter supported boom were used to simultaneously acquire these photographs of the lake at a scale of approximately 1:1897. This stereo pair was taken on September 30, 1995, for a low-altitude photogrammetric project to monitor riparian and channel habitat conditions on Sealaska Corporation lands. The area covered in stereo is 1.7 acres. The larger image size of this fixed-base large scale aerial photography results in more accurate stream and forest measurements than can be obtained from high altitude photography. The scale in this methodology is determined by the relationship of distance in space between two parallel cameras' axes, the cameras' focal lengths, and camera format and is independent of ground control or flying height. This independence increases the usefulness of this method because large areas of forest and stream occur on rough terrain in remote areas where it would be difficult to establish ground control. Funding provided by Sealaska Corporation, Juneau, Alaska. Photographs by R. Grotefendt. (See "Fixed-Base Large Scale Aerial Photography Applied to Individual Tree Dimensions, Forest Plot Volumes, Riparian Buffer Strips, and Marine Mammals" by Richard Grotefendt, et al. on page 270.)

iii

DEDICATION

The proceedings for 1996 are dedicated to André J. Coisman, Washington Office Engineering Geometronics Program Manager. The Forest Service and the remote sensing and geospatial community lost a dear friend and ardent supporter when André was killed on June 24th in a motor vehicle accident. André spent his career in the Forest Service working in mapping sciences. His work included assignments in Regional Geometronics staffs in Milwaukee, San Francisco, Albuquerque, and Denver. While in the Rocky Mountain Region in Denver, André served with distinction as the Geometronics Group Leader. In July, 1994, André joined the Washington Office Engineering staff as the Geometronics Program Manager for the Forest Service. In this capacity, André provided national leadership in geometronics policy development and interagency coordination. André was a true professional. He loved his work and his unflagging optimism touched us all. André embraced the future; he built bridges, not walls—he knew no other way.

PREFACE

The goal of the Sixth Biennial USDA Forest Service Remote Sensing Applications Conference was to provide a forum for resource managers and specialists to share information and experiences regarding practical applications of remote sensing and related technologies. This conference has evolved and expanded to include the participation of people from regions, forests and districts throughout the Forest Service, and other land management agencies, universities and private industry. The excellent quality and diverse range of the technical presentations, poster displays and exhibits contributed to the success of the conference.

In addition, many people contributed to the success of the conference by assisting in the planning and organization of the conference activities, and I wish to acknowledge their efforts:

Dave Wolf	Program Committee, Moderator
Jerry Greer	Proceedings Editor
Gail Shaw	RSAC Secretary—Conference Coordination
Karen Nabity	Exhibits & Displays
Jenny Alban	Exhibits & Displays
Mike Duncan	Graphics Support
Mike Hamby	Moderator
Ebeth McMullen	Moderator
Roger Crystal	Moderator
Paul Greenfield	Moderator
Dave Schultz	Moderator
Ron Skillings	Moderator
Liz Wegenka	Conference Registration
Melinda Walker	Conference Registration
Carol Brady	Conference Registration
Larry Jensen	Graphics Support
Jan Johnson	Conference Audio Visual
Vicky Varner	Conference Audio Visual
Mike Hoppus	Conference Audio Visual
Don Evans	Conference Audio Visual
Paul Ishikawa	Conference Arrangements
Chuck Dull	WO ENG Remote Sensing Coordinator
André Coisman	WO ENG Geometronics Program Manager
Robin Carroll	Manager, Geometronics Service Center
Bryon Foss	WO ENG Assistant Director of Engineering

Tom Bobbe
Manager
Remote Sensing Applications Center

TABLE OF CONTENTS

Opening Remarks and Welcome to the Sixth Biennial Forest
Service Remote Sensing Applications Conference
    T. Bobbe 1

Welcome to the Sixth Biennial Conference on Remote Sensing
Applications for the Forest Service
    E. Estil 3

Washington Office Perspective
    J. McDougle 6

Remote Sensing in the USDA Forest Service, "Where We Stand"
    C. Dull 9

The U.S. Geological Survey's Land Cover
Characterization Program
    T. Loveland & N.Van Driel 17

The National Biological Service/National Park Service
Vegetation Mapping Program—Overview and Update
    M. Nyquist & M. Story 26

An Overview of the Bureau of Land Management's
National Applied Resource Sciences Center and Geographic
Sciences Laboratory
    J. Turner 31

Project 615—Implementation and Add-On Technology
    S. Bain 36

Geomatics at Colorado State University
    R. Hoffer & D. Dean 37

Remote Sensing Data Purchase for Ecosystem Management
    H. Lachowski, S. Bain, T. Wirth & P. Maus 46

Assessment of Remote Sensing/GIS Technologies to Improve
National Wetlands Inventory Maps
    B. Wilen & G. Smith 50

Exploratory Use of Synthetic Aperture Radar (SAR):
Conclusions from Forest, Wetland, and Geomorphic Mapping
    C. Williams & J. Hampshire 65

The Southwest Colorado Vegetation Classification Project: An
Example of Interagency Cooperation
    M. Walker, A. Cook, J. Ferguson & S. Noble 74

Applications of Remote Sensing on the Sandhills of the
Nebraska National Forest
    V. Emly, R. Sprentall, H. Lachowski, V. Varner & P. Maus 80

Technological Building Blocks for Deriving Timberland Suitability Answers
K. Mayeski & F. Krueger 87

Integration of Remote Sensing with Vegetation Mapping, Integrated Resource Inventories, and GIS Database Projects
J. Milliken & R. Warbington 98

Forest Biometrics from Space
T. Hill 107

The Implementation of GIS to Assess the Role of Insects and Pathogens in Forest Succession
L. Lewis & L. Stipe 120

Spatially Explicit Ecological Inventories for Ecosystem Management Planning Using Gradient Modeling and Remote Sensing
R. Keane, C. McNicoll, K. Schmidt & J. Garner 135

Imagery Used in Support of GIS on the Stikine Area of the Tongass National Forest
J. Schramek 146

Development of a Geographic Information System for the Chugach National Forest
C. Markon & B. Williams 155

Large Area Change Detection Using Satellite Imagery: Southern Sierra Change Detection Project
L. Levien, C. Bell & B. Maurizi 164

Monitoring Aspen Decline Using Remote Sensing and GIS
T. Wirth, P. Maus, J. Powell, H. Lachowski, K. Suzuki, J. McNamara, P. Riordan & R. Brohman 174

Using Satellite Imagery to Detect and Monitor Forest Change
K. Green & B. Costentino 184

Evaluating the Utility of Partitioning Remotely Sensed Data
B. Cooke 91

Efforts to Improve Vegetation Maps from Landsat™ Imagery Through Subpixel Analysis
J. Spruce, W. Graham & D. Gibas-Tracy 196

High Resolution Commercial Satellite Imagery and Data Exploitation for Forestry Applications
M. Bullock, P. DeWolf & S. Wagner 202

Digital Aerial Photography Used in River and Riparian Corridor Analysis—Blackfoot River Study
D. Johnson & J. Raine 212

Evaluation of a Color Infrared Digital Camera System for
Forest Health Protection Applications
A. Knapp & M. Hoppus 213

Bread Making and Designing Resource Inventories: The GIS
Connection (The Whole Loaf)
G. Lund, W. Wallace & W. Wigton 220

Technology Transfer, an Important Role: GPS in Uruara, Brazil
C. Rodriguez-Pedraza 230

Laser System Remote Sensing Applications
J. Moll 237

Submeter GPS Positioning Over Long Baselines
C. Sumpter & L. Gross 241

Using GPS in Wildland Fire Management
D. Mangan 251

Integrating Global Positioning System (GPS) Data into
Cartographic Feature Files (CFF)
C. Brady 257

A Real-Time GPS System for Monitoring Forestry Operations
W. Michalson, J. Single, M. Spadazzi & M. Wehr 264

Fixed-Base Large Scale Aerial Photography Applied to
Individual Tree Dimensions, Forest Plot Volumes, Riparian
Buffer Strips, and Marine Mammals
R. Grotefendt, B. Wilson, N. Peterson, R. Fairbanks, D. Rugh,
D. Withrow, S. Veress & D. Martin 270

Use of Historical Aerial Photos to Evaluate Stream
Channel Migration
J. Barry, S. Reutebuch & T. Robison 286

1995 Scanned Aerial Photography of the Kisatchie National
Forest in Louisiana
C. O'Neil, L. Handley, S. Hartley, J. Johnston, B. Coffland
& L. Schoelerman 293

Using "Pseudo" Digital Ortho Photography or the Good, the
Bad and the Ugly
K. Winterberger 300

Accuracy of USGS Digital Elevation Models in Forested Areas
of Oregon and Washington
S. Reutebuch & W. Carson 310

UTOOLS and UVIEW: Analysis and Visualization Software
R. McGaughey & A. Ager 319

An Ecological Approach to Assess Vegetation Change after
Large Scale Fires on the Payette National Forest
S. Boudreau & P. Maus 330

Resources at Risk: The Boise NF Fire-Based Hazard/
Risk Assessment
T. Burton, D. Dether, J. Erickson, J. Frost, L. Morelan, W. Rush,
J. Thornton, C. Weiland & L. Neuenschwander 340

Technical Upgrades to the USDA Forest Service Firefly
Infrared Mapping System
R. Wicks 349

The Application of Forward Looking Infrared Imaging Systems
in Wildland Fire Suppression
W. Krausmann & P. Hicks 355

Integrating GIS and BEHAVE for Forest Fire
Behavior Modeling
J. Campbell, D. Weinstein & M. Finney 363

Application of Thermal Infrared (FLIR) and Visible
Videography to the Monitoring and Restoration of Salmonid
Habitat in the Pacific Northwest
N. Poage, C. Torgersen, D. Norton, M. Flood & B. McIntosh 376

New Technologies for Infrared Remote Sensing and
Fire Mapping
J. Hoffman & R. Grush 380

Development and Utility of a Four-Channel Scanner for
Wildland Fire Research and Applications
V. Ambrosia, J. Brass, R. Higgins & E. Hildum 390

Poster Session Abstracts 400

Final Agenda 419

Session Photographs 426

Commercial Exhibits 433

Register of People Attending 435

AUTHOR INDEX

Ager, Alan 319
Alban, Jenny (poster) 405, 413
Ambrosia, Vincent 390
Bain, Stan 36, 46
Barry, Jeff 286
Beiso, Debbie (poster) 400
Bell, Cynthia 164
Bobbe, Tom 1
Bond, Doug (poster) 400
Boudreau, Susan 330
Brady, Carol 257
Brass, James 390
Brohman, Ron 174, 410
Bullock, Michael 202
Burton, T. A. 340
Campbell, Jeff 363
Carson, Ward 310
Coffland, Bruce 293
Collins, Laurel (poster) 409
Cook, Allen 74, 401
Cooke, Bill 191
Costentino, Brian 184
Crump, Roger (poster) 401
Cutler, Richard (poster) 410
Dean, Denis 37
Dether, D. M. 340
DeWolf, Paul 202
Dull, Charles 9, 415
Edwards, Larry (poster) 402
Edwards, Thomas (poster) 410
Emly, Virginia 80
Erickson, J. R. 340
Estil, Elizabeth 3
Faintich, Marshall (poster) 404
Fairbanks, R. L. 270
Ferguson, James 74
Finney, Mark 363
Flood, M. 376
Frost, J. P. 340
Garner, Janice 135
Gibas-Tracy, Dawn 196
Graham, William 196
Green, Kass 184
Gross, Leslie 241
Grotefendt, Richard 270
Grush, Ronald 380

Hall, Ronald (poster) 404
Hampshire, John 65
Handley, Lawrence 293
Hann, Wendel (poster) 406
Harcksen, Kathleen (poster) 400
Hartley, Steve 293
Hicks, Phillip 355
Higgins, Robert 390
Hildum, Edward 390
Hill, Chris (poster) 405
Hill, Timothy 107
Hoffer, Roger 37
Hoffman, James 380
Hoppus, Mike 213, 405, 408
Hose, Mark (poster) 401
Ishikawa, Paul (poster) 405, 413
Johnson, Dale 212
Johnston, James 293
Keane, Robert 135, 406
Kennedy, Robert (poster) 406
Klein, Cherie (poster) 407
Knapp, Andrew 213, 408
Krausmann, William 355
Krueger, Faye 87
Lachowski, Henry 46, 80, 174, 408
Levien, Lisa 164
Lewis, Lowell 120
Linden, David (poster) 414
Long, Donald (poster) 406
Loveland, Thomas 17
Lund, Gyde 220
Mangan, Dick 251
Markon, Carl 155
Martin, D. J. 270
Maurizi, Barbara 164
Maus, Paul 46, 80, 174, 330, 408, 411
Mayeski, Kim 87
McDougle, Janice 6
McGaughey, Robert 319
McIntosh, B. 376
McKean, Jim (poster) 409
McNamara, Jim 174, 410
McNicoll, Cecilia 135
Menakis, James (poster) 406
Michalson, William 264
Milliken, Jeff 98
Moisen, Gretchen (poster) 410
Moll, Jeffry 237

Morelan, L. Z. 340
Morrison, Mike (poster) 411
Myers, Jeff (poster) 412
Myhre, Dick (poster) 413
Myneni, Krishna (poster) 401
Neuenschwander, L. F. 340
Neufeld, David (poster) 413
Nichols, Frederic (poster) 413
Noble, Suzanne 74
Norton, D. 376
Nyquist, Maurice 26
O'Neil, Calvin 293
Paschke, Jeanine (poster) 414
Pastrone, Mark (poster) 405
Pertica, Alex (poster) 415
Peterson, N. P. 270
Poage, N. 376
Powell, Jay 174, 410, 415
Pywell, Ross (poster) 413
Raine, Julie 212
Reilly, Edward (poster) 416
Reutebuch, Stephen 286, 310
Riordan, Pat 174, 410
Robison, Thomas 286
Rodriguez-Pedraza, Carlos 230
Rugh, D. J. 270
Rush, W. R. 340
Schmidt, Kirsten 135
Schoelerman, Lynn 293
Schramek, James 146
Single, Joshua 264
Smith, Glenn 50
Spadazzi, Michael 264
Sprentall, Robert 80
Spruce, Joseph 196
Stipe, Lawrence 120
Story, Michael 26
Sumpter, Carl 241
Suzuki, Kevin 174, 410
Thornton, J. L. 340
Torgersen, C. 376
Turner, James 31
Van Driel, Nick 17
Varner, Vicky 80, 408, 411
Veress, S. A. 270
Wagner, Steve 202
Walker, Melinda 74
Wallace, Wanda 220

Warbington, Ralph 98
Ward, Denny (poster) 415
Wehr, Michael 264
Weiland, C. A. 340
Weinstein, David 363
Wheatley, Carl (poster) 416
Wicks, Ronald 349
Wigton, William 220
Wilen, Bill 50
Williams, Bruce 155
Williams, Cynthia 65
Wilson, B. 270
Winterberger, Ken 300
Wirth, Tim 46, 174, 408, 410
Withrow, D. E. 270
Wolf, Dave (poster) 418

OPENING REMARKS AND WELCOME TO THE SIXTH BIENNIAL FOREST SERVICE REMOTE SENSING APPLICATIONS CONFERENCE

Tom Bobbe
Conference Chair
Remote Sensing Applications Center
Salt Lake City, Utah 84119

Welcome to the Sixth Biennial Forest Service Remote Sensing Applications Conference. In spite of the current budget climate, it is encouraging to see such a good turn out that includes people from a broad cross section of the remote sensing and geospatial community. From the Forest Service we have people from Research, State & Private Forestry, the Washington Office, Regional, Forest, and District Offices representing all regions of the Forest Service. In addition we have several different agencies, including the Bureau of Land Management, US Geological Survey, National Biological Service, US Fish & Wildlife Service, NASA, US ARMY Missile Command, and several universities, as well as private industry are contributing to the conference.

The theme of the conference is "People in Partnership with Technology". During this week you will hear technical presentations and view poster displays and exhibits that feature practical applications in utilizing remote sensing, GIS, image processing, and GPS technologies. These applications address a wide variety of natural resource and ecosystem management issues. This is the third time I have had the pleasure of serving as the chairperson for this conference. I think this is the best combination of presentations and poster displays we have seen so far. The topics listed on the agenda cover a wide range of technology and applications. Of course all of the credit for the success of the conference goes to the people who are taking the time to present their work and experiences with us.

Looking back at previous conferences has helped confirm what I feel is a significant trend. An increasing number of resource specialists and managers are becoming the practitioners and users of remote sensing and GIS technology. This is a very positive sign that we are forming a valuable partnership with technology. Recent developments in image processing and GIS software and hardware now provide new ways for resource specialists and managers at the field level to acquire, and analyze data more efficiently and at a lower cost. However we must recognize the critical need to provide adequate training and support so these technologies are used properly.

We have witnessed many significant advances in electronics and computer technology that provides new capabilities to the remote sensing, and geospatial community. We are now exploring the use of new advanced airborne and satellite

1

sensor systems. And we have also found new methods to utilize existing remote sensing systems that have been around for many years.

We are also forming partnerships with other agencies to share data, ideas and combine efforts to implement ecosystem management. Several presentations listed on the agenda will describe cooperative efforts in applying remote sensing and GIS for ecosystem management.

Perhaps the most valuable aspect of this conference is not listed on the agenda. In this room we have a very unique collection of people covering a diverse range of resource disciplines, technical expertise and backgrounds. Use the week as an opportunity to network and exchange ideas with others who share a common interest, or who may have an idea on how to approach a problem you are facing.

At this time I need to take a few moments to acknowledge several people who helped make this conference possible. First, I appreciate the support Region 2 provided by agreeing to cohost the conference here in the Denver area. Dave Wolf from R2 Geometronics helped a great deal in making local arrangements and preparing the agenda.

Several people from Robin Carroll's staff at the Geometronics Service Center helped in a number of ways. Liz Wegenka and Carol Brady organized and handled much of the registration activities. Karen Nabity worked very hard in setting up the vendor exhibits.

Just about everyone from my staff at the Remote Sensing Applications Center was involved in the preparations for the conference, but I want to specifically mention Gail Shaw who helped with registration and other countless details. Jenny Alban spent a great deal of time and effort organizing the poster display area.

I also want to mention Jerry Greer. Jerry served as the conference chair for the first three conferences. Without Jerry's efforts this conference would have never got off the ground. Jerry is once again serving as the conference proceedings editor, as he has done for all the previous conferences. The proceedings will be published by the ASPRS.

And finally I thank the next two speakers on the agenda for taking the time from their busy schedules to be wish us. Elizabeth Estill, the Regional Forester from R2 will be sharing with us her perspectives on what is happening in the Rocky Mountain Region. Associate Deputy Chief Janice McDougle will provide an overview of what is happening nationally. I am very pleased that Elizabeth and Janice can be with us today.

Again I thank all of you for coming and helping make this conference a success.

Welcome to the Sixth Biennial Conference on Remote Sensing Applications for the Forest Service

Elizabeth Estil
Regional Forester, R2
Lakewood, Colorado

A New Beginning for Remote Sensing in the Forest Service

In my welcome to you, I would like to take this opportunity to make some comments about the role of technology such as remote sensing in the Forest Service's current operating environment, and pose to you some of what I see as the challenge and responsibility facing those of you working in this field.

I would be amiss in my responsibilities as a line officer if I didn't publicly acknowledge that we almost need a new beginning for remote sensing in the Forest Service.

In the past, some of the technologies which make up the field of remote sensing were frankly oversold. Claims for the utility of these technologies to the basic work of the Forest Service were overstated. In spite of much good and valiant effort, some units of the organization ended up with little or nothing to show for their investment in people and time. Interpretation of the first generations of satellite images never could provide the information claimed by some proponents. These older image products were inherently limited, overpriced (at least for our purposes), and sometimes constructed for purposes not really germane to our Agency.

It is important today to acknowledge these "facts" as being part of the heritage of remote sensing in the Forest Service, and to be cautioned by them, because... these things occurred in more placid times when the Agency had less at stake, and we had more freedom to experiment and make mistakes.

Now, the situation for remote sensing is changing in two very important ways:

First, the base technologies of remote sensing (satellite and airborne imagery, global positioning, and scientific visualization) are maturing. A new generation of products of considerably grater use to land managers may be about to emerge. In particular, the SIR-C space borne imaging radar systems, the GOES Sounder 8 and 9 series of satellites, and the new earth observation applications being developed by NOAA's Hydrology and Environmental Modeling Offices are of interest to us.

These seem to hold a lot of promise for the Forest Service with respect to developing better information about the status of our land, atmosphere and waters, and the physical and social processes changing them.

And second, the environment for remote sensing in the Forest Service is changing. For one thing, we are finding we have needs for information about ecosystems at scales above those of the District and Forest levels. We need contextual information about ecosystem provinces and regions. We need better information about things such as snow deposition patterns, soil temperatures, subterrain water courses, and the large-scale physical processes which shape our ecosystems. To carry out our mission to understand, protect and utilize ecosystems, we must *temporize* in the matter of resource information.

There is no substitute for ground-acquired, detailed information about the structures of plant communities, faunal habitat requirements, and riparian areas. The Forest Service must develop high quality and detailed information about the characteristics of our ecosystems directly related to the scale at which biotic processes occur. This is generally at what we now call "the stand level" and smaller. While the powers of remotely sensed information technology are truly remarkable, not only is there "no reasonable substitute" for the kind of information we must eventually have, there is "no substitute" for this kind on information ... period. So, our first priority must continue to be the development of these ground-based information systems.

Our responsibilities to make decisions about the management of ecosystems cannot be postponed for the five to seven years our Agency needs to develop these information systems.

Remotely sensed or acquired data is our single best interim information source. I hope you, at this conference today, will understand both the urgency this situation confers on your work, and also the responsibility to be judicious in your work.

Challenge and Responsibility

So, as always in the case of advanced technologies in the Forest Service, you are working in a highly volatile atmospheric mix of challenge, hardship, risk and responsibility.

If I were to try to summarize and pick out the most important of the particular challenges and responsibilities facing you who are working in remote sensing, I would list the following five points.

First, in the time of declining budgets and staff, increasing workloads, and need for higher quality analysis, you absolutely must focus on development of products and tools for *Today's Work*.

There are a couple of parts to this point. You need to consciously separate "work" from "experimentation", and make Today's Work the first priority. You need to do this so that we create enough slack in the system to permit experimentation and learning. We no longer have any such "slack". If we don't create it by focusing on the Agency's mission, there well be no "advanced technologies" in the Forest Service.

As we pare back our personnel and increase our workloads, we will have less and less opportunity to assume the user to be a highly skilled professional. Therefore, products and tools you create must be made more readily understandable by the ordinary person. They must be much more refined before they are delivered to Forests and Districts.

The products you create also must be integrated with our existing and developing information system. They must complement and supplement the other systems we are trying to bring on-line, such as GIS and integrated resource information sets.

Second, you must continually work on behalf of the Agency to separate fact from fiction. Remote sensing is an increasingly powerful and mesmerizing science. It is fun, seductive, and glamorous! Keep your wits about you!

Third, if the Agency is to exercise judgment, each of you must work very hard, sometimes at personal expense, to keep up with the rapidly advancing changes in the technologies which make up "remote sensing". The Agency needs each of you to be the best people around.

Forth, we need good demonstrations of truly useful applications. And I suggest to you that this requires a particular kind of partnership. You who are technologists must neither destain nor be afraid of those of us who are line officers. You must cultivate us, practice plain English and plain thought, and listen to the needs, responsibilities, timeliness, short attention spans and other burdens we bear. Truly useful applications, those things which go beyond "experiment" and "demonstration" are likely to come only out of this kind of synthesis of stewardship and technology.

Fifth, and finally, we must build close working partnerships with the other federal agencies whose primary mission it is to acquire and develop earth observation products. Such as, the EROS Data Center; the National Geophysical Data Center here in Boulder; NOAA's Hydrology and Environmental Modeling offices; and the Jet Propulsion Laboratory.

The Forest Service should always have a place for new technologies which will improve its ability to carry out its mission and better serve the American public. I appreciate being invited to this conference and I wish you the best. I hope each of you will find your week to be of great value. With the proper tools of science, we can better approach the ultimate challenge of protecting the resources and serving the people.

Washington Office Perspective

Janice McDougle
Associate Deputy Chief
National Forest System

Welcome to the Sixth Biennial USDA Forest Service Remote Sensing Applications Conference. I am very pleased to be part of this very important conference and have the opportunity to share a Washington Office perspective on the role of remote sensing and related technologies to help meet the Forest Service mission of caring for the land and serving people. You are a select group that has the expertise, leadership and experience to utilize this technology in the Forest Service.

We are facing another challenging period in the Forest Service with restructuring and downsizing. Managers must deal with increasingly complex resource issues during a time of decreasing budgets and fewer people. The need to acquire and utilize current resource information, in a timely and efficient manner is even more critical. Remote sensing along with other geospatial technologies are already contributing to Forest Service planning and decision making processes, but we must continue to explore new methods to collect and manage resource information more efficiently.

The Forest Service has been using traditional remote sensing methods for the detection and characterization of many natural resource phenomena since the 1930's. The Forest Service has pioneered many techniques in utilizing aerial photography and other remote sensing systems for many applications such as: completing forest inventories, mapping insect and diseases outbreaks, monitoring riparian and range habitat, mapping fires and fuels, and creating GIS data bases. The Forest Service has also played a major role in recent developments to utilize remote sensing and digital image processing to support ecosystem management and analysis.

Remote sensing systems offer a unique capability to assist in managing our National Forests by providing resource information within large geographic areas, in a short time and in a cost effective manner. Accurate and current resource information is vital to our land management and decision making process. With the recent contract award to IBM for GIS hardware and software, the Forest Service will have completed the first stage of acquiring advanced state-of-the-art computing technology. This award will also provide the foundation for add on technology with the next procurement for remote sensing technology under the Project 615 program. This next phase will provide image processing capability for remotely sensed data. The combination of remote sensing and image processing will provide a valuable capability to collect and manage resource data more effectively. Virtually everyone in the Forest Service will have the capability to view and manipulate imagery in some way.

Users of digital imagery will range from novice, who will simply use imagery as displays or backdrops for GIS to advance users who complete vegetation and ecological unit mapping. Improved access and lower costs of imagery will increase the number of users of remote sensing data. Because of the importance of remote sensing technology to ecosystem management issues, it is critical that the Forest Service streamline the process to provide imagery to users who need it. The Forest Service Interregional Ecosystem Management Coordination Group has recently implemented a service-wide program to procure existing satellite imagery. This procurement will provide the Forest Service with AVHRR, Landsat Multi Spectral Scanner (MSS) triplicates, and Landsat TM imagery, for all lands in the continental United States at a significant cost savings. This imagery will provide a synoptic view of large geographic areas at scales consistent with the Forest Service Hierarchy of Ecological Units.

The Remote Sensing Applications Center (RSAC), located in Salt Lake City, Utah, is a Forest Service facility with lead responsibility to provide technology evaluation, development, and training support in the use of remote sensing, image processing, and Global Positioning Systems (GPS) for Resource Management. RSAC is managing the Forest Service satellite imagery data procurement and is available to provide technical support and training to all Forest Service staffs in its use.

Managing all the data from satellites, digital cameras, aerial photography, video, GPS, digital orthophotos, and other sensors will be challenging. We may think we have a lot of data now, but it will minor compared to the amount of information that will be collected, analyzed, and shared in the next 6-8 years. How will we manage this information? How will we track this information? How will we share this information efficiently? The Forest Service will rely on you, the experts in remote sensing to provide technical leadership on these issues.

To provide for the most efficient and appropriate use and applications of GIS, remote sensing and other technologies the Forest Service is beginning to create and support technical centers, "Centers of Excellence". The foundations for such centers exist in some Regions, research centers, at RSAC and other sites within the Forest Service. The Forest Service must support the "Centers of Excellence" concept now and in the future to ensure the efficient and appropriate use of remote sensing and other technology.

The complexity of managing National Forest lands is increasing, and will continue to increase, as we place additional or conflicting demands on the land. The IBM award is a major step forward in providing the needed tools to help cope with these and many other complex management issues. GIS will provide the foundation to integrate other supporting technology such as remote sensing, image processing, and global positioning systems. The Forest Service, a major innovator, will continue to meet the diverse requirements of the agency for integrating technologies in a well designed and organized approach. However, integration of these technologies will not come easily. Technology development, technology transfer and training must continue. The Forest Service will continue to look to you for agency guidance and support. You the user will have the greatest responsibility, in learning and apply these new technologies.

In closing, I want you to know the importance of your work is to the Forest Service. Use this week to share experiences and knowledge that each of you have gained. This conference provides a unique opportunity to share valuable information and experiences. I know the Forest Service is in good hands with professionals such as yourselves. With your guidance with new technologies the Forest Service with continue to meet our mission in "Caring for the Land and Serving People".

REMOTE SENSING IN THE USDA FOREST SERVICE, "WHERE WE STAND"

Charles W. Dull
National Remote Sensing Program Manager
USDA Forest Service
Washington, DC 20250

Introduction

The sixth Forest Service Remote Sensing Applications Work Conference was a tremendous opportunity to discuss and learn about new ecosystem management applications for remote sensing and related advanced technologies. Each of the past remote sensing work conferences have been memorable occasions. At the first conference in 1986 held at NASA Ames, Moffett Field, the entire conference was invited to watch the deployment of a U2. Two years later, in 1988, at Bay St Louis, MS, we visited the test facility for the Saturn launch vehicles. In 1990, the conference was hosted in Tucson, AZ, with emphasis on arid lands remote sensing applications. In 1992 at Orlando, FL, we visited the Kennedy Space Flight Center. The last conference was in Portland in 1994, with a tour of and review of remote sensing applications to support ecosystem restoration on Mt. St Helens. But most memorable, each of the conferences had outstanding programs and provided an opportunity to meet with remote sensing specialist and natural resource managers to discuss issues related to ecosystem management applications for remote sensing.

I have entitled this paper "Where we stand" after an essay by the same name about conservation issues and the Forest Service written by Gifford Pinchot shortly after the turn of the century. Gifford Pinchot is considered the father of forestry in the United States and was the first Chief of the Forest Service. He was a very forward looking person and innovative in the use of advanced technology of his time. He was the first to us a typewriter and dictaphone in the government, and initiated the use of horizontal filing systems (advanced technology for his time). He recognized the need for mapping and survey skills in his forester's field guide. If he were here today, I think he would be proud of how we are using remote sensing, GIS, GPS, and other advanced technologies to support conservation issues and ecosystem management. Our current chief has stated on many occasions, the need to take full advantage of advanced technologies to help us do a better job of managing our natural resources.

It's hard to determine "where we stand" in the Federal government this year. We have undergone right-sizing, downsizing, re-engineering, and reorganization. The forty agencies under the Interior appropriations have worked under 13 continuing resolutions. But we all realize that the budget process is not just about numbers or dollars, but about policy...and environmental policy is increasingly important in establishing budget appropriations. However, resource managers have every right to be very optimistic that remote sensing and associated technologies are advancing at a rapid rate and have much to offer conservation leaders. The current administration has supported, with a high level of interests, the greater utilization of advanced technologies and the sharing of information .

REMOTE SENSING - PROVIDING A VIEW FROM ABOVE FOR FOREST PLANNING

Ecosystem Management has been embraced as the operating philosophy of the Forest Service. Indeed it represents a treat to many land owners and resource managers. But upon further inspection, most people are accepting ecosystem management as a good philosophy and good business...good business economically, environmentally, and socially. There is a worldwide movement toward integrating human, biological, and physical dimensions of natural resource management.

Remote sensing has a great deal to offer to help "view" natural resources from a landscape to eco-regional level. The Forest Service is working in partnership with other land management agencies on broad area assessments to identify current conditions, trends, and future planning considerations. The broad area assessments will make our forest plan revision process more efficient and cost effective. Within the next five years, the Forest Service has a legal mandate to complete a majority of national forest plan revisions. The broad area or eco-regional assessments will help make this process more efficient. A broad framework does exist to support the various levels of planning...from the RPA Program, Forest Plans, and Project level plans. Just as we have established a hierarchy of ecological units, there is a corresponding information hierarchy. All of it is interconnected and dependent on each other.

A multi-resolution imagery data base is now available to support the broad area assessments and other resource issues. Working with a federal consortium of imagery users, the Forest Service now has access to full CONUS coverage of multi-date imagery from AVHRR and Landsat MSS. Leveraging our dollars with this federal consortium consisting of the GAP Program, EPA/EMAP,

NOAA/C-CAP, USGS Water Resources, and the EDC we also have domestic coverage with Landsat TM data, with the prospect of multi-date coverage soon.

Digital imagery purchases by the Forest Service have ranged from $400,000 in 1991 to $207,000 in 1995. The Forest Service is actually acquiring more data, but taking advantage of partnerships and cooperative agreements with other federal and state agencies to reduce the cost of imagery acquisition. In FY96 and FY97, the Inter-Regional Ecosystem Management Coordinating Group (IREMCG) provided $500,000 to fund the procurement of the multi-resolution satellite data sets. This activity does not preclude the procurement of additional newer imagery or the addition of other image sources from Region or Forest use.

INSTITUTIONALIZING REMOTE SENSING IN THE FOREST SERVICE

With the availability of imagery to support the broad area assessments and planning process and new sources of data emerging soon, the institutional capability for managing remote sensing in the Forest Service is materializing. Hardware and software provided by the Project 615, IBM contract has provided a standard platform to integrate remote sensing into GIS. The Forest Service will proceed with the process of awarding a follow-on contract for image processing and hardware enhancements. People will be trained to use these tools on a much broader level. Training will be available through the IBM contract, the new image processing contract, and on special applications coordinated by the Remote Sensing Applications Center (RSAC) in cooperation with the Regions, Research Stations and private industry.

NEW EARTH OBSERVING SATELLITES

As of March 1996, there are 47 current, planned, or proposed earth observing satellites scheduled for launch or to be available by the year 2010. Costs for satellite data and the hardware and software to process it are falling. However, decisions on how to gather, process, store and access data collected by these new satellites won't be easy. Already, governments, research institutions, and private industry are struggling with how to make the expected influx of satellite data more accessible and usable to more users and establish imagery acquisition and distribution policies.

Land management agencies have been asked about their information requirements to support system designs for several of the proposed systems to be launched by the United States. But we have much work to do to match land management information requirements with system capabilities. These new systems will offer much greater spatial and spectral resolutions with more frequent coverage. We can also expect more timely data delivery through the use of portable ground stations, use of the internet, and new image browse and ordering capabilities.

These new systems are truly international in scope. Eleven countries are planning satellites with resolutions equal to or better than the Landsat series. The United States leads in launches of experimental systems and the only commercial systems proposed are being led by U. S. firms. With one exception, all the commercial systems proposed to date are planning high resolution coverage of relatively small areas. Most do not have the capability to collect global data sets for global change research. All of the proposed commercial systems appear to be well within the technological state-of-the-art. The major challenges they face include securing sufficient financing to deploy their satellites and ground systems and finding a sufficient market to establish a viable business.

FOREST SERVICE AERIAL PHOTOGRAPHY PROGRAMS

Although there is a wealth of proposed new and existing satellite data available, aerial photography is still relied upon by the Forest Service for most mapping and resource management operations.

The Forest Service has operated an annual aerial photography acquisition program at approximately $4-5 million over the past five years. Contract resource aerial photography at scales of 1:12,000; 1:15,840; or 1:24,000 still represent the major expenditures, with contracts administered through the Aerial Photography Field Office (APFO) in Salt Lake City, UT. Project contract photography administered through the Regional Geometronics Offices accounted for 12% of the aerial photo costs. The Forest Service continues to operate force account aerial photo operations in Atlanta, Ogden, Ft Collins and Portland. NASA is helping to acquire photo coverage using an ER-2 in Alaska, and the National Aerial Photography Program (NAPP) accounted for 12% of last years funding for aerial photography.

The NAPP provides 1:40,000 scale, quarter quad centered, B&W or CIR aerial photo coverage of the United States on a five year cycle. Funding levels for the NAPP program have historically remained level, with the exception of the BLM withdrawing from the program last year. The NAPP still relies on state cost share contributions to help fund the program. Total federal funding amounted to $3.8 million last year.

This year, 1996, was the end of the second, five year cycle of the NAPP. An effort is underway to make the NAPP and the National Digital Orthophotography (NDOP) more operationally and programmatically integrated. The development and implementation of a fixed 7-year plan for NAPP acquisition has been implemented by the NAPP Steering Committee. The new 7-year cycle will consider Federal requirements for aerial photography to support NDOP requirements, Federal requirements for aerial photography to support other non-DOQ programs, and where known, potential opportunities for cooperative activities with states.

Going to a 7-year plan, and maintaining Federal funding levels combined with potential state cooperative funding, offers the greatest opportunity for NAPP to achieve full national coverage within the proposed cycle. Both NAPP and NDOP Steering Committees recognized that a stable program of acquisition is critical to meeting NDOP requirements for aerial photography and must be a reliable tool to conduct budget planning for cooperative state activities.

Significant progress in producing digital orthophotos in the Unites States has been achieved through the cooperative efforts of the Forest Service, Natural Resources Conservation Service (NRCS), Farm Services Agency (FSA), U. S. Geological Survey (USGS), and cooperating states. As of May 1996, over 51,000 quarter quads have been completed or are in work, covering 24% of the conterminous U.S. The Forest Service is producing DOQ's at the Geometronics Service Center (GSC). The GSC is undertaking an aggressive DOQ production program to produce DOQ's to the national standard. In a cooperative agreement with the USGS, for every DOQ produced by the GSC, the USGS will provide a DOQ over National Forests, in a one to one exchange agreement.

Resource aerial photography continues to be the primary source for land management information in the Forest Service. The APFO awarded 22 new contracts for the Forest Service in FY95. A total of 38 contracts were administered for 51 project areas at a total cost of $1.3 million. The aerial photography will cover 56,000 square miles of National Forest and adjacent lands. We continue to see good completion rates for our contracts due to continued review of contract specifications and interaction with APFO and contractors.

13

MAPPING PROGRAMS IN THE FOREST SERVICE

Nationally significant geospatial data management activities are ongoing at the GSC in cooperation with the Regional Geometronics Groups, RSAC, and the GIS Center of Excellence in Ogden, UT. Not only are the GSC and the Regional Geometronics Groups meeting the DOQ production requirements of the agency, but they continue to revise and produce the primary base series (PBS) and secondary base series (SBS) maps covering 20% of the United States from which the majority of land management decisions rely upon in our agency. The digital collection of this information is then produced as cartographic feature files (CFF's) for the vertical integration and development of GIS databases on National Forests. Digital elevation models (DEM's) are also produced at GSC to support DOQ production, terrain analysis, and image map production.

GLOBAL POSITIONING SYSTEM (GPS) APPLICATIONS IN THE FOREST SERVICE

The Forest Service currently operates approximately 1500 resource grade GPS receivers, available on all Ranger Districts and Research Stations. We continue to support a base station network, comprised of 24 stations strategically located to cover 95% of all National Forest lands to provide post processed differentially corrected GPS data.

We have strived to continue a GPS technical services project, located at the Missoula Technology Development Center (MTDC), working in cooperation with RSAC, and Regional GPS coordinators. Last year, each WO staff director was briefed on the significance of GPS technology and the applications to their programs. The GPS technical service program is funded equally by each resource staff in the Washington Office. This allows the agency to continue to test and evaluate GPS equipment with industry at three Forest Service test sites (under a forest canopy) for resource applications. It also supports the facilities to key the P(Y) code receivers for access to the Precise Positioning Service. This is truly a strong commitment by Forest Service management to continued the advancement of this technology in the agency.

GPS has also been used extensively for cadastral surveys, with approximately 50 survey grade receivers purchased by the agency. Land surveys, control surveys, corner search, right-of-way surveys, and establishment of high accuracy reference networks, use real-time kinematic,, on-the-fly and post processed methods and PPS technologies for high precision surveys. These methods have greatly increased the efficiency of survey operations within the Forest Service.

14

In addition to the GPS activities within the agency, we continue to work with the GPS Interagency Advisory Council to express the requirements of the Forest Service and USDA to support the development of new GPS policies. On March 29, 1996 a Presidential Decision Directive was issued to broaden the use of and promote the GPS industry, both domestically and internationally. This directive will allow selective availability to be turned off within the next ten years to provide more precise real time capabilities. A continuously operated reference station system (CORS) is now in operation with 73 stations providing correctors for differential processing. One of the most significant new developments in GPS will be the use of system correctors rather than station correctors for more precise positioning.

The Federal Government is selling blocks of radio frequencies which have the potential to interfere with GPS reception. It is imperative that the agencies involved in the spectrum analysis recognize the impacts of this potential problem. The next block of GPS satellites scheduled for operation as soon as 2001, may have a second frequency to support civil applications providing better positioning. Funding to support the continued operation of the GPS constellation will continue to be under review. With civil, commercial and international GPS applications projected to exceed military applications by a ration of 8:1 by the end of the 1990's, the government is looking at the National Highway, Maritime, and Aviation trust funds to support future operations.

SUPPORT TO FIRE OPERATIONS

We have made excellent progress in remote sensing support to wildfire detection, suppression, and resource restoration and rehabilitation following major fires. After the large fires of 1994, RSAC has been working with the BLM, industry, and Regional remote sensing specialist to support the National Interagency Fire Center (NIFC) to upgrade the airborne scanning (firefly) systems, and make greater utilization of FLIR and GPS to reduce losses resulting from fire and increase fire fighter safety.

REMOTE SENSING AND INTERNATIONAL FORESTRY

Remote sensing continues to support International Forestry operations. The International Forestry budget and staff was severely cut last year. But remote sensing and related technologies will continue to play a vital role to support the sustainable forestry programs and global conservation leadership of the Forest

15

Service. Although severely reduced, we are concentrating our efforts in Russia, Indonesia, East Africa, Mexico, Brazil and Eastern Europe.

SUMMARY

Resource managers aware of the applications for advanced technologies, have good reason to get excited with all the advances in remote sensing and related technologies (be they current, planned, or proposed). Today, we have new sources of imagery and geospatial data becoming available at a rapid rate and at reduced cost. Hardware and software is available and leaping forward in capability and capacity to manage and analyze large volumes of information. Training opportunities are available for those wishing to apply these new technologies. Resource managers will transform data into information, information into knowledge, and knowledge into wisdom for the wise use of our nations natural resources.

THE U.S. GEOLOGICAL SURVEY'S
LAND COVER CHARACTERIZATION PROGRAM

Thomas R. Loveland
and
Nick Van Driel
U.S. Geological Survey
EROS Data Center
Sioux Falls, South Dakota 57198

ABSTRACT

The U. S. Geological Survey's (USGS) National Mapping
Division has started a Land Cover Characterization Program
in response to the need for land cover and vegetation data
for inventory, monitoring, modeling, and management in the
public and private sectors. The program is built on the
experiences of the interagency Multiresolution Land
Characterization consortium. The land cover program's
general goal is a multiscale, multipurpose land
characteristics data base. Customer needs, source data, and
analytic techniques form the basis of the program's four
components. The small-scale component uses satellite images
acquired by the National Oceanic and Atmospheric
Administration's advanced very high resolution radiometer.
Multidate satellite images have been analyzed and combined
with ancillary data to produce a multilevel, geographically
referenced land cover data base for global change research.
Landsat thematic mapper data are the source for the
intermediate-scale component that delivers processed
satellite data and ancillary data sets to cooperators. The
USGS will synthesize a national land cover characterization
from cooperators' results. Current activity in the
large-scale component uses digital orthophoto quadrangles as
a source for land cover and land use interpretations in
selected urban areas for the National Water Quality
Assessment Program. A special projects component will
accommodate cooperative activities in areas where standard
products are not satisfactory. The USGS Land Cover
Characterization Program is compatible with current concepts
of government operations, the changing needs of the land use
and land cover data users, and the technological tools with
which the data are applied.

HISTORICAL PERSPECTIVE

The U.S. Geological Survey (USGS) has a long heritage of
leadership and innovation in land use and land cover
mapping. The USGS Anderson system (Anderson and others,
1976) defined the principles for land use and land cover
mapping that have been the model both nationally and
internationally for more than 20 years. USGS scientists
successfully applied these principles by mapping the United
States. Recently, the USGS demonstrated the utility of
multipurpose land cover characteristics data bases, which
build on the previous USGS efforts (Loveland and others,

1991; Reed and others, 1994). The land characterization
approach interprets multiresolution and multitemporal data
into flexible data bases describing landscape types,
processes, and conditions. The results of the research
phase have been broadly accepted within the national and
international environmental assessment community (Steyaert
and others, 1994).

The Land Cover Characterization Program (LCCP) is founded on
the premise that the Nation's needs for land cover and land
use data are diverse and increasingly sophisticated. The
range of projects, programs, and organizations that use land
cover data to meet their planning, management, development,
and assessment objectives has expanded significantly. The
reasons for this are numerous, and include the improved
capabilities provided by geographic information systems,
better and more data-intensive analytic models, and
increasing requirements for improved information for
decision making.

MULTIRESOLUTION LAND CHARACTERIZATION

The interagency Multiresolution Land Characterization (MRLC)
consortium has been an important contributor to the form and
objectives of the LCCP. The MRLC consortium is comprised of
several federal programs and organizations with
responsibilities to produce or use large-area coverage land
cover data. The original consortium members, the National
Biological Service Gap Analysis Project, the National
Oceanic and Atmospheric Administration Coastal Change
Analysis Program, the Environmental Protection Agency
Ecological Monitoring and Assessment Program, and the EROS
Data Center and National Water Quality Assessment Program of
the USGS, all share a need for national land cover data.
The U.S. Forest Service (USFS) is a recent new member to the
MRLC consortium. The founding philosophy of the MRLC
consortium involves sharing the intellectual, technical, and
financial development of multiscale land characteristics
data and information. This cooperative effort builds on the
strengths of the respective programs and serves the national
interest more effectively than could any individual program.

The goal of the MRLC consortium is to provide land
characteristics data across a range of spatial and temporal
scales, coupled with mechanisms for monitoring, targeting,
and assessing environmental changes. MRLC objectives
include the development of a data base framework that leads
to: (1) a global 1-km land characteristics data base; (2) a
national 30-m land characteristics data base covering the
United States; and (3) a multiresolution system for
monitoring synoptic environmental processes and targeting
significant areas of change.

As a member of MRLC, the USFS has access to the consortium's
national Landsat Thematic Mapper (TM) data set. This data
set consists of peak growing season coverage for the 430

conterminous U.S. path/rows, and multitemporal coverage for an additional 95 path/rows that are primarily in areas with deciduous forest cover. The scenes are all during the 1991-1993 period. The data are all radiometrically and geometrically corrected. Specific processing standards include:

- Using USGS 1:100,000-scale digital line graphs (DLG) representing transportation and hydrography for geographic control. The 1:100,000-scale DLG's provide a consistent and detailed source of ground control for the entire United States.

- Terrain correction processing to remove distortions caused by relief (the georeferenced digital elevation model is delivered as part of the standard data scene-based data set).

- Resampling to a 30-m Universal Transverse Mercator grid using cubic convolution resampling methods.

- Scene clustering based on the Spectrum clustering technique (Benjamin, 1995). The clustered product consists of 240 classes and is generated from the six Landsat TM reflective channels.

Implicit in this approach is the opportunity for each project to specify alternate projection and resampling parameters. This is essential to ensure that specific program technical requirements are not compromised.

In addition, the USFS is receiving copies of the conterminous U.S. Advanced Very High Resolution Radiometer (AVHRR) composite data set for 1989 to the present (Eidenshink, 1992), and the 1973, 1986, and 1992 Landsat Multispectral Scanner (MSS) data produced as part of the North America Landscape Characterization project (U.S, Environmental Protection Agency, 1993).

GUIDING PRINCIPLES

The LCCP builds on the heritage and success of previous USGS land use and land cover programs and projects. It will be compatible with current concepts of government operations, the changing needs of the land use and land cover data users, and the technological tools with which the data are applied. The program is founded on the following guiding principles:

Land characteristics data bases must be compatible with previous USGS land use and land cover products and compatible with Federal Geographic Data Committee (FGDC) standards. This is essential for providing the continuity that allows analysis of land use and cover change.

The USGS will actively solicit interagency partnerships to

involve other Federal and State agencies in all aspects of the program, including funding, planning, mapping, and applications. This program will assist Federal agencies in combining their efforts to characterize the landscape. No single organization can afford the cost of a national land characterization program.

Because no single product satisfies all users or their applications, the program will be based on a flexible data base strategy that facilitates both standardized categorized land use and land cover, and land characteristics describing landscape processes and dynamics.

Multiresolution products will be designed for use at the local, regional, and national levels.

The program will be user driven and implemented on a cooperative basis. This will assure allocation of efforts to pressing national priorities.

The program will provide cyclic data that permit analysis of landscape change.

The LCCP is based on an integrated strategy of research, applications, production, and data management.

PROJECT COMPONENTS

One land cover product for the Nation would simply be inadequate. The LCCP thus consists of four components, each with the following unique but complementary land cover, land use, and ancillary data products:

1) Small-scale (1:2,000,000) national and global land cover characteristics data produced periodically from coarse resolution (e.g., 1-km) remotely sensed data.

2) Intermediate-scale (1:100,000) national land cover characteristics data produced on a cyclic basis from intermediate-scale remotely sensed data, such as data from the Landsat thematic mapper (TM).

3) Large-scale (1:24,000) land use and land cover data for metropolitan areas and other areas experiencing rapid growth. Generally, this component will use the digital orthophoto quadrangle data as a mapping source.

4) Special projects will be conducted, as needed, to produce unique land cover characteristics data for situations in which the first three standard products do not satisfy user requirements.

As stated in the guiding principles, all four components will be developed through partnerships that include public and private participants. The following sections summarize the key aspects of each component. Note, however, that product definitions, timing, and other specific features will be thoroughly reviewed and refined during the first year of the program.

Small-Scale Component

The small-scale component will be a land cover characteristics raster data base with a grid cell size of 1km^2. It will be designed for use in ongoing national operational programs that require coarse resolution and nationally consistent data. This data set will be produced on a 10-year cycle corresponding to the national population census. An experimental 1990 data base produced by the USGS is currently available (Loveland and others, 1993). The next national update will be in the year 2000.

Initially, the data set will be produced from advanced very high resolution radiometer (AVHRR) satellite data. Data from advanced sensors, such as the moderate resolution imaging spectrometer, may be used in the future. This product includes the following:

Seasonal land cover regions as the spatial component. They represent common mosaics of land cover, phenology, and landscape biophysical processes. See Loveland and others (1995) for a description of seasonal land cover regions.

Attributes describing each region, including land cover types, vegetation components, phenology, spectral measures, and site characteristics (political boundaries, hydrology, elevation, general soils, climate, and ecoregions).

Derived thematic maps that are based on a translation of the seasonal land cover regions into common land cover classification legends; these include the USGS Anderson system (Anderson and others, 1976), National Terrestrial Land Cover (Jennings, 1995), Biosphere-Atmosphere Transfer scheme (Dickinson, 1986), and Simple Biosphere Model (Sellers and others, 1986).

Source data used in the development of the data base, including all nonproprietary satellite data and related earth science data sets.

A global land cover characteristics data base also will be produced as part of the LCCP. This effort is under way; a global 1-km data base produced from 1992-1993 AVHRR data will be completed by 1997. This product is intended for use in continental and global scale studies.

Intermediate-Scale Component

The intermediate-scale component will be the primary detailed national land cover product. The efforts of the MRLC consortium will be the foundation of this component. The data base, developed from Landsat TM satellite data or an equivalent source, will be a raster product with 30-m resolution. The data set is intended for use in regional (e.g., State, multicounty, ecoregions) land management, planning, and environmental assessment applications.

The general philosophy for the generation of the intermediate-scale land cover product relies on collaboration rather than the standardization of methods and classification schemes. Each MRLC participant, for example, is taking the lead in developing land cover data sets responsive to their program's specifications. The classifications are based on relatively similar but not identical classification legends. For example, GAP classifications emphasize natural terrestrial vegetation while NAWQA requires detailed agricultural categories. The different land cover classifications developed by each program are shared between the programs in order to accelerate each other's mapping efforts and to provide consistency checks. The individual project data sets will be synthesized into a national data base suitable for a broader range of applications and programs.

The overall all strategy will be implemented in two phases:

Phase 1: Every 5 years, a national preprocessed Landsat TM data set will be released by the USGS to United States Government cooperators. This data set will be georeferenced and will include appropriate ancillary data.

Phase 2: Every 10 years, the USGS will synthesize a national land cover characteristics data base. This product will be generated on the basis of interpretations made from a consistent, preprocessed Landsat TM data base. The primary sources of these interpretations will be projects conducted by State and Federal agencies and the private sector. For example, the National Biological Service (NBS) Gap analysis program, with State-level land cover mapping projects in most States, will be a key source of such land cover data. Although the USGS will not dictate a standardized land cover legend for these projects, it will develop a set of specifications regarding minimum documentation standards. To facilitate access to the project-level data sets derived from projects, the USGS will maintain an archive of land cover products developed from the national TM data base. As national land cover interpretations are completed, the USGS will synthesize the project-level interpretations into a nationally consistent classification legend, add appropriate attributes, and distribute the data base to anyone. This component is based on the current interagency land cover mapping initiative coordinated by the Multiresolution Land Characterization (MRLC) consortium (Loveland and Shaw, 1995).

Large-Scale Component

The large-scale component will emphasize land use and land

cover. Keeping with the approach developed by Anderson and others (1976), the component will use land cover as a surrogate for land use. The large-scale data set, which will be in vector format, will be developed from the interpretation of digital orthophoto quadrangle data. The data set will be consistent with the 1:24,000-scale USGS quadrangle maps. Unless special circumstances arise, source data will be used that can be released with the land use and land cover interpretations. This data set will be produced for metropolitan areas and other rapidly growing parts of the country. The target is to produce land use and land cover data every 10 years for all standard metropolitan statistical areas. The classification legend will be similar to the USGS Anderson system so that comparisons with the previous USGS land use and land cover data set will be possible. However, refinement of the Anderson system is likely. The LCCP will work closely with the proposed FGDC Land Cover Subcommittee and with the National Water Quality Assessment Program to define the final classification legend.

Special Projects

In cases where standard products will not satisfy user requirements, cooperative activities will be established to generate the necessary products. The ways in which special projects are conducted will be determined on a case-by-case basis.

Coordination

Many organizations and individuals do land cover and vegetation mapping, often without knowledge of each other's work. However, many of these activities are known to analysts and managers in the LCCP, and several initiatives are under way to provide coordination at state, regional, and national levels. Specifically, the Department of the Interior (DOI) has created a Land Cover Working Group under the Interior Geographic Data Committee (IGDC). This working group, with membership from all DOI bureaus, will gather and publish information on each bureau's land cover and vegetation mapping plans, current activities, and related information sources. This information will be made available through the World Wide Web (WWW). In fiscal year 1997, the IGDC Land Cover Working Group plans to expand its membership to other agencies under the FGDC. Because the FGDC includes 14 agencies, this working group will provide a coordination forum for all Federal land cover and vegetation mapping activities. Some of these Federal activities include coordination components at the national, regional, and state levels; these components will be used and expanded where appropriate. In addition, the LCCP will follow the successful MRLC Program example of coordination at the project level among agency programs with requirements for land cover data. MRLC participants include the USGS, NBS, Environmental Protection Agency, National Oceanic and

23

Atmospheric Administration, and U.S. Forest Service.

Outreach Activities

Communication of the LCCP's activities and goals will be the program's outreach focus in fiscal year 1996. Information will be delivered through existing land cover and vegetation mapping projects and cooperatives, on the WWW under the USGS National Spatial Data Infrastructure node, through committees at bureau, department, and agency levels, and by presentations at professional meetings and conferences. These activities will announce the program's products and schedules and will encourage partnerships among individuals and groups with similar goals.

Standards

The USGS will work with the FGDC to establish a land cover working group. The initial tasks of the working group will be to share information about Federal plans and activities, develop a national land cover legend for the intermediate-scale synthesized data set, assist the FGDC land cover classification subcommittee, and write documentation standards for Federal land cover projects.

The USGS will lead an effort to develop a crosswalk or conversion process for existing land cover classifications. With this conversion capability, valuable maps and data can be used as historic references in landscape monitoring activities.

CONCLUSION

Land cover, land use, and vegetation mapping data are essential elements in a wide array of government and private sector activities, including inventory, management, monitoring, and modeling. The time and expense required to develop land cover and vegetation data demand efficient collection and analysis procedures that allow multiple use of a single product. The LCCP reflects the USGS commitment to providing high quality land cover data for effective stewardship of the Nation's resources.

REFERENCES

Anderson, J.R., Hardy, E.E., Roach, J.T., and Witmer, R.E., 1976, A Land Use and Land Cover Classification System for Use with Remote Sensor Data, U.S. Geological Survey Professional Paper 964, Reston, VA: U.S. Geological Survey.

Dickinson, R.E., Henderson-Sellers, A., Kennedy, P.J., and Wilson, M.F., 1986, Biosphere-Atmosphere Transfer Scheme (BATS) for the NCAR Community Climate Model, NCAR Technical Note NCAR/TN -275+STR, Boulder, CO.

Eidenshink, J.C., 1992, The 1990 Conterminous U.S. AVHRR

Data Set, Photogrammetric Engineering and Remote Sensing, v. 58, no. 6, p. 809-813.

Jennings, M.D., 1996, Nomenclature and Mapping Units for Gap Analysis Land Cover Data, in Technologies for Biodiversity Gap Analysis, Proceedings of the ASPRS/GAP Symposium, Charlotte, NC (in press).

Loveland, T.R., Merchant, J.W., Ohlen, D.O., and Brown, J.F., 1991, Development of a Land Cover Characteristics Data Base for the Conterminous U.S., Photogrammetric Engineering and Remote Sensing, American Society for Photogrammetry and Remote Sensing, Falls Church, VA, v. 57, no. 11, p. 1453-1463.

Loveland, T.R., Ohlen, D.O., Brown, J.F., Reed, B.C., Merchant, J.W., and Steyaert, L.T., 1993, Prototype 1990 Conterminous United States Land Cover Characteristics Data Set, CD-ROM, USGS CD-ROM Set 9307.

Loveland, T.R. and Shaw, D.M., 1996, Multiresolution Land Characterization: Building Collaborative Partnerships, in Technologies for Biodiversity Gap Analysis, Proceedings of the ASPRS/GAP Symposium, Charlotte, NC (in press).

Reed, B.C., Loveland, T.R., Steyaert, L.T., Brown, J.F., Merchant, J.W., and Ohlen, D.O., 1994, Designing Global Land Cover Databases to Maximize Utility: The U.S. Prototype, in W.K. Michener, J.W. Brunt, and S.G. Stafford, editors, Environmental Information Management and Analysis: Ecosystem to Global Scales, London, Francis and Taylor, Ltd., p. 299 - 314.

Sellers, P.J., Mintz, Y., Sud, Y.C., and Dalcher, A., 1986, A Simple Biosphere Model (SiB) for Use Within General Circulation Models, Journal of Atmospheric Science, v. 43, p. 505-531.

Steyaert, L.T., Loveland, T.R., Brown, J.F., and Reed, B.C., 1994, Integration of Environmental Simulation Models with Satellite Remote Sensing and Geographic Information Systems Technologies: case studies, in Pecora 12 Symposium, Land Information from Space Based Systems, Sioux Falls, SD, August 1993, Proceedings: Falls Church, VA, American Society of Photogrammetry and Remote Sensing, p. 407 - 417.

U.S. Environmental Protection Agency, North America Landscape Characterization Research Plan, EPA/600/R-93/135, EPA Office of Research and Development, Washington, D.C., 419 p.

THE NATIONAL BIOLOGICAL SERVICE/NATIONAL PARK SERVICE VEGETATION MAPPING PROGRAM - OVERVIEW AND UPDATE

Maurice O. Nyquist and Michael H. Story
National Biological Service
Technology Transfer Center
Denver Federal Center, Bldg. 810, Suite 8000
P.O. Box 25387
Denver, CO. 80225-0387

ABSTRACT

This paper describes the process that is being used to develop and implement a program for mapping vegetation cover types in approximately 250 units of the National Park Service. This program is being carried out through the use of contract sources due to the size of the effort, which entails all or portions of approximately 4000 1:24,000 scale quads. The major focus of this effort is to obtain uniform, consistent, baseline data on the composition and distribution of vegetation cover types for each of the National Park units (excluding Alaska) in the Inventory and Monitoring Program. These data will be used for various purposes at the park, regional, and national levels. The data for each park must therefore be consistent in detail and accuracy, and available in a format that allows for ready transfer to various GIS within the Park Service and elsewhere.

Development of these vegetation data requires several coordinated phases. These includes the development of a hierarchical classification methodology and scheme that meets NPS needs while also being useful to other programs (eg. GAP, EMAP, NAWQA, etc.), surveys of the parks to determine the existence and appropriateness of available vegetation data, definition and acquisition of necessary aerial photography, pre-classification field work (survey) to review and refine the classification scheme for each park, interpretation of photography for vegetation cover type information, conversion of the vegetation information to digital GIS and DBMS files, vegetation community description and documentation, field verification of the classification and mapping of each park as an assessment of accuracy, and transferring the data and technology to the NPS. Requirements for the geospatial and attribute accuracy are defined to assure that the vegetation data meet quality requirements of the NPS, FGDC and National Map Accuracy Standards.

The vegetation information (metadata and digital data) will become part of the National Biological Information Infrastructure (NBII)/National Spatial Data Infrastructure (NSDI) through the Federal Geographic Data Committee (FGDC) "Clearinghouse" concept. Overview documents, standards, protocols and other information about the program can be accessed on the Vegetation Mapping Program WWW-site at URL <http://www.nbs.gov/npsveg/intro.html>.

INTRODUCTION

This paper discusses a major nationwide program for producing maps of the vegetation of a majority of the units managed by the National Park Service. First we will discuss how the program fits into the NPS Inventory and Monitoring (I&M) Program, the type of requirements for such a program, and how we plan to meet those requirements. We will lastly discuss the specific activities to date.

This effort will be accomplished through the cooperative efforts of the National Park Service and the National Biological Service. Originally the NPS GIS Division became involved in the program as technical consultants because the GIS Division had been responsible for similar activities in the past. The GIS Division was subsequently transferred to the NBS as the Technology Transfer Center and continues to provide technical guidance, program management and funding.

The Vegetation Mapping Program is one of many 'parallel' efforts within the NPS I&M Program. To better understand the requirements of this vegetation mapping effort it will help to take a look at the NPS I&M program and its other efforts. One major goal of the program is to provide baseline data to all 250 I&M designated parks. Although there will be other park specific information required, the following digital data themes are to be acquired for all I&M parks: Digital Line Graphs (DLG)-(Hypsography, Hydrography, Political Boundaries, Administrative Boundaries, Roads and Trails), Digital Elevation Models (DEM), Digital Orthophoto Quads (DOQ), Soils, Geology, Vegetation, Bibliographies, Species Lists, Air Quality Inventory and Water Quality Inventory.

Vegetation mapping is only one of many such efforts in I&M Program. Vegetation mapping is a dualistic concept comprised of two components: vegetation and mapping. But why is vegetation mapping so important to the NPS I&M program?

Why Vegetation:

> Because vegetation is an integrated expression of the environment over time ... Vegetation = f (flora + fauna + climate + topography + soils + fire + perturbations + time)

> Because Vegetation information is required for so many NPS applications - resources management, research, operations, interpretation and planning/compliance

> Because vegetation information provides predictive capabilities necessary for ecosystem management (e.g. most plant associations have a narrower ecological amplitude than most of their component species), as well as other levels of management (e.g. useful for plant and animal species management).

27

Why Mapping:

> Because of the need to describe the whole park and environs; mapping defines the distribution of park resources in space and time.

> Because mapping of resources provides the framework to tie all the other resource components together.

> Because comprehensive mapped information will aid in determining data voids and will help identify more effective and efficient data gathering strategies to fill data voids.

Hence Vegetation Mapping:

> Provides the synergistic blend of the beneficial attributes of vegetation and mapping sciences that enables the consistent and useful characterization of a fundamental park resource in time and space.

PROGRAM PURPOSE AND GOALS

The purpose of the vegetation mapping program therefore is:

- to fulfill a high priority requirement of NPS I&M Program
- to begin long term vegetation monitoring program
- to provide data for short term applications related to:
    - resource management
    - research
    - planning and compliance
    - interpretation
    - operations

The goals of the vegetation mapping program are the comprehensive mapping of NPS vegetation resources in a way that:
(1) has a nationally consistent, hierarchical, useful, classification scheme,
(2) has a level of detail (i.e. both spatial resolution and classification level) useful to park management, as well as other program applications,
(3) is highly accurate (i.e. both spatial and thematic), and
(4) meets the professional standards of the scientific community and applicable standards of the FGDC/IGDC.

In other words, the goals of the program are to provide vegetation data and information that are uniform, consistent, professional, accurate and useful. Differences between this National (Servicewide) program and others that NPS has undertaken in the past include both the scope and scale. For example, results will be used at local, regional, and national scales, participation of multiple agencies and contractors, and requires coordination at multiple levels, digital products, completeness and consistency in detail and accuracy.

APPROACH

These differences require a different approach than has been used in the past. The general approaches that have been taken to meet the goals of the Vegetation Mapping Program are:

- develop standard national vegetation classification
- develop and use effective inventory and mapping protocols
- produce digital vegetation maps of 250 parks and environs
- develop database for field vegetation documentation
- acquire suitable aerial photography
- verify and assure requirements compliance
- produce and serve metadata for all products
- use contracted services to implement most aspects of program

Past efforts have been park specific and have therefore focused on specific requirements of one or more park programs such as fire fuel modeling, habitat assessment, planning, etc. Those efforts would therefore utilize a classification system that was specific to that park and that program. The idea of standards as requirements follows throughout this program to assure consistency. Also by standardizing on products and not procedures permits flexibility to utilize new technologies or approaches as they evolve over the life of the program, while maintaining fixed requirements for the contractors to meet.

REQUIREMENTS

Some of the requirements for the vegetation mapping project as specified in the contract documents are:

"Projects will be in accordance with the NPS overall management policies, standards, and guidelines. Work will conform to standards developed by the Federal Geographic Data Committee (FGDC) and is expected to meet professional standards of the scientific community."

"Uniformity in classification methodology over the entire park system is critical to achieve the overall inventory and monitoring goals of this initiative. All parks must be mapped at the same level of classification detail (i.e. plant association/cover type) and must be able to be re-aggregated to the FGDC vegetation classification scheme."

"A minimum mapping unit of 0.5 hectare or smaller will be used with a thematic classification accuracy of >80% for each vegetation class. All cartographic products will meet National Map Accuracy Standards and all digital products will have metadata files meeting FGDC standards."

PRODUCTS

The results of the efforts of this program will be deliverable products that will include:

- digital file of vegetation in prescribed GIS format;
- digital metadata file for each digital file delivered;
- textual description of vegetation classes;
- classification accuracy verification (contingency table);
- spatial accuracy verification and calculated rmse (root mean

square error);
- field data entered into an SQL based DBMS;
- analog field notes and site descriptions;
- location of field sites including labeling on aerial photographs used in the project;
- annotated field site photographs and interim interpretations;
- all photography, maps, and supplies initially furnished or paid for by the government;
- hard copy of the vegetation map for each park, tiled, edge-matched, and scaled to the "park-special" (USGS topographic) map; otherwise at some specifically prescribed scale; and hard copy plot of the vegetation map for each park on individual 1:24,000 quads.

PROGRESS TO DATE

A Standardized National Vegetation Classification Scheme has been developed and it is being reviewed and modified by the FGDC Vegetation Subcommittee for eventual adoption as a federal standard. Protocols for field methods and accuracy assessment have been developed. These protocols and the vegetation classification scheme are being tested through a series of prototype projects in parks located in different ecosystems with the attendant variability in vegetation complexity and having a range of differences in size, topographic relief and accessibility. These prototype projects are in various stages of completion but current results indicate the classification system is reliable and that the protocols are reasonable. Currently, a benchmarking phase of the program is beginning, which will test replicability of the products and compare various protocol efficiencies. Meanwhile, several small to medium sized parks will have completed vegetation data bases, including the other program products mentioned above.

The major challenge of the program is to balance, in a cost effective and efficient manner, the requirements at the park level with requirements at the regional and national levels. The approach we have taken is to develop standards that can meet the needs at regional and national levels and are flexible enough to meet the needs at the local park level (i.e. hierarchical, nested standards and data collection at the elemental level). To keep informed about further program progress and availability of products continue to check the Vegetation Mapping Homepage at URL <http://www.nbs.gov/npsveg/intro.html>.

An Overview of the Bureau of Land Management's National Applied Resource Sciences Center and Geographic Sciences Laboratory

by James Turner
Geographic Sciences Laboratory Manager

Like the U.S. Forest Service, the Department of the Interior's Bureau of Land Management (BLM) is endeavoring to keep up with the rapid pace of change that characterizes contemporary society. The Bureau is faced with finding new ways to accomplish its land management mission while improving customer service and operating within limits imposed by shrinking budgets and reductions in personnel. To meet this challenge, the BLM continues to modify its organizational structure so that more employees can be placed "on-the-ground" to resolve important land management issues.

A national support structure has been created to provide BLM field personnel with the highest level of scientific and technical assistance. In 1994, the Secretary of the Interior authorized the establishment of the National Applied Resource Sciences Center (NARSC). NARSC was assigned a pivotal role in supporting on-the-ground management activities in the areas of biological resources, engineering, lands and minerals, library services, mapping sciences, photo imaging, physical sciences, communications, and technology transfer. The Center's focus is to provide scarce skills and specialized equipment that augment the work of field resource professionals in accomplishing the Bureau's mission. The Center also serves as a brokerage for bringing skills together, as a link between policy and operations, and as a member of field teams to sustain the long-term health and productivity of public lands and resources.

The NARSC Organization

The blueprint for the Center's organizational structure was developed by internal employee teams with the primary objective of being as responsive as possible to customer needs. The management structure includes a Director, who maintains extensive customer contacts to help define high-priority Bureau requirements; five Administrative Group Supervisors who are responsible for all aspects of personnel management and employee development; and three Work Flow Coordinators, who administer NARSC's resources to meet customer needs, ensure that open communications are maintained with customers, and track the progress of more than 800 active projects. The Work Flow Coordinators and Group Supervisors also work together to establish teams, allocate personnel resources, manage project schedules, and monitor customer satisfaction.

Center personnel are organized into five administrative groups, each consisting of approximately 22 employees who are mentored by a supervisor. The administrative groups are organized around the Center's five functional roles: Resource Analysis and Assessment; Architecture and Engineering; Information and Communications; Geographic Sciences; and Natural Resource Sciences. Although employees are assigned to thematic groups, NARSC products and services are provided without regard to artificial organizational boundaries through the establishment of multi-disciplinary project teams.

Teams are formed if customer requests involve a high degree of complexity, are of a critical or sensitive nature, or must be completed as quickly as possible. After accepting the request, the best available talent is assembled under a Team Lead to provide the required product or service. Typically, project teams consist of individuals from one or more of the administrative groups, and they may include field personnel or individuals from other agencies and the private sector.

The NARSC structure was designed to be responsive to the changing needs of the Bureau of Land Management and other partner agencies and organizations -- providing quality support and products in a timely manner. Ecosystem-based management requires that the BLM adopt a different approach to managing the public lands and resources, working across traditional boundaries in cooperation with other agencies and the public-at-large. Without question, new and better ways must be found to accomplish the Agency's mission while reducing costs, improving the way we do business, and maximizing service to customers. The employees of the National Applied Resource Sciences Center are actively engaged in accomplishing these goals.

Primary Roles

BLM Headquarters has given NARSC scientists, engineers, and technicians key roles in the following areas:

Scientific and Technical Consulting - Providing assistance and advice on natural resource management issues; developing and providing training in coordination with the BLM's National Training Center; identifying research needs and assessing research results; providing scarce skills not found elsewhere in the Agency; and, serving as an agent to bring scientific and technical skills together to assist customers. Examples of work in this role include riparian ecological site training; in-stream flow assessment reports; soil survey enhancement demonstration efforts; and toxicology, geohydrology, and remote sensing support.

Information Exchange - Collecting and organizing scientific and technical publications; synthesizing applied resource science; and disseminating information. Examples of this role include preparation of technical notes and references, literature searches and reviews, writing and editorial support, graphics services, audio/visual productions, and technical assistance brokerage service.

Production - Providing products and services supporting on-the-ground resource management activities and supplying tools or specialized equipment to support the use of science and technology in making sound resource management decisions. Examples of this role include the administrative status (surface and subsurface ownership) map series; special management maps; hazardous materials site evaluations; digital elevation models; and architectural and engineering drawings, specifications, and cost estimates.

Analysis and Assessment - Responding to customers' requests to analyze and assess land management issues, and providing tools and expertise to integrate social, economic, and environmental aspects of ecosystem-based management. Examples include rangeland health standards and guidelines, natural resource damage assessment, and the Colorado River salinity control initiative.

Systems - Designing resource systems for use in assessing, documenting, and monitoring ecosystem conditions; promoting data sharing and acquisition of interagency resource information systems and applications; and coordinating scientific systems design and maintenance, data management, and geospatial projects. Examples of work in this role include the River Management Modeling System, Integrated Landscape Analysis, user representation for the Wildlife and Fisheries Information System (WFIS) and the Grazing Authorization and Billing System (GABS), and GIS applications.

Geographic Sciences Laboratory

The Geographic Sciences Laboratory, which is an integral part of the NARSC organizational design, is intended to bring together specialized equipment and scarce skills in a collaborative setting that promotes innovation. Many of the projects undertaken by NARSC specialists are performed in laboratory work areas, where specialists from within and outside the Agency are able to interact in an atmosphere that is conducive to teamwork and enhanced productivity.

Sophisticated computer hardware and software, precision measuring equipment, and unique data archives provide a resource that customers can use to complete projects or test new ideas and technological applications in ecosystem-based management. The resources of the lab are available to all BLM specialists and their counterparts in other agencies and organizations and can be scheduled through the Geographic Sciences Laboratory Manager. The following laboratory components form the basis for the Geographic Sciences Laboratory:

Aerial Photography Archive and Processing Laboratory (APAPL)

The Bureau's Aerial Photography Archive and Processing Laboratory was established in Denver in the early 1970s to facilitate access to aerial photography acquired by BLM field offices. Although satisfying the need for easier access to the photography holdings was the primary objective for establishing the laboratory, there was also a requirement to control and safeguard the aerial photography -- to ensure its viability as an irreplaceable public record. Both of these requirements are currently being met by the APAPL through the application of archival storage and protection techniques and the in-house reproduction of original project film.

Although the APAPL was originally established to provide aerial photography archival and reproduction services to BLM field offices, the customer base has expanded over the years to include state and local governments, other Federal agencies, colleges and universities, businesses, and private citizens. Customers currently have at their disposal approximately 400,000 frames of aerial photography covering much of the western United States. The photography dates from the late 1960s and was acquired using various camera and film combinations, including black and white, natural color, and color infrared. Copies of the photography can be obtained in a variety of formats either through in-house reproduction facilities or contract laboratories.

Cartography Laboratory

NARSC mapping science professionals and technicians utilize the resources of the Cartography Laboratory to produce both standard and specialized mapping products. Administrative maps such as the 1:100,000 Surface and Minerals Ownership map assist field office specialists and the public-at-large in locating and identifying public lands, while specialized products such as wilderness, recreation, and off-highway vehicle maps are designed and produced to meets specific program requirements.

Although traditional cartographic techniques are still employed to a large extent, the Laboratory is being modernized to provide services such as high-resolution scanning, precision digital plotting, and automated design and map construction. The laboratory is equipped with an extensive complement of hardware, including IBM® RISC 6000 workstations, IBM-compatible and MacIntosh computers, a Tangent® Scanner; large-format digitizing tablets, and a Kongsberg® flat-bed plotter. Software includes ERDAS Imagine® (ver. 8.2), ARC/INFO® (rel. 7.03), the INFORMIX® relational data base management system, TechBase®, Microstation®, and AutoCAD®. †

Image Processing Laboratory

Resource specialists are able to use the Image Processing Laboratory to perform a variety of tasks, including vegetation classification, feature mapping and extraction, data verification and updating, visualization, and three-dimensional data modeling. The Laboratory also serves as the repository for the Bureau's archive of satellite imagery, which includes data from the Landsat Multi-Spectral Scanner (MSS) and Thematic Mapper (TM), and SPOT sensors.

Similar in configuration to the other laboratories, the Image Processing Laboratory is equipped with a full suite of hardware and software capabilities available to assist in analyzing and interpreting image-based data. Hardware includes IBM® RISC 6000 workstations, IBM-compatible personal computers, large-format digitizing tablets, Hewlett-Packard 650C DesignJet plotters, and a Xerox® electrostatic plotter. Software includes ERDAS Imagine (ver. 8.2), OrthoMAX®, ARC/INFO (rel. 7.03), INFORMIX®, TechBase®, EarthVision®, and AutoCAD®.

GIS Laboratory

The GIS Laboratory provides a range of tools that NARSC specialists can use to develop GIS applications and assist field offices in resolving problems based on spatial data analysis and interpretation. ERDAS Imagine (ver. 8.2), ARC/INFO (rel. 7.03), INFORMIX®, TechBase®, and AutoCAD® software applications are loaded on a UNIX cluster that resides in the lab. These applications can be used in conjunction with complementary applications in other laboratories to tailor solutions directly to field office needs.

Photogrammetry Laboratory

NARSC cartographers and photogrammetrists use the capabilities of the Photogrammetry Lab to produce a variety of products, including digital elevation models that support field GIS requirements and large-scale site maps that are used by Bureau engineers and architects. NARSC staff are also regularly consulted by government solicitors to assist with litigation involving the definition of cadastral boundaries.

In addition to three Zeiss Planicomp P3 analytical stereo plotters, the Photogrammetry Lab is also equipped with IBM® RISC 6000 workstations and IBM-compatible personal computers. Software includes ERDAS Imagine® (ver. 8.2), ARC/INFO® (rel. 7.03), INFORMIX®, TechBase®, Line Trace Plus, and AutoCAD®.

Requesting Assistance

Personnel from any Federal agency can contact NARSC for assistance. The assistance can range from a brief telephone consultation to a multi-year, team-based partnership. In most cases, initial consultation is provided at no cost. Costs associated with projects or partnership agreements will usually be shared among cooperating organizations. If you are interested in obtaining more information about NARSC or the Geographic Sciences Laboratory, please contact one of the Work Flow Coordinators at (303) 236-6454 or visit the NARSC World Wide Web site on the Internet at http://www.blm.gov/narsc/.

† *The use of trade names does not constitute an endorsement by the Bureau of Land Management or the Department of the Interior.*

Project 615 - Implementation and Add-On Technology

Stan Bain
USDA Forest Service, Remote Sensing Application Center
2222 West 2300 South, Salt Lake City, UT 84119

The Forest Service is making progress in the development of a service-wide Integrated Information Management System with GIS capabilities to store, retrieve, and present spatially referenced information. Under this initiative, known as *Project 615*, we have awarded the first contact to the IBM Corporation. This award begins the pilot phase. In this phase, we address several concerns of Congress, test hardware and software configurations and develop plans for full implementation of the new IBM platform. Fifty four pilot sites have been selected and IBM platform have been installed so we may address pilot phase questions and concerns. A number of other sites also have received IBM equipment but are not part of the pilot phase testing. Later this year, the results of the pilot test and the implementation plans will be presented to Congress prior to nationwide implementation of the IBM based platform.

Under the *Project 615* program, we are pursuing a second procurement of add-on technology. We are adding image processing and remote sensing related technology to the Forest Service Integrated Information Management System. This effort has been in progress for the past couple of years. We have been identifying functional requirements and developing procurement documentation. Both Forest Service users and the vendor community have been and will continue to participation through the Request For Information (RFI) and Request For Comments (RFC) stages of the procurement process.

With a recent personnel change, a Memorandum of Understanding (MOU) is being developed between the Information Systems and Technology (IS&T) group and the Remote Sensing Application Center (RSAC). This MOU will shift the remote sensing procurement activities to RSAC, even though it is still part of the *Project 615* program for which IS&T retains responsibility. The MOU will facilitate coordination between IS&T and RSAC staffs as we acquire additional state-of-the-art remote sensing and image processing technology.

GEOMATICS AT COLORADO STATE UNIVERSITY

Roger M. Hoffer and Denis J. Dean
Department of Forest Sciences
Colorado State University
Fort Collins, Colorado 80523

ABSTRACT

With the continuing escalation of interest and activity in GIS throughout the United States and the world, there has been a significant increase in demand for a variety of educational opportunities. Among the colleges of natural resources throughout the country, one of the largest and best equipped remote sensing and GIS programs has been developed at Colorado State University (CSU). At present, five faculty teach a variety of remote sensing, GIS and GPS courses to over 400 graduate and undergraduate students each year, and an additional 1,300 students take classes that are prerequisites or background for the core courses. A minor in Spatial Information Management is being pursued by about 50 undergraduates. In addition, there are 40 graduate students in the geomatics program.

This paper describes some of the teaching, research and service activities at CSU. Some of the implications of the rapid growth of geomatics will be discussed, both from the perspective of institutions of higher learning and from the perspective of natural resource agencies such as the US Forest Service.

INTRODUCTION

Remote sensing has been around in one form or another for many decades, but the recent widespread growth of Geographic Information System (GIS) technology seems to have generated a renewed interest in this area. Because so much of the digital data that is utilized in GIS originally comes from remote sensing, this renewed interest in the characteristics of remotely sensed data is very understandable.

Another development that has spurred renewed interest in remote sensing is the widespread availability of the Global Positioning System (or GPS). GPS technology is providing us with the capability to easily and inexpensively locate positions on the earth's surface with accuracies that were previously just a dream. This capability has obvious impacts on our ability to georeference and validate remotely sensed data, which again makes such data less expensive and more accessible.

Remote Sensing, GIS, and GPS involve somewhat overlapping but nevertheless distinct sciences and technologies; yet they are clearly interrelated. All three of these technologies are primarily concerned with recording and analyzing the locations and attributes of the spatial features of the earth's surface or near-surface environment. *Geomatics* is a term that has been coined to describe the combination of these spatial sciences. A basic definition of geomatics is *the science of collecting, manipulating, and analyzing spatially explicit data*. This definition encompasses all aspects of remote sensing, including the use of aerial photos for traditional photo interpretation and photogrammetry, the digital processing of data gathered by thermal or multispectral scanners and radar systems borne by either air or spacecraft, and the gathering and interpretation of data obtained by various other instrument systems. Geomatics also encompasses all aspects of developing and utilizing GIS. Developing and maintaining the locational and associated attribute databases that are an inherent part of GIS certainly falls under the definition of geomatics, as do the techniques and processes used to develop new information from these databases via GIS's cartographic analysis capabilities. Finally, geomatics includes all aspects of utilizing GPS technologies, which are, after all, designed to allow GPS users to spatially locate points with a high degree of accuracy.

We were very pleased to be asked by the organizers of this Conference to discuss the remote sensing, GIS, and GPS--the geomatics--activities at Colorado State University (CSU). We would also like to describe some of the challenges faced by CSU and other institutions of higher learning in keeping pace in this rapidly changing field. This paper will briefly describe the geomatics program at CSU; we hope that this description will illustrate some of the strengths, opportunities and problems facing what we believe is a successful academic program.

GOALS AND PURPOSES OF THE PROGRAM

Geomatic activities at CSU are centered in the Remote Sensing and GIS Program (RS/GIS Program), which was established as part of the College of Natural Resources in 1988. CSU is a land grant university (i.e., it was established under the auspices of the land grant act of 1864). Like all land grant universities, CSU is required to perform teaching, research and service activities. The RS/GIS Program has attempted to develop a strong presence in all three of these areas.

The goals of the RS/GIS Program reflect this three-part emphasis. The Program's goals are: (1) to provide an outstanding educational experience in the field of geomatics for graduate and undergraduate students, both in individual courses and in complete programs of study; (2) to develop a pertinent and innovative research program that will contribute to the advancement of human knowledge and mankind's ability to effectively and efficiently utilize spatial data relating to the natural and cultural features of the earth's surface or near-surface environment; (3) to develop and conduct cooperative applications-oriented projects with appropriate local, state, or federal agencies and/or the private sector that will expand knowledge of the utility of the existing technologies and that will provide students with meaningful learning experiences; and (4) to serve the scientific, academic, government, community and private sectors of society through a variety of activities that will advance the understanding of the principles and applications of remote sensing and GIS technologies, and that will help to create a better local, national and international environment in which to live.

HISTORY AND CURRENT STATUS

Since its founding in 1988, the RS/GIS Program has developed a national and international reputation for excellence. We believe that this reputation is based on the quality of the program's students and faculty, the excellent array of pertinent courses available at CSU, our access to some of the best computer facilities available at any educational institution in the world, and our innovative research program designed to accommodate both applications-oriented projects and basic research efforts.

Perhaps the best way to describe the RS/GIS Program is to divide its activities into the teaching, research and service areas that make up any land grant university. However, to describe the program in this way runs the risk of implying that these three areas are separate from one another. It must be remembered that the program is designed to be integrated; almost all of our activities may be characterized as being primarily oriented toward one of these areas, but each activity undoubtedly has significant impacts in multiple areas. It is important to keep these interrelationships in mind as the separate components of the program are described.

Teaching

The teaching program includes a variety of both graduate and undergraduate courses. Eight GIS core courses (three undergraduate and five graduate) and five remote sensing core courses (two undergraduate and three graduate) are currently offered. Five faculty members teach these core courses to more than 400 graduate and undergraduate students each year. An additional 1,300 students (primarily undergraduates) take classes that are prerequisites or background for the core courses.

The demand for these core courses has grown rapidly and far exceeds our current capabilities. For example, the basic undergraduate Introduction to Geographic Information Systems class has a ceiling of 75 students per semester, but still fills to capacity early during registration each semester. Many students recognize the beneficial aspects of this course in helping them find employment when they graduate-- approximately half of the recent natural resource job announcements that we have seen at CSU have specifically stated a requirement for applicants to have knowledge of GIS and remote sensing.

The interest in GIS is so high that many students who complete the initial Introduction class have a great interest in additional studies. For this reason, an integrated follow-up course was designed and instituted in 1995. This class builds on the basic concepts and theories of spatial data management and cartographic analysis developed in the introductory class, and it gives students more hands-on experience with the software packages most commonly used in real-world GIS applications.

To provide a more complete program to meet the demand for appropriately educated undergraduate students, a minor in Spatial Information Management Systems was approved and instituted in 1993. This minor requires 21 credits of course work in GIS, remote sensing, geography, computer programming and statistics. The minor is particularly well suited to the approximately 200 students currently majoring in Natural Resource Management. It is also becoming a popular minor for students in several other majors. At present, there are over 50 students officially enrolled in the minor. Students who have graduated with the Spatial Information Management Systems minor have been in high demand for private industry, local, state and federal government agency positions.

At the graduate level, students can pursue either MS or PhD degrees emphasizing geomatics. At present, there are 40 graduate students in the Remote Sensing and GIS Program (30 MS and 10 PhD) who have a wide array of backgrounds and interests.

These students came to CSU from 23 different undergraduate universities. In addition to these 40 students directly enrolled in the RS/GIS Program, many other graduate students who are not enrolled in the Program but have thesis or dissertation projects that utilize remote sensing and/or GIS take some or all of Program's courses. Finally, the international stature of the Program is reflected in the number of international students who have enrolled; over the last six years, graduate students have entered the program from at least seven countries.

A list of the core RS/GIS courses currently available from the College of Natural Resources is shown in Table 1. In addition to these College of Natural Resource courses, there are also several courses involving geomatics that are offered through the College of Engineering, the Department of Atmospheric Sciences, and the Department of Anthropology.

Research

The Remote Sensing and GIS Program faculty and graduate students pursue a broad variety of research activities. Essentially all of the research projects are conducted cooperatively by faculty and graduate students, often with technician-level assistance provided by undergraduates. Many of these research projects are funded by natural resource management organizations and thus have a decided real-world, applications-driven orientation. For many of our MS students, projects of this sort are an excellent way of obtaining valuable research experience, meeting the requirements of their graduate programs, and receiving a livable wage. Some recent examples of projects such as this include conducting satellite image classifications of various watersheds across Colorado, performing cost-effectiveness analyses of established ways of mosaicing and interpreting airborne video imagery, and comparing the information content of different types of remotely sensed data.

For almost all of our PhD students and those MS students who express an interest in extending their education beyond their MS program, simple application-oriented research projects are often insufficient to provide the intellectual challenge their graduate program warrants. The graduate programs of these students are often extended to include more basic research topics that may not have direct or immediate benefits in the same sense as does a classified satellite image of an area of interest. However, since these projects investigate more fundamental aspects of geomatics, in time their results could be far more wide-reaching than those produced by the applications-oriented projects. These projects are sometimes funded by research organizations such as NASA or the research division of the US Forest Service, but more often they are unfunded or are performed as unfunded extensions of applied research projects. Some examples of project falling into this category include the development of new techniques for shadow removal from aerial photographs, investigations of ways of using integrated remote sensing and GIS techniques to model non-point source groundwater contamination, and explorations of ways of combining geomatics and artificial intelligence techniques to produces new and unique visualizations of spatial data.

Service

CSU faculty and students (both graduate and undergrad) participate in a wide variety of service and outreach activities. We receive many requests for demonstrations of remote sensing and GIS technologies from individuals and groups throughout the United States and the world. We try to accommodate these requests through tours and demos conducted at CSU, and by sending written documentation to those individuals unable to visit the CSU campus.

Faculty and students have been active in several cooperative research and service projects designed to assist local, state or federal government agencies, as well as private groups, in the use of geomatic technologies. Many of these projects involve helping organizations with existing geomatic capabilities utilize their systems in new and innovative ways. Other projects involve helping organizations with little or no geomatic experience to incorporate geomatic techniques into their operations. An excellent example of this latter case involved a community planning project in a neighborhood of Estes Park. This project involved a class working with a neighborhood citizen's group, planners from the town of Estes Park, community liaison officers from Rocky Mountain National Park, professional planners from a commercial planning firm, and many others. The goal of this project was to help the community utilize geomatic techniques in developing a land use plan for a visually sensitive area bordering Rocky Mountain National Park. After much discussion and input from all of the concerned parties, a series of manageable questions of interest were developed and a strategy for answering these questions was outlined. The students finalized this solution strategy and then implemented it, providing the community with valuable information needed in the development of their land use plan. Since the completion of this project, the community has approached the Larimer County zoning board and petitioned to have their plan codified as part of the County zoning plan.

Another major cooperative project involves the Colorado Rockies Regional Cooperative, or CORRC. CORRC is a consortium of federal, state and local governmental agencies, nonprofit groups, private for-profit firms, and two universities. All of these organizations share an interest in land management in Northern Colorado. One aspect of the cooperation between these groups fostered by CORRC involves the sharing of spatial data. As part of the CSU contribution to CORRC, the RS/GIS Program maintains a spatial data library covering much of north central Colorado. This data library consists of a variety of cartographic data sets including data layers such as topography, hydrography, vegetative cover, transportation infrastructure, political boundaries, and so on. All of the data in the library has been donated voluntarily by CORRC members, and is made publicly available via the world wide web through the computer facilities of CSU. It is hoped that in the future this data base can be further developed to include larger portions of Colorado and a more extensive set of data layers, thereby allowing it to become a major focal point of activity for teaching, research, and outreach in the future.

Numerous other service related activities are underway at CSU. Unfortunately, it is often difficult to determine if an activity could best be characterized as teaching, research or service. For example, we have an ongoing cooperative arrangement with IBM corporation. This co-op involves having a CSU student who is familiar with how geomatic techniques are implemented in natural resource management work part-time at the IBM facility in Boulder that is delivering both the US Forest Service's and the BLM's new office automation equipment. This activity certainly benefits the student educationally, so it may be considered a form of teaching. However, it also involves a substantial research component, because the students involved (and CSU) have typically worked with IBM in developing new and innovative ways of incorporating IBM products into natural resource management activities. Finally, this could also be called a service activity, because it involves a cooperative arrangement between CSU and an outside organization that is mutually beneficial. Thus, deciding what exactly constitutes a service activity is not always as clearcut as one might expect. However, no matter where you place the lines between teaching, research and service, it is clear that CSU has a very active and diverse service program.

Facilities

The levels of teaching, research and service just described would be impossible to maintain without the outstanding computer facilities available in the College of Natural Resources. At the present time, there are three primary facilities available to faculty and students in the RS/GIS Program. These are the Computer Learning Lab, GIS Technology Lab, and the Advanced Technology Lab. The Computer Learning Lab consists of forty-five networked Pentiums and 486 PCs running several remote sensing and GIS software packages, as well as a large array of general purpose software. This is our primary undergraduate teaching facility, and is used not only by RS/GIS students but also by many other students from throughout the College of Natural Resources.

The GIS Technology Lab has fifteen 486 PCs with access to all the same software available in the Computer Learning Lab, but in addition these machines run the PC ARC/INFO GIS software package. These machines are also equipped with tablet digitizers. This facility is also heavily used for undergraduate teaching. However, in addition to regular undergraduate instruction the GIS Tech Lab is often used in our service program as the primary center for our non-academic short courses. These courses allow professionals from land management organizations to come to CSU for week-long training sessions to help them learn how to utilize GIS technologies in their professional lives.

The Advanced Technology Lab is a truly unique facility, consisting of eighteen SUN SPARC UNIX workstations. These machines run ARC/INFO, ERDAS, and GRASS software. There are many peripheral devices in this lab, including a number of large format digitizers, scanners, plotters, tape drives, and so on. This facility is used for graduate level instruction and serves as our primary research facility.

In addition to these primary facilities, there are also a number of satellite labs located throughout the College, each containing a handful of workstations and/or PCs. The satellite lab that is used exclusively by graduate students in the RS/GIS program currently contains two UNIX workstations, three PCs, and a variety of peripherals including a large-format digitizer and a high resolution, large-format color inkjet plotter. This lab, like many of the other satellite labs scattered around the College, is expected to grow as time goes by.

This suite of computer facilities, and the Advanced Technology Lab in particular, are widely recognized and acknowledged as being the best computer hardware and software facilities for remote sensing and GIS teaching and research in any college of natural resources in the country. All of these facilities are funded through three primary sources. Basic overhead costs and salaries for professional support staff are covered by general operating funds of the College. Equipment purchase, upgrade and maintenance are covered by research funds and monies raised through a special "technology fee" paid as part of each student's tuition. Together, these relatively substantial and reliable sources of funding (in 1995, the total level of funding provided by these three sources exceeded $450,000) have been invaluable in allowing CSU to develop and maintain its outstanding facilities.

The Results

The demand for graduates from the RS/GIS Program is very high. Over the last six years, all of our graduating MS and PhD students have found employment in the geomatics field. Several recent graduates have had two or three job offers from which to choose, and some have received excellent job offers well in advance of finishing their degree. Program faculty often receive calls from leaders in corporations and agencies throughout the country inquiring as to the availability of students about to finish their degree. Comments are frequently made concerning the competence and effectiveness of employees who are RS/GIS Program graduates.

The RS/GIS research programming is also producing very successful results. Program faculty and students have made presentations at numerous geomatic conferences throughout the United States and the world. Many of these presentations are invited, and virtually all of them involve descriptions of research projects of interest. In addition, Program faculty and students have had considerable success at publishing research results in a wide range of refereed scientific journals. Perhaps the most telling mark of the success of the RS/GIS research program is our continued ability to secure research funding. Currently, the faculty of the RS/GIS program serve as principle investigators on 13 research projects with total budgets exceeding $900,000. If program efforts were not seen as successful, it seems unlikely that we would be able to maintain this level of support from outside research organizations.

Finally, the RS/GIS service program is very large and visible. Once again, the fact that we continue to receive many service requests from numerous organizations indicates that the service activities of the program are seen as successful by individuals and organizations seeking geomatics support.

OPPORTUNITIES AND PITFALLS OF THE PRESENT AND THE FUTURE

Spatial information technology has developed rapidly during the past decade. Worldwide revenues just for GIS software are projected to exceed $563 million in 1995. The number of GIS workstations in the United States has grown from 99,000 in 1989 to a projected 700,000 in 1997 (Hemenway, 1995). The number of users is expected to far outstrip this seven-fold increase in hardware facilities. GIS technology provides the user with the ability to manipulate and analyze any data that has a spatial component, and since virtually all data describing the physical world contains a spatial component, this technology can be applied to an extremely wide-ranging set of disciplines. Over the past seven years, it has been estimated that the world-wide annual rate of growth in GIS has been approximately 17% per year, and this growth is expected to continue at an even faster rate in the future.

From the above figures, it seems clear that GIS as well as other areas of geomatics are growing rapidly. This will result in a continuing and increasing demand for well educated students, both undergraduate and graduate. This demand poses both great opprotunities and great challenges to institutions of higher learning such as CSU. On the positive side, this great demand virtually ensures that there will be some level of political support for programs like ours. However, the great challenge facing organizations like ours is translating this support into the tangible results needed to maintain up-to-date programs in such a rapidly changing field. It is not only necessary for programs of higher learning to keep abreast of developments in computer hardware and software (a very expensive proposition), but also to keep faculty and staff up-to-date with the state of the art. In an environment of constant demands for classes, "publish or perish" research deadlines, and seemingly never-ending requests for service activities, finding the time needed to remain up-to-date can be difficult.

Perhaps the greatest challenge facing educators in the geomatics field is gaining acceptance within the halls of academia itself. Academic institutions are, by their very nature, cautious and conservative when it comes to admitting new disciplines into their ranks. This natural conservatism places great pressures on geomatics programs to demonstrate that they are not simply service providers and that they represent a new discipline of science worthy of the same respect accorded to more established disciplines. If the political support from outside academia generated by the economic forces driving geomatic technologies could be harnessed and turned into support within academia for the science of geomatics, many of the problems facing geomatic programs such as ours could be solved.

CONCLUSIONS

In summary, remote sensing, GIS, and GPS -- that is, geomatics -- are widespread, important sciences throughout the world, having tremendous economic and social significance. Spatial data management will have a increasingly important impact on natural resource management, urban and county land-use planning, utility company operations, mineral exploration, land management, watershed management and hydrology, epidemiology, and all other disciplines requiring use of spatial data. **The education of knowledgeable people to effectively and efficiently utilize the full potential of geomatics capabilities is vital**. It is people who make tools such as GIS, remote sensing and GPS functional and meaningful.

LITERATURE CITED

Hemenway, D. 1995. Personal Communication.

Table 1:   List of courses.

Remote Sensing and GIS Courses Available At
Colorado State University

Course Designation and Title[1]	Credit[2]	Level[3]

Core Remote Sensing and GIS Courses (and Instructor)

Course Designation and Title		Credit	Level
NR 322 (F)	Introduction to Geographic Information Systems (Dean)	3	B
NR 323 (F)	Remote Sensing of Natural Resources (Hoffer)	3	U
NR 422 (F)	GIS Applications in Resource Management (Coleman)	4	B
NR 493 (F)	Seminar on Remote Sensing and GIS Applications (Hoffer)	1	U
NR 503 (F)	Remote Sensing for Resource Management (Hoffer)	3	G
NR 504 (F)	Computer Analysis of Remote Sensor Data (Hoffer)	3	B
NR 505 (F)	Principles of Geographic Information Systems (Laituri)	3	B
NR/ST523 (F/ST)	Quantitative Spatial Analysis (Reich)	3	G
NR 621 (F)	Advanced Concepts in GIS (Dean)	3	B
NR 793 (F)	Remote Sensing and GIS Seminar (Hoffer)	1	G
XX 495 or 695	Special Problems in Remote Sensing/GIS (All)	V	B
XX 698	Remote Sensing/GIS Research (MS) (All)	V	G
XX 699	Thesis in Remote Sensing/GIS (All)	V	G
XX 798	Remote Sensing/GIS Research (Ph.D.) (All)	V	G
XX 799	Dissertation in Remote Sensing/GIS (All)	V	G

Other Related Courses

Course Designation and Title		Credit	Level
GR 100 (ER)	Introduction to Geography	3	U
F 201	Forest Mapping	3	U
GR 210 (ER)	Physical Geography	3	U
NR 420(F/FW)	Integrated Resource Management	4	U
NR 541 (LA)	Regional Resource Planning Techniques	3	B
LA 510	Advanced Analysis Methods	3	B
LA 560	Structure of Landscape Patterns	3	B
F695	Research Methods in Forest Science	2	G
EE 653	Image Processing and Pattern Recognition	3	G
AT 737	Satellite Observation of Atmosphere & Earth	3	G
FW 371	Wildlife Management Techniques	4	B

KEY

NR	=	College of Natural Resources
		(Departmental Affiliation of Instructor shown in parentheses)
GR	=	Geography Courses, College of Natural Resources
ER	=	Department of Earth Resources, College of Natural Resources
AT	=	Dept. of Atmospheric Sciences, College of Engineering
EE	=	Dept. of Electrical Engineering, College of Engineering
F	=	Dept. of Forest Sciences, College of Natural Resources
FW	=	Dept. of Fishery and Wildlife Biology, College of Natural Resources
LA	=	Dept. of Landscape Architecture, College of Agricultural Sciences
XX	=	Any Department

[1]Department and College offering course. (See key above)

[2]Credit = Semester Credit Hours; V = Variable Credit.

[3]G = Graduate; B = Both graduate and undergraduate; U = Undergraduate

REMOTE SENSING DATA PURCHASE FOR ECOSYSTEM MANAGEMENT

Henry Lachowski
Stan Bain
Tim Wirth
Paul Maus

USDA Forest Service
Remote Sensing Applications Center
Salt Lake City, Utah 84119

ABSTRACT

The USDA Forest Service guiding principles for ecosystem management
recognize that ecosystems are constantly changing and must be viewed
over long periods of time. Ecosystems must also be viewed from a
variety of scales, from the very large, to the very small. These basic
principles are being addressed with the help of remote sensing and
other geospatial technologies. Resource managers depend on remote
sensing to provide data for large area assessments, for forest plan
preparation and monitoring, and for project implementation. The Forest
Service Interregional Ecosystem Management Coordination Group (IREMCG)
has provided funding for the purchase of remotely sensed data to
support field implementation of ecosystem management. The purpose is
to provide uniform and complete coverage for the lower 48 states, at
three levels, to assist in large area assessments, and other ecosystem
management requirements. The products come from existing US Geological
Survey EROS Data Center archives, and consist of the following:
weather satellite AVHRR coverage, Landsat MSS triplicates from 1970s,
1980s, and 1990s, and Landsat TM from early 1990s. The purchase is
being done during 1996 and 1997. The coordination of this purchase and
related activities for the IREMCG is being done by the Remote Sensing
Applications Center and by Regional Coordinators.

THE INTERREGIONAL ECOSYSTEM MANAGEMENT COORDINATION GROUP (IREMCG)

The Interregional Ecosystem Management Coordination Group consists of
Deputy Regional Foresters and Assistant Research Station Directors.
The group was formed in 1993 to promote and work towards "a
consistentphilosophy and approach to ecosystem management, and
ensurecoordination of planning, information management, human
dimension, and scientific analysis activities in support of ecosystem
management in the regions and research programs". The charter of the
IREMCG specifies among other items to promote consistency for
information management through coordinated mapping techniques, and
consistency in the National Hierarchy of Ecological Units. Another
part of the charter specifies a coordinated implementation and revision
of forest plans, through the use of most appropriate scientific
knowledge and technologies. The IREMCG sponsors and provides funding
for activities at the national level that support this charter.

PURPOSE OF THE REMOTE SENSING DATA PURCHASE

One of IREMCG requirements is to provide framework for addressing information at multiple scales and levels in support of ecosystem management. Remotely sensed data and related products can partially satisfy this requirement. Data at multiple scales and resolutions provide a wealth of information on the location, extent, and condition of natural resources. It provides a basis for current vegetation, for modelling, and for change detection. The purpose of this national purchase is to provide a general and consistent framework across administrative boundaries that can support multi-scale requirements of ecosystem management.

Products to be purchased are currently being used by other federal agencies, and are available through USGS EROS Data Center in Sioux Falls, SD. These products, at three levels and scales are as follows:

IMAGE TYPE	SCALE	RESOLUTION
Weather Satellite Imagery (AVHRR)	1:1,000,000	1 km
Landsat MSS Triplicates (1973, 1986 & 1991)	1:250,000	80 m (1 acre)
Landsat TM (1990s)	1:100,000	30 m (0.25 acres)

The procurement is phased over two years to meet the anticipated needs for large area assessments, and other spatial analysis requirements. The data and products generation and delivery is being coordinated through Regional Coordinators and the Remote Sensing Applications Center in Salt Lake City. All the products are georeferenced to facilitate efficient use and to allow input into geographic information systems.

DESCRIPTION OF DATA AND PRODUCTS

Advanced Very High Resolution Radiometer (AVHRR)

AVHRR data is obtained from the Department of Commerce, NOAA weather satellites. It has relatively course spatial resolution of 1.1 km x 1.1 km pixel size with five spectral bands. Broad area coverage is obtained with a swath width of 2,700 kilometers. Temporal resolution is excellent, with repeat coverage over a given area every 12 hours. Major ecosystem management applications for this data include: vegetation discrimination, vegetation biomass, snow/ice discrimination, vegetation/crop stress, and geothermal mapping. This imagery is currently used by the Forest Service in association with USGS, on development of a global vegetation data set and with the United Nations on verification for a global vegetation characterization data base.

The image data and following products derived from AVHRR imagery are available:

- Land Cover Data Base for the United States
- Seasonal Land Cover Regions of the United States
- Forest Cover Types of the United States

47

Landsat Multispectral Scanner (MSS)

MSS imagery has a spatial resolution of 80 m, spectral resolution in 4 bands, covers the same geographic location every 16 days and covers an area of 185 x 170 km per scene. The Landsat program gathered digital MSS data from 1972 through 1992. The result is a twenty year time span of data that will support evaluations of change in landscapes or land cover.

MSS data is available through the EROS Data Center working in cooperation with the NASA Landsat Pathfinder Program. Pathfinder efforts are focused on evaluation of global change using available remote sensing technologies. The North American Landscape Characterization (NALC) Project, a component of the Landsat Pathfinder Program, is developing an archive of MSS data and production of three-date georeferenced data sets, and placed into a UTM map projection. Pixels are resampled into a 60 x 60 meter size format for compatability with the 30 m x 30 m Landsat Thematic Mapper (TM) data resolution.

The standard NALC imagery set is as follows:

> Triplicate images georeferenced and coregistered data from the 1991, 1986, and 1973 composite images
>
> Digital Terrain Model data

Landsat Thematic Mapper (TM)

TM imagery has 30 m spatial resolution with 7 spectral bands. The Landsat satellite covers the same geographic location every 16 days, each scene covering 185 x 170 km. The characteristics and quality of Landsat data, and the ability to collect new data directly comparable to that in the archive, make this a unique resource to address a broad range of issues in ecosystem management. There are 430 Landsat Thematic Mapper scenes that cover the entire lower 48 states. This is similar for the Landsat Multispectral Scanner described above.

COORDINATION OF DATA PURCHASE

This purchase of remote sensing products is being done through the USGS EROS Data Center in Sioux Falls, South Dakota. EROS is the depository for a variety of geospatial data, including satellite imagery. The procurement is based on a partnership with five federal agencies. This substantially lowers acquisition cost.

Each Forest Service Region has a coordinator (point of contact) to coordinate the acquisition within the Region, and for Research Stations, State and Private Forestry and other units. RSAC provides general information about data, products, and ordering procedures, and coordinates the purchase with the EROS Data Center.

To utilize data and products, users will need a computer system that allows display and manipulation of raster and vector data, with appropriate input devices. There are a number of image analysis systems that can meet this need. The IBM equipment in the Forest Service has some capability to display and manipulate image data. Image analysis will need to be done on image analysis systems, and

properly trained personnel are needed in order to fully utilize remote
sensing data.

There are numerous applications of the remote sensing data being
purchased. To assist in applications, RSAC has created a CD containing
basic Arc/Info layers of boundaries and other ancillary data for the
entire country, at an approximate scale of 1:2 million. The CD
contains the following coverages:

    Forest Service boundaries
    Ecoregions
    Landsat path-row locations
    Latitude-longitude (2-degree)
    Forest vegetation (from AVHRR)
    State and county boundaries
    Hydrology.

The National Hierarchy of Ecological Units, adopted recently by the
Forest Service, provides a convenient framework for categorizing land
units, and for applications of remote sensing and other technologies.
Figure 1 lists the main ecological units and possible remote sensing
imagery sources that can match the mapping requirements at those
levels. The challenge to the ecosystem managers and to the remote
sensing community is to test these linkages, and to provide assistance
in field applications.

The Ecological Unit Hierarchy and Remote Sensing Imagery

Ecological Unit	Imagery Source
⊛ Ecoregion (1,000,000's to 10,000's of square miles)	AVHRR
⊘ Subregion (1,000's to 10's of square miles)	Landsat MSS
⊛ Landscape (1,000's to 100's of acres)	Landsat TM
⊛ Land Unit (100's to less than 10 acres)	Aerial Photography Other Airborne Systems

Figure 1. The National Hierarchy of Ecological Units and remote
sensing imagery sources.

ASSESSMENT OF REMOTE SENSING/GIS TECHNOLOGIES TO IMPROVE NATIONAL WETLANDS INVENTORY MAPS

Bill O. Wilen, Ph.D.
U.S. Fish and Wildlife Service
National Wetlands Inventory
Washington, D.C. 20240

Glenn S. Smith
U.S. Fish and Wildlife Service
National Wetlands Inventory
Region 5
Hadley, Massachusetts 01035

ABSTRACT

The National Wetlands Inventory (NWI) produces two very different kinds of information through remote sensing. First, detailed wetland maps are produced to support site-specific decisions. Secondly, national statistics on the current status and trends of wetlands are developed to provide information supporting the development or alteration of Federal programs and policies. The national scope and required level of detail dictated that a remote sensing tool, combined with field work, be used to conduct the project. High altitude aerial photography and satellite imagery were investigated as possible data sources. The results showed that high altitude color infrared photography provided the spectral and spatial resolution needed to obtain the required classification accuracy. The identification of temporarily and certain seasonally flooded forested wetlands remains problematic. Medium- to high-resolution multiband satellite and airborne imagery of visible, near-infrared, mid-infrared, thermal, passive microwave, and radar spectral regions were investigated using advanced GIS software packages to solve this problem.

BACKGROUND

The National Wetlands Inventory of the U.S. Fish and Wildlife Service was developed to generate information on the characteristics, extent, and status of the Nation's wetlands and deepwater habitats (Wilen and Bates 1995). The Emergency Wetlands Resources Act (16 U.S.C. 3931), as amended by P.L. 102-440, requires the National Wetlands Inventory to complete maps for the conterminous United States by September 30, 1998; to update the report on wetlands status and trends on a 10-year cycle; to produce wetland maps of Alaska by September 30, 2000; to produce a digital database for the United States by September 30, 2004; and to archive and make final maps and digitized data available for distribution.

MAPPING

The National Wetlands Inventory has produced (final and draft) maps for 87 percent of the conterminous United States and 29.5 percent of Alaska. The current budget will allow for mapping only 1 percent each year. The mandated (1998) date for completion of mapping wetlands of the lower 48 States will not

50

be met until at least 2011. The current budget will likewise allow for mapping only 1 percent of Alaska each year. The mandated date (2000) for completion of mapping the wetlands of Alaska will be missed by decades.

DIGITAL DATABASE

The NWI has completed 24 percent of the digital database. Statewide databases have been built for 10 States and initiated in 5 additional States. Digitized wetland data are also available for portions of 35 other States. Continued development of a digital wetlands database will have to be on a user-pays basis. This will set back the completion date of 2004 mandated by the Emergency Wetlands Resources Act of 1986. The actual completion date cannot be estimated.

WETLANDS STATUS AND TRENDS

Map making is not the only function the National Wetlands Inventory conducts through remote sensing. The NWI has produced three reports to Congress on the Status and Trends of the Nation's wetlands. In 1990, the Service produced the latest update of the status and trends report to Congress as mandated by the Emergency Wetlands Resources Act of 1986. Future national updates are to be completed on a 10-year cycle in the years 2000, 2010, 2020, and beyond. Current funding may not allow completion of the report due to Congress in 2000.

DISTRIBUTION OF PRODUCTS

National Wetlands Inventory maps and digital data are distributed widely throughout the country and the world. The principal venues for distribution are the 32 State-run distribution centers throughout the Nation; U.S. Geological Survey centers at 1-800-USA-MAPS; the Library of Congress and the Federal Depository Library System; and most recently the NWI Home Page on the Internet. The URL address for the Home Page is: http://www.nwi.fws.gov.

REMOTE SENSING OF FORESTED WETLANDS*

The National Wetlands Inventory has found that leaf-off, color infrared aerial photography from the early spring is best for detecting deciduous forested wetlands. Evergreen forested wetlands are a bigger problem because dense evergreen stands of the same species can occur both in the wetlands and adjacent uplands. At times, height of the evergreen canopy may reflect a difference in wetness. Wetland evergreens may be somewhat reduced in height. Wet evergreens may show signs of chlorosis due to water stress. Saturated soils or understory wetland signatures may be evident in canopy openings. The photointerpreter uses landscape and topographic position, soils information, and extensive field work to identify subtle photo signatures of evergreen and the drier deciduous forested wetlands.

*A detailed discussion on the use of high-altitude aerial photography for inventorying forested wetlands in the United States is found in (Tiner 1990).

The problem is further compounded by altered hydrology. Dams, levees, channelization projects, failed drainage, and stream-water diversions often prevent or impair normal seasonal flooding. Photointerpreters must determine whether the hydrology has been altered to the extent that the area is no longer a wetland. If an area no longer has wetland hydrology, it is a historic wetland and not mapped. If the water level has been artificially lowered, but it maintains sufficient hydrology to be classified as wetland, it is given the special modifier of Partly Drained (Cowardin et al. 1979). Generally, the modifier is only applied if there is a visible system of ditches or channels. In the southern United States, wetlands may be used for pine plantations. Some of these wetlands are effectively drained, others partly drained, and others undrained. Separating former wetlands that have been effectively drained and are no longer mapped from those that have been partly drained is often problematic.

Some types of evergreen forested wetlands and temporarily flooded deciduous forested wetlands are difficult to identify even in the field. It can require extensive soil sampling to determine the limits of hydric soils. Despite the best efforts of photointerpreters, often only general wetland boundaries with omission errors can be produced. Through remote sensing the National Wetland Inventory has identified 51.7 million acres of forested wetlands or approximately 2.6 percent of the surface of the conterminous United States (Dahl and Johnson 1991). Nearly half of the wetlands are forested. The extent of the omission errors has not been estimated. Additional details on remote sensing of wetlands are provided in Wilen and Pywell (1992) and Wilen and Bates (1995).

INCREASING THE SCALE OF AERIAL PHOTOGRAPHY

Two important studies address this topic (Tiner and Smith 1992 and MacConnell et al. 1992). Tiner and Smith (1992) investigated scales from 1:58,000 to 1:12,000. MacConnell et al. investigated scales from 1:58,000 to 1:4,800. The larger the scale of the photography, the larger the area of the wetland signature on the photograph and the larger the canopy openings that allow the photointerpreter to see more of the saturated soil or understory wetland vegetation (Figure 1). The larger the scale the greater the stereoscopic exaggeration and thus the better the view the interpreter has of landscape and topographic position.

Generally, with increasing photo scale, wetland/upland boundaries are more refined and distinct, smaller polygons are identified, and forested wetlands are easier to identify but are still difficult. Forested wetland/upland boundaries are easier to delineate at 1:12,000 than 1:5,760 photography because, at 1:5,760 and larger scales the photointerpreter could not see the forested wetland for the trees. At 1:12,000 the interpreter was able to see the forested wetland boundary and not draw the boundary line on a tree-to-tree basis (MacConnell et al. 1992).

The amount of effort required to produce a standard National Wetlands Inventory map increases dramatically with increased scale (Table 1). The second consideration is how much of the increased detail can be effectively displayed at a scale of the standard NWI map which is 1:24,000 (see Figure 1). Lastly, and most importantly, forest wetlands larger, and in some cases much larger, than the minimum mapping unit are still not detected. The problem with identifying forested wetlands is spectral resolution, not spatial resolution. Increasing the scale of photographs only increases the spatial resolution.

Figure 1. Pen width and relative sizes of polygons at different scales of photography (MacConnell et al. 1992)

SCALE Representative Fraction	Verbal	PEN LINE WIDTH ON THE GROUND 3x0 (.25mm)	4x0 (.18mm)*
1:4,800	1"=400'	4.0'	2.8'
1:7,200	1"=600'	6.0'	4.2'
1:12,000	1"=1,000'	10.0'	7.0'
1:24,000	1"=2,000'	20.0'	14.0'
1:40,000	1"=3,333'	33.3'	23.3'
1:58,000	1"=4,833'	48.3'	33.8'

*Practical limit of fine pen point size for photo annotation which is wide enough on the larger scale photos to be easily discernible for transfer to base maps. 4x0 pens are normally used by NWI on 1:58,000 photos.

Table 1.   Summary of labor required at various steps (excluding ground-truthing and Regional Quality Control) to prepare a large-scale wetland map from different scales of photography. The symbol "x" is used to indicate ratios within a particular step, so 2x takes twice as much time as "x". Hours of labor for "x" in each step is designated in parentheses (Tiner and Smith 1992).

Scale of Aerial Photography (# photos)

Step	1:58K (3)	1:36K (6)	1:24K (12)	1:12K (42)
Data Preparation	0.5x	x (1 hr.)	2x	7x
Photointerpretation	0.9x	x (10.25 hrs.)	2.1x	4.4x
Quality Control (National PIQC)	x	x (0.5 hrs.)	2.5x	6x
Map Preparation	0.5x	x (28 hrs.)	2.2x	3.3x

EVALUATION OF OTHER SENSORS

As early as 1979, the National Wetlands Inventory evaluated 80-meter resolution multispectral scanner (MSS) data for its value in identifying and mapping wetlands, and over the years has looked at the new sensors as they have come on line.

President's Domestic Policy Council's Wetlands Task Force

The President's Wetlands Task Force requested that the Federal Geographic Data Committee's Wetland Subcommittee report on the application of satellite data for mapping and monitoring of wetlands. On January 14 and 15, 1992, the subcommittee held a meeting to discuss the current application of satellite data for mapping and monitoring of wetlands. The subcommittee invited top-level technical experts from the following organizations to address a preset list of questions and describe their experiences: Earth Observation Satellite Company; SPOT Image Corporation; Ducks Unlimited; U.S. Environmental Protection Agency, Environmental Monitoring Systems Laboratory; National Oceanic and Atmospheric Administration, Coast Watch, Change Analysis Program; Earth Satellite Corporation; U.S. Geological Survey, Earth Resources Observation System Data Center; and Maryland Department of Natural Resources.

The Wetlands Subcommittee reported that the detail and reliability of information derived from satellite data have steadily improved. These improvements include advancements in spatial and spectral resolution, georeferencing, and digital image processing techniques, along with growing experience using satellite data. Significant strides have been made in integrating ancillary data, such as soils and digital elevation models, into the classification of satellite data. This integration is dependent upon the use of geographic information system (GIS) technology. Stream gauging data and rainfall data are now being used to select the best scenes for wetland identification. Even with these improvements, satellite data can not match the accuracy of areal extent, classification detail, or reliability that can be extracted from conventional aerial photography using manual photo-interpretation techniques, such as those used by the U.S. Fish and Wildlife Service's National Wetlands Inventory Project. However, for some regions, satellite remote sensing may be the most cost-effective means for conducting reconnaissance wetland surveys.

The power of satellite imagery lies in its ability to be easily integrated with all other sources of data in a GIS, contributing to the accuracy of the GIS. The U.S. Department of Agriculture's Natural Resources Conservation Service (NRCS) believes that satellite technology can help to classify certain administrative classes of wetlands legislated by the Farm Bills of 1985, 1990, and 1996. Many other resource managers have complained that, in practical application, the promise of space-based remote sensing has not measured up to National Wetlands Inventory's actual performance. The subcommittee believes satellite data, when used in conjunction with NWI digital data produced through the use of aerial photography, can provide a tool for monitoring water levels in wetlands and monitoring the cover change of adjacent uplands. Synergistic effects created by combining both satellite data and NWI digital data have greater value than using either data source alone. Such data sets have the potential to be synoptic and accurate (Federal Geographic Data Committee 1992).

The President's Wetlands Task Force also directed the Wetlands Subcommittee of the Federal Geographic Data Committee to complete reconciliation and integration of all Federal agency wetland inventory activities. The Federal Geographic Data Committee's Wetlands Subcommittee developed a strategic interagency approach (FGDC, Wetlands Subcommittee 1994).

A working group was formed with representatives from the U.S. Department of the Interior (U.S. Fish and Wildlife Service (FWS) and U.S. Geological Survey (USGS)), the U.S. Department of Agriculture (Natural Resources Conservation Service (NRCS)), the U.S. Department of Commerce (National Oceanic and Atmospheric Administration (NOAA)), the Environmental Protection Agency, and the Maryland Department of Natural Resources.

Pilot Study. The working group began a pilot study to better understand the issues and problems associated with the data comparison task. Wicomico County, Maryland, was selected as the pilot because: (1) wetland data and other spatial data in digital form were available from the various government agencies, (2) the county's proximity to the Washington, D.C., area facilitated field analysis where necessary, and (3) the county has an abundance of forested wetlands, which are generally recognized as the most difficult wetland type to map.

Description of Wicomico County, Maryland.* The study area is on the Atlantic Coastal Plain of Maryland's eastern shore. Lithologically, this part of the Coastal Plain is composed of marine units of varying thicknesses. Clay, sand, and shells are the major deposits. Characteristically, the surface is of low elevation, usually between 0 and 25 feet. The low elevation and the broad smoothness of the region cause the streams to have a gentle gradient and the stream incision is minimal. The amount of stream incision directly affects the level of the water table; thus, the water table is high throughout the county. The low elevation and smoothness of the land surface combined with a gentle stream gradient and high water table result in a well-developed floodplain and extensive areas of wetlands. Small changes in elevation, microtopography, or parent material will determine whether a given site is wetland or upland.

Results. The study provided clear evidence that there were significant disagreements in wetland delineation among the various government wetland data sets and that some data sets (e.g., the National Wetlands Inventory maps) had significant omission errors while other data sets had significant commission errors. It was not possible to determine the accuracy of the data sets due to the lack of a standard to measure against. It was clear that no agency had solved the problem of mapping temporarily flooded deciduous forested or evergreen forested wetlands. Everyone either overmapped or undermapped these wetland types (Shapiro 1995).

Field of Dreams

Nine data overlays for Wicomico County were assembled in a GIS (Table 2). Because the scale, content, resolution, format, and collection methods for each of the data sets varied, the U.S. Geological Survey designed a GIS analysis interface (Sechrist 1995). It allowed the operator to view, manipulate and compare the data sets.

Table 2. Wicomico County Data Overlays

1. USGS, 1:100,000-scale digital line graphs
2. USGS, 1:250,000-scale land use/cover vector data
3. FWS, 1:24,000-scale wetland vector data
4. NOAA, 30-meter resolution raster wetland and upland data
5. NRCS, Natural Resources Inventory, point wetland and upland data
6. NRCS, 1:24,000-scale raster swampbuster wetland
7. NRCS, 1:20,000-scale soil vector data
8. MD, Department of Natural Resources (DNR), 4-foot resolution color infrared digital orthophoto quarter quadrangle images (DOQQ)
9. MD (DNR), wetland vector data registered to 1:12,000 scale DOQQ

*This description of Wicomico County, Maryland, is from an unpublished paper by Tera Paul, U.S. Geological Survey.

To this data set an independent contractor provided field verified data for 130 sites. A group of agency scientists made wetland determinations at 100-foot intervals on 11 transects across wetland and upland boundaries. Lastly, 10 wetland/upland boundaries were established in the field and surveyed. Groups of five ground water wells were placed across each of these surveyed wetland/upland boundary lines and their positions were also surveyed.

The problems all parties had identifying forested wetlands and the amount of data collected and entered into a GIS result in a kind of "field of dreams" that attracted several other groups to attack the problem of mapping forested wetlands with a variety of sensors.

Passive Airborne Microwave

The emission of radiation from the Earth's surface at microwave wavelengths is dependent on many environmental factors, including soil temperature, surface roughness, vegetation water content, soil water content, bulk density and soil texture. It is possible to estimate soil moisture content and water table depth by measuring the intensity of emitted microwave radiation in different wavelengths. Clouds and rain are the main sources of interference at wavelengths shorter than 5 mm. Radar, TV installations, and galactic and ionospheric radiation interfere with reception of wavelengths over 30 cm. Between those wavelengths, radiation is most sensitive to the water content in the soil.

To predict those variables from emitted radiation, it is necessary to collect data from 2 or 3 radiometers operating at different wavelengths between 0.5 and 30 cm. The portion of soil water content indicated by microwave radiation variability is the free water content, or water that is not bound, chemically or physically, to the soil particles. Free water is expressed as grams per cubic centimeter, which is roughly equivalent to percent of total volume (assuming a density of 1 g/cc). Saturation occurs when all the free space in the soil is occupied by water. Emitted microwave radiation was measured from an aircraft using radiometers equipped with antennas operating at different wavelengths. Two antennae (18 cm and 6 cm) were mounted externally to the bottom of the aircraft.

The tests conducted by Photo Science and Geoinformatic were sufficient to prove the technical feasibility of using airborne microwave radiometry for mapping soil moisture characteristics. The results of the tests, with respect to the potential of the technology for wetlands mapping, were both encouraging and disappointing. The spatial resolution of the data is clearly not sufficient for this approach to provide improvements over other existing methods of wetlands mapping for general inventory purposes. However, there is enough empirical correlation between the soil moisture characteristics as interpreted from the radiometer data, known wetlands and the orthophoto image to suggest that there may be applications for the technology to meet certain specific mapping purposes (PhotoScience 1993). The Department of

Transportation is investigating its use to attempt to separate uplands from areas that contain wetlands or potentially have wetlands.

Airborne Terrestrial Applications Sensor (ATLAS), Airborne Synthetic Aperture Radar (AirSAR), and Shuttle Imaging Radar - C (SIR-C)*

NASA's Commercial Remote Sensing Program in cooperation with the U.S. Environmental Protection Agency, U.S. Geological Survey, U.S. Fish and Wildlife Service, EarthSat Corporation, and the University of Colorado at Colorado Springs undertook a project to verify and validate the utility of commercially available satellite and airborne imagery processing techniques to accomplish less-expensive, more-reliable wetland maps. The project team had all the ancillary data collected for the earlier study plus 1:8,000 scale color infrared photography acquired over the study area. The remote sensing data collected is detailed in Table 3.

Table 3. Remote sensing data types used in wetland classification analysis

Test Data	Date of Acquisition	Total/Kinds of Channels	Spatial Resolution
ATLAS Daytime	4/5/95	15 channels in VIS, NIR, MIR, TIR[1]	2.5 m
ATLAS Pre-Dawn	4/6/95	15 channels in VIS, NIR, MIR, TIR	2.5 m
SIR-C	4/12/94	8 channels 4 polarities C and L bands	12.5 m
AirSAR	6/2/95	12 channels 4 polarities C, L, and P bands	9.0 m

[1]VIS, NIR, MIR, and TIR refer to visible, near infrared, mid-infrared, and thermal infrared spectral regions.

The characteristics of the SIR-C and AirSAR data are provided in Table 4. A description of the multi-sensor data stack is presented in Table 5.

---

*This description of the project is derived from an unpublished paper entitled "An Assessment of Remote Sensing/GIS Technologies for Delineation of Wetlands" prepared by the Commercial Remote Sensing Program Office of the National Aeronautics and Space Administration and Lookheed Stennis Operations both located at the John C. Stennis Space Center, Mississippi.

Table 4. Characteristics of radar (SIR-C and AirSAR) data

	SIR-C	AirSAR
Acquisition date	4/12/1994	6/2/1995
Projection	Ground range	Ground range
Wavelength (cm)	C band (5.7 cm) L band (24 cm) X band (3 m)[1]	C band (5.7 cm) L band (24 cm) P band (68 cm)
Line spacing (m)	12.5	8.23
Pixel spacing (m)	12.5	8.23
Incidence angle(^{0})	24.53	21.34
Polarizations	HH, HV, VH, VV[2]	HH, HV, VH, VV

[1]X band was acquired with C and L bands but not included in the data set from JPL.

[2]HH, HV, VH, and VV refer to horizontal send and receive, horizontal send and vertical receive, vertical send and horizontal receive, and vertical send and receive.

Table 5. Description of multi-sensor data stack

Number	Description	Center Wavelength (μm)
1	ATLAS band 1 (day mission, green band)	0.466 μm
2	ATLAS band 4 (day mission, red band)	0.637 μm
3	ATLAS band 6 (day mission, near infrared band)	0.773 μm
4	ATLAS band 11 (day mission, thermal band)	9.2 μm
5	ATLAS band 12 (day mission, thermal band)	9.9 μm
6	ATLAS band 11 (night mission, thermal band)	9.2 μm
7	ATLAS band 12 (night mission, thermal band)	9.9 μm
8	AirSAR PHH	68 cm
9	AirSAR PHV	68 cm
10	AirSAR PVV	68 cm
11	AirSAR LHH	24 cm
12	AirSAR CTP	5.7 cm
13	ATLAS day-night thermal difference	9.9 μm

<u>Analyses</u>. Data analysis was divided into two major components: delineation of wetland cover types and identification of general wetland boundaries. Traditional remote sensing classification algorithms (supervised and unsupervised) and neural network analyses were applied to the data sets to study cover type and wetness mapping capabilities of AirSAR, ATLAS, and SIR-C data. In addition, two alternative techniques (cluster busting and a hybrid analysis) were conducted. AirSAR, SIR-C, plus ATLAS visible, near infrared, and thermal (daytime and predawn) data were used as the primary input for the wetness analysis, while cover type analysis was performed on the multisensor data stack, ATLAS visible through near infrared, AIRSAR and SIR-C data.

<u>Conclusions</u>. The project yielded some encouraging results, particularly the AirSAR imagery, but the goal of producing more reliable wetland maps was not achieved. AirSAR was the best overall data source for detecting wetland versus upland areas. It is anticipated that a three component scattering model (surface scatter, canopy scatter, and double bounce scatter) could be helpful in reducing wetland omission errors by separating clearcut wetlands from uplands (Freeman and Durden 1992). Results suggest that ATLAS 2.5-meter spatial resolution tends to measure reflectance/radiance from individual trees instead of from larger habitat patches. The resampling of the ATLAS data from 2.5 to 9 meters enabled mixed covertypes to be more visually apparent on color infrared composite image displays. Suggesting again that spectral not spatial resolution is the limiting factor in remotely sensing forested wetlands.

Results of cluster separability analysis of the multisensor ATLAS/AirSAR data stack suggest that AirSAR data contributes more to spectral separability than ATLAS day and pre-dawn data. Comparison of ATLAS daytime thermal and pre-dawn thermal data types indicates daytime thermal data tended to have increased separability of wetness over pre-dawn thermal data. Neural network classifications corroborated this result. Neural network approach to wetness classification yielded similar results compared with traditional unsupervised classification techniques. Additional research is being conducted at the University of Colorado, Colorado Springs, on the use of neural nets on these data sets.

Wetland/upland classification analyses performed on AirSAR, SIR-C, ATLAS daytime thermal, and ATLAS pre-dawn thermal data layers indicate these data types have potential for identifying wetness in vegetated cover but do not contain the information content needed to discriminate wetland cover types to the level of detail used by the National Wetlands Inventory. However, both the ATLAS daytime and AirSAR data appear to have sufficient information to derive general land cover maps suitable for updating NWI cover-type change. Other studies have shown the promise of using satellites to monitor changes and losses of wetlands (EarthSat 1993).

<u>National Cooperative Highway Research Board (NCHRB)</u>

The NCHRB has funded a project entitled "Remote Sensing and Other Technologies for the Identification and Classification of Wetlands." The contractor for the project is Normandeau Associates. The goal is to define methodologies for efficiently and effectively identifying, classifying, and locating wetlands within potential highway corridors and alignments beyond the detail and accuracy of National Wetlands Inventory maps. The problem is that highway corridors and alignments have been selected based on National Wetlands Inventory maps to find after field work that some forested wetlands were unmapped.

In one case, after field investigations it was determined that an unselected corridor in fact had fewer wetlands than the one selected using National Wetlands Inventory maps. The contractor expects an evaluation of new technologies and geographical information systems in combination with other existing information resources will be required. The contractor has visited the Wicomico field of dreams and is aware of what has already been investigated.

Additional Studies

Two additional studies are underway; the first is using very high resolution digitized multitemporal color infrared photography and the second is using National Technical Means.

DISCUSSION

Why can't we find a solution to the problem of photointerpreting some types of forested wetlands? Radiometric data are often hard to evaluate because there are numerous variables that prevent correlation between radiometric response and ground phenomena. They include the amount of energy reaching the sensor, illumination of the object being sensed, atmospheric variables, differences in reflectance due to season and growth stage of vegetation, sensor ability to capture and record the data, etc.

Remote sensing specialists can deal with these problems, but the photointerpreters have misled those trying to help us by constantly referring to "signatures" in photointerpretation. I have used the term in this paper, but the term "signature" is misleading. It implies a distinctive mark that is unique and consistent. Wetland signatures on color infrared film are not unique or consistent. Signatures vary according to time of day (affecting shadow and glare), season (including degree of shadow, endless stages of leaf-out and aging, short-term and long-term weather patterns, combinations of weather patterns, snow cover, ice cover and water cover that is sometimes so extensive or deep early in the year, that you can't delineate wetlands), emulsion (which can be predominantly blue, green, pink, red, or purple in tone or the tone might be washed out or too dark). Photointerpreters must deal with these variations in signatures. It takes expertise, experience, and the power of reasoning to transcend these variations. If we can't solve it, what can we do to reduce the problem?

o   Film development and duplication are two areas where there appears to be potential for major technological advancements. Often there is a significant difference between whether the duplicated photos are processed as a single print or as an entire roll. It appears that color infrared film and duplicates can be developed to enhance or mask wetlands. The reasons for the problems introduced in the development and duplication process are not fully understood. Important insights into some of these problems are provided in Hershey and Befort (1995).

o Soils maps are an important source of collateral data for mapping wetlands. Checking for hydric soils is presently a cumbersome and time-consuming process. A product that relates soils data to the USGS quad map would save time in the photointerpretation process. The photointerpreter could easily relate the land-surface form features (contours) in relation to the hydric soils units. Even having the normal NRCS soils map in a quad style format would be helpful.

o Due to the number of variables, subtleties and exceptions that go into every photointerpreted delineation and classification, it appears that any solution must involve a human. The solution needs to be computer assisted where the computer assists the human.

o Surface water often blends in with shadows on color infrared film. A technique needs to be developed to separate shadows from open water. On the edges of color infrared photographs you can see the glare of the sun as it is reflected off water in open wetlands. This same phenomenon can be experienced from an airplane. AirSAR L band radar with a HH (horizontal send and receive) polarization revealed drainage patterns due to the strong double-bounce return. Spectral reflectance from the smooth, highly reflective water surface is bounced back toward the receiver by vertically oriented trunks. A sensor that takes advantage of these phenomena would be useful for identifying flooded forested wetlands.

CONCLUSION

The problem of photointerpreting certain types of forested wetlands has not been solved. We have learned that the problem is spectral not spatial resolution. In fact, at some point too much spatial resolution worsens the problem. Potential new data sources such as the Airborne Terrestrial Applications Sensor (ATLAS), Airborne Synthetic Aperture Radar (AirSAR), Shuttle Imaging Radar-C (SIR-C), Airborne Multisensor Pod System (AMPS) and the civilian use of National Technical Means are becoming increasingly accessible for evaluation. In addition, several new high resolution remote sensing satellites are just over the horizon. The full potential of the existing sensors has not been adequately explored. A Geographic Information System (GIS) provides the tools to georeference, quantitatively compare, analyze, visualize, tabulate, and produce composite maps necessary for evaluation. Without a GIS, the tasks of quantitatively comparing various maps and remote sensing data would be impractical. Before the remote sensing satellite builders can help us solve our problems, we must learn to be able to clearly communicate with them. Terms like photo signatures cause confusion. Signature implies something distinctive, unique and consistent and, in reality, photo signatures are not. They vary according to a multitude of variables already discussed.

Our investigations have shown that wetland mapping omission errors result from photointerpreters not field checking what appear to be apparent upland "signatures." Wetland "signatures" and confusing "signatures" are investigated but apparently obvious upland "signatures" are ignored. The lesson is that some percentage of the areas with a potential wetland landscape and topographic positions need to be field checked even though they have what appears to be an obvious upland "signature." Note how many times I have used the misleading term "signature." In order to assess accuracy, you need a standard to compare against. Comparing data sets do not result in an accuracy assessment. They result in a comparison.

A systematic analysis of existing data sources needs to be conducted before their full potential can be appreciated. Testing should be undertaken to determine the most effective resolution for various applications. Builders of satellites must be careful that they do not collect data at such fine resolutions, e.g., one or two meters, that it adds to rather than reduces classification confusion when automated classification approaches are utilized.

REFERENCES

Cowardin, L.M., V. Carter, F.C. Golet and E.T. LaRoe 1979, Classification of Wetlands and Deepwater Habitats of the United States: USDI Fish and Wildlife Service, Washington, D.C., FWS/OBS-79/31: 103 pp. http://www.nwi.fws.gov/classman.html

Dahl, T.E. and C.E. Johnson 1991, Status and Trends of Wetlands in the Conterminous United States, Mid-1970's to Mid-1980's: U.S. Department of the Interior, Fish and Wildlife Service, Washington, D.C.: 28 pp.

EarthSat, Inc. 1993, Cross-correlation Analysis of Wetland Change: Rockville, Maryland: 40 pp.

Federal Geographic Data Committee 1992, Application of Satellite Data for Mapping and Monitoring Wetlands - Fact Finding Report; Technical Report 1: Wetlands Subcommittee, FGDC. Washington, D.C.: 32 pp. plus Appendices.

Federal Geographic Data Committee, Wetlands Subcommittee 1994, Strategic Interagency Approach to Developing a National Digital Wetlands Data Base (second approximation): Federal Geographic Data Committee, Washington, D.C. ftp://www.fgdc.gov/pub/general/subcommittees/wetlands/ interagency.data.base.approach.8294.txt

Freeman, A. and S. Durden 1992, A Three-component Scattering Model to Describe Polarimetric SAR Data, Radar Polarimetry, SPIE Vol. 1748: pp. 213-224.

Hershey, R.R. and W.A. Befort 1995, Aerial Photo Guide to New England Forest Cover Types: Gen. Tech. Rep. NE-195. Radnor, PA: US Department of Agriculture, Forest Service, Northeastern Forest Experiment Station: 70 pp.

MacConnell, W.M., J. Stone, D. Goodwin, D. Swartout and C. Costello 1992, Recording Wetland Delineations on Property Records. The Massachusetts Experience 1972 to 1992: Department of Forestry and Wildlife Management, University of Massachusetts, Amherst, Massachusetts: 60 pp.

PhotoScience, Inc. 1993, Report on Passive Airborne Microwave Radiometer Technology and Potential Applications for Wetlands Mapping: Gaithersburg, Maryland: 23 pp.

Sechrist, D.R. and B.O. Wilen 1995, Wetlands Evaluation: A Geographic Information System Analysis of Multi Agency Wetlands Data for Wicomico County, Maryland: Selected Papers in the Applied Computer Sciences, V 4: American Society for Photogrammetry and Remote Sensing/Congress on Surveying and Mapping, Annual Convention and Expositions Technical Papers, Vol. 2: pp. 348-366.

Shapiro, C. 1995, Coordination and Integration of Wetland Data for Status and Trends and Inventory Estimates: Federal Geographic Data Committee Wetlands Subcommittee, Technical Report 2: 210 pp.

Tiner, R.W. 1990, Use of High-Altitude Aerial Photography for Inventorying Forested Wetlands in the United States: Forest Ecology and Management, 33/34: pp. 593-604.

Tiner, R.W. and G.S. Smith 1992, Comparisons of Four Scales of Color Infrared Photography for Wetland Mapping in Maryland: National Wetlands Inventory, Fish and Wildlife Service, U.S. Department of the Interior.

Wilen, B.O. and M.K. Bates 1995, The U.S. Fish and Wildlife Service's National Wetlands Inventory Project: Vegetatio, Klumer Academic Publishers, Belgium, Vol. 118: pp. 153-169.

Wilen, B.O. and H.R. Pywell 1992, Remote Sensing of the Nation's Wetlands, National Wetlands Inventory: Proceedings; Fourth Biennial Forest Service Remote Sensing Applications Conference, 6-10 April 1992, Orlando, Florida.

EXPLORATORY USE OF SYNTHETIC APERTURE RADAR (SAR): CONCLUSIONS FROM FOREST, WETLAND, AND GEOMORPHIC MAPPING

Cynthia L. Williams
Institute of Northern Forestry
Cooperative Research Unit
PNW Research Station
USDA Forest Service
308 Tanana Drive
Fairbanks, AK 99775

John Hampshire
Dept. of Electrical & Computer Engineering
Carnegie Mellon University
Pittsburgh, PA 15213-3890

ABSTRACT

In interior Alaska and elsewhere, SAR has proven to be an important remote sensing tool for mapping forest successional stage, biomass, and river geomorphology. However, the wide variety and complexity of the analytical methods used at different sites have made difficult any predictions of the likely utility of SAR at new sites. We use fast, uniform, interactive image analysis techniques to examine the following questions: Can SAR imagery be useful for quick, exploratory analyses of unfamiliar sites? To what extent can we generalize across sites about the utility of each of the three bands and four polarizations available from AIRSAR imagery? Image interpretation and exploratory classification of AIRSAR imagery from six unfamiliar boreal, temperate, and tropical sites are useful for mapping of major landscape units and unusual habitats. Different types of ecological information dominate the AIRSAR imagery at our sites from Alaska, Canada, Michigan, and Peru. Although comparisons of knowledge-based classifications and exploratory classifications confirm the importance of ground-based data for mapping, exploratory classification can very quickly delineate major landforms, map wetlands, and distinguish most major vegetation types. At one boreal site, we distinguish eight different structural wetland types with exploratory classifications. We present generalizations about the utility of specific AIRSAR bands for study of forest, wetlands, and geomorphology.

INTRODUCTION

Resource managers, ecologists, and other field scientists must consider a variety of practical considerations when choosing among sources of remote sensing imagery. Managers prefer sensors which have been of proven utility for the specific uses of interest and those sensors which are of predictable utility at new sites. It is sometimes important that a sensor have potential for revealing something new about a site, rather than merely extrapolating from known sites to larger areas. Sometimes the effectiveness of a remote sensor is dependent upon having a certain amount of prior knowledge about the site.

The research we are describing here began with research at Bonanza Creek Experimental Forest and the Bonanza Creek LTER site in interior Alaska. The uses of Synthetic Aperture Radar (SAR) for ecological and natural resource applications have been under evaluation for some time. Since 1988, investigators from the Institute of Northern Forestry and the Jet Propulsion Laboratory have collaborated on many studies using SAR for ecological research. The early studies at Bonanza Creek and elsewhere typically have involved intensive study of a few sites and intensive analysis to demonstrate that specific ecological parameters can be extracted from radar backscatter. But, until now we haven't been able to say how generalizable our findings are to other sites and other ecosystems.

This ongoing research demonstrates the value of using cross-site comparisons of SAR imagery both for evaluating SAR imagery and its potential at new sites and for learning something new about individual landscapes. We use cross-site comparisons and rapid, uniform methods to address the following questions: What types of information emerge from different SAR bands when classifications are based on no prior knowledge of land cover? If we do not manipulate the imagery into estimation of any particular parameters, what types of information dominate the images? Can we generalize about the utility of different SAR bands across sites, particularly with respect to forests, wetlands and geomorphology? In a practical sense, these are the types of information potentially available to users before visiting new sites, so these are the ways in which SAR can contribute to the planning of field sampling and biological inventories. Our goal is to enable users to choose among bands and to potentially choose among sources of SAR imagery which use different sensor configurations.

BACKGROUND

SAR is an active microwave remote sensor which can utilize four different bands or wavelengths of radiation. SAR imaging systems can transmit and receive signals of different polarizations; backscatter intensity and polarization is dependent on environmental conditions and surface geometry. A major advantage over optical sensors is that SAR imaging is not influenced by sun angle, darkness, or cloud cover. SAR imagery is currently available from a variety of satellite and aircraft systems. Aircraft-borne AIRSAR imagery uses C-band (about 6 cm), L-band (15-30 cm), and P-band (30-100 cm).

Capabilities of SAR have been documented in a variety of ecosystems and for a wide variety of purposes, including mapping of perpetually cloud-covered regions (Pope et al., 1994), land cover classification (deGrandi et al., 1994; Dobson et al., 1995a), measurement of biomass (Dobson et al., 1995b; Kasischke et al., 1994), mapping inundation and flooding events (Melack et al., 1994; Hess et al., 1995), monitoring of freeze-thaw events (Rignot and Way, 1994), crop inventory (Lemoine et al., 1994), estimation of soil moisture (Lin et al., 1994; Dobson and Ulaby, 1986), and a variety of ice, earth science, and marine applications (Evans, 1995) . On the Tanana River floodplain in interior Alaska, we have used SAR for classification of successional stages, biomass estimates, mapping of flooding, detection of freeze-thaw transitions, and mapping of structural wetland types (Williams et al., 1995; Rignot et al., 1994 a,b,c).

METHODS

Our objective was to determine what types of information emerge from different SAR bands, both with direct image interpretation and with image classifications based on no prior knowledge of land cover. We used visual interpretation and classification of multi-band, multi-polarization AIRSAR images from six different sites in North and South America. This experimental approach relies on eosMapper software, under development by Hampshire, and its unusually rapid viewing and classification capacities. The steps in classification are as follows: Twenty minutes are spent on each image defining its visually distinctive components, or training areas. For each training type, the distribution of backscatter for each polarization is fit to log normal, Nakagami-M, Rayleigh, or Rician parametric models; model fit reflects both color and texture of the training areas. The parametric models are then applied to the original image at the scale chosen by the investigator. We limited the total time spent generating each classification to thirty minutes. After classification, we used field checking or inspection of land cover maps or aerial photography to identify the classified features. For each site we summarized the information derived from each band. By comparing and summarizing across sites we could then predict the likely utility of this imagery to other users at other sites.

STUDY SITES

BNZ: This boreal site along the floodplain of the Tanana River in interior Alaska includes the Bonanza Creek LTER site. Successional processes in floodplain forests at BNZ have been the focus of more than 30 years of ecological research. The Tanana River is a glacial fed river, in transition at BNZ from a braided to a meandering river.
TFL: This boreal site encompasses the Tanana River floodplain and the Tanana Flats wetlands complex fed by groundwater from the Alaska Range to the south. It includes a wide variety of both wetland and forest types, maintained by hydrologic processes, river depositional processes, and fire.
TOTAT: This boreal site extends from the Tanana River (75 km downriver from BNZ) south along the Totatlanika River. Aerial photos indicate forest and wetland complexes different from those at TFL.
MI: This temperate site on the upper peninsula of Michigan includes forests in the Hiawatha National Forest. Forest stands here have been intensively managed in a variety of ways. Uses of SAR imagery at this site have been intensively studied by Ulaby and others at the University of Michigan.
SFEN: This boreal site encompasses the BOREAS South Fen site in Saskatchewan, Canada. It also includes both a wide variety of boreal wetlands and forests with different logging histories.

MANU: This tropical site in Manu National Park, Peru includes a variety of tropical forest types along a river similar in structure to the Tanana River in Alaska.

RESULTS

C-Band Imagery

All of the sites included in this study are forested to some extent, but C-band AIRSAR imagery provides little information about these forests. For the most heavily forested sites, MI and MANU, C-band provides the least information, clearly separating only vegetation and water. Structural and compositional differences among forest stands are not revealed at BNZ, TFL, TOTAT, or SFEN, even though these sites contain complex mosaics of stand types. However, at most sites C-band does delineate clearings, whether caused by logging, wildfire, hydrologic conditions or vegetational succession. At BNZ and TFL the 10-year old Rosie Creek burn is seen most clearly with C-band, although its backscatter in some places resembles wetlands. At SFEN, clearcuts with varying amounts of regrowth are separated, again with some confusion between wetland and clearing. Understanding of specific landscapes aids in distinguishing the different types of clearings.

 At most sites C-band SAR imagery shows fine-scale geomorphic details by scattering off the top surfaces of vegetation. These details do not show in classifications. At BNZ, TFL, and TOTAT cross-cutting depositional relationships of the river floodplain are very clearly seen from the imagery. At these sites and at SFEN, watercourses and drainages are delineated. At BNZ, C-band shows stand management units and old forested dune systems.

C-band classifications clearly separate open water, whether lake, river, or wetland, from vegetation. At BNZ, the classification provides a detailed map of floodplain wetlands. At MI open water is confused in some places with pavement. At TOTAT and TFL, C-band imagery highlights a total of 5 unique wetland areas; these areas are not distinguishable on aerial photography as different from other wetlands in the scenes.

In summary, fine-scale detail dominates C-band imagery. Depending on the site, this can include geomorphic details revealing landscape history, current watercourses and drainages, distribution of open water wetlands, or management history. Because C-band does not penetrate vegetation canopies, it does not provide structural and compositional information about forests. Much of the information C-band imagery provides will be lost in classification procedures. Interpretation of C-band imagery may require more knowledge of landscape function that for the other bands.

L-Band Imagery

L-band imagery provides much greater detail about forests at these sites than does C-band. Classifications also reflect more of this information with L-band than with C-band. At TFL, BNZ, and TOTAT high biomass white spruce and balsam poplar stands on the floodplain are clearly separated from low biomass black spruce stands. At TFL, mixed white spruce-balsam poplar stands are clearly distinguished from pure white spruce; white spruce stands also separate out into several density or biomass categories. Pure balsam poplar stands are visually distinct on the image, although this distinction was not included in the training areas. Alder stands are classified with low density white spruce. Shrublands on the older floodplain are clearly separable from forests, and wetland areas of the Tanana Flats are distinct from the shrubby and forested islands. At the SFEN site, shrublands are clearly separated from most forests, although conifer/aspen mixtures are separated from shrublands only by landscape position. Also at the SFEN site, four different kinds of wooded wetlands are classified; these belong to one category on Forestry Canada maps. Clearcuts at BNZ and SFEN are clearly distinguished at L-band. Differences in regrowth are apparent at both sites. However, the Rosie Creek fire is less conspicuous, and appears similar to the recent floodplain.

At BNZ and TFL, L-band classifications clearly stratify the landscape into upland, old floodplain, recent floodplain, and wetland complex. TOTAT is a more diverse landscape and its stratification is not as clear at L-band as at P-band. At SFEN major landscape units are more apparent than with C-band; again, P-band is even more useful for this. At all these sites, high biomass forests are very clearly separated from lower biomass vegetation. Large alluvial fans at BNZ are most clearly delineated with L-band, as are old forested dunes. However, these do not dominate the classifications.

At TOTAT and TFL, classifications separate watercourses identified with C-band into those bounded by woody and non-woody vegetation. At TOTAT one additional distinct wetland is identifiable. The L-band SAR penetrates more deeply through wetland vegetation to reveal inundation patterns in the three wetland types visible with C-band. At SFEN, this penetration through vegetation highlights the wetlands much more than at C-band, but they are still only distinguishable from clearcuts by size and shape.

At most of our sites, L-band imagery is dominated by both vegetation structure and geomorphic features. The usefulness of this imagery varies a great deal from site to site depending on both the specific vegetation and the scale of geomorphic features. The fine scale geomorphic detail evident in C-band imagery is lost at L-band. Narrow watercourses may be confused with shrublands, floodplain history is less apparent, and open water is no longer separable from vegetated wetland. New wetlands can be identified, and inundation information becomes available for previously identified wetlands.

P-band imagery

P-band imagery also provides a great deal of information about forest composition and structure at these sites, and much of this information is different from that revealed with L-band. At BNZ and TFL, alder stands are distinguished from other woody vegetation; at these sites most other forest detail appears at L-band. At SFEN, P-band classification seems more strongly tied to biomass than to species composition. Still, jack pine, larch/black spruce/jack pine, aspen/black spruce/jack pine, aspen/jack pine, wooded wetlands, and open wetlands are distinguishable with varying degrees of accuracy. At the densely forested MI site, P-band allows discrimination of 13 landscape units, corresponding primarily to vegetation types defined by composition and structure. At Manu, the other densely forested site, 8 landscape units were classified. These units correspond to river and to major vegetation units, including *Cecropia* forest, palm forests, deciduous forest, evergreen forest, open forests with many dead trees and lianas, and early floodplain regrowth.

At the SFEN site, the landforms underlying the vegetation appear most clearly at P-band. At TFL the stratification of the landscape into wetlands complex, floodplain, older floodplain, and uplands dominates the classification. Large alluvial fans also dominate the TFL classification. P-band imagery highlights differences between the wetland complexes at TOTAT and TFL. At MANU and MI, information about landforms is mostly obscured at P-band by vegetation.

P-band SAR penetrates furthest into vegetation; as a consequence additional kinds of wetlands are distinguished. At all sites, wetlands appear larger at P-band because any water underlying fringe vegetation is detected. At TFL and TOTAT, floating *Menyanthes* mat vegetation is evident, although it was not detected as wetland with other bands. Previously identified wetlands can be divided into those with standing water and those without. At TOTAT, the distinctions between those wetlands and watercourses bordered with woody vegetation and those unbordered are most dramatic at P-band.

Depending on the site, P-band imagery may be dominated by vegetation or landform information. Vegetation biomass appears as an important factor in the classifications for all these sites. This imagery can provide important structural information about wetlands, and allows identification of wetland types, especially floating mats, not detected at shorter wavelengths.

GENERALIZATIONS FOR LANDSCAPE MAPPING WITH SAR

Forest Vegetation:
L and P band, alone or in combination, give the greatest information about forest structure and composition. The combination frequently gives greater discriminatory power.

Geomorphology:
The scale of geomorphic processes of interest, if known, determines the most useful radar wavelengths. C-band (short wavelength) best shows floodplain depositional relationships and watercourses. At the longer wavelength L and P bands, larger geomorphic features such as alluvial fans or dune systems can dominate images and classifications.

Wetlands:
C, L, and P band images each contribute uniquely to mapping of wetlands. Together, they can be used to distinguish as many as 8 different types of wetlands in a single scene. The utility of the three bands will vary depending on the wetlands present in a particular landscape.

DISCUSSION

Evaluation of SAR imagery for ecological use requires both quantitative accuracy assessment for classifications and investigations of the generality of results across a wide variety of landscapes. This investigation demonstrates that different types of information (about wetlands, canopy vegetation, or geology, for instance) can dominate imagery from different sites. Microwave backscatter is strongly dependent on the size, orientation, and moisture status of the components of a landscape. These components are rarely well understood in advance. For this reason, the information to be derived by use of any one SAR sensor configuration may not always be predictable. This unpredictability can be viewed either as a problem, hindering production of global scale products, or as an opportunity to gather otherwise unavailable new information about each new location. The value of SAR has already been demonstrated for a wide variety of ecological uses. Our investigation demonstrates the further usefulness of SAR for assessment of unfamiliar sites, and it begins analysis of the extent to which utility of SAR can be predicted.

Our cross-site comparisons do allow us to formulate generalizations about the utility of different SAR bands. In addition our results show for both familiar and unfamiliar sites the importance of interactive analysis with direct involvement of resource managers, ecologists, or others knowledgeable about landscapes and vegetation.

CONCLUSIONS

Cross-site comparisons of AIRSAR imagery enable us to make predictions about the utility of different SAR bands for study of wetlands, forests, and geomorphology. At some sites one band is clearly most valuable on first analysis. For heavily forested sites, this is usually P-band. At other sites, most dramatically at wetland-dominated sites, each band provides a unique contribution to landscape classification. We demonstrate the value of SAR imagery as an exploratory tool for unfamiliar sites. The imagery and the techniques we use are valuable tools for rapid preliminary mapping and for landscape stratification prior to field sampling, biodiversity inventory, or detailed mapping.

Our analyses also highlight the value of fast, flexible, interactive methods of SAR image analysis both for exploratory analyses and for more detailed classifications and mapping. The techniques we have used can be used with little training and without any knowledge of a specific site or even of landscapes in general. They allow rapid comparisons of sites. These techniques allow direct, interactive involvement of resource managers, ecologists, and other field scientists in image classification and interpretation. This dramatically increases the usefulness of SAR imagery, because knowledge of landscape history and dynamics, and of vegetation composition and structure can be immediately incorporated into analysis. This increases the potential for gaining new insights from SAR rather than just using it as a tool for extrapolation.

ACKNOWLEDGMENTS

We thank Reiner Zimmermann for assistance in interpretation of the Manu National Park classification. We thank Leland Pierce for assistance with interpretation of the Michigan classification.

REFERENCES

deGrandi, G, G. LeMoine, H. de Groof , C. Lavalle, A. Sieber. 1994. Fully polarimetric classification of the Black Forest MAESTRO 1 AIRSAR data. Int. J. Remote Sensing 15: 2755-2775.

Dobson, M.C., F.T. Ulaby, and L.E. Pierce. 1995a. Land cover classification and estimation of terrain attributes using synthetic aperture radar. Rem. Sens. Environ. 51: 199-214.

Dobson, M.C., F.T. Ulaby, L.E. Pierce, K. Bergen, and K. Sarabandi. 1995b. Land cover classification and biomass estimation with SIR-C/X-SAR. . IEEE Trans. Geosci. Rem. Sens. (in press).

Dobson, M.C. and F.T. Ulaby. 1986. Preliminary evaluation of the SIR-B response to soil moisture, surface roughness, and crop canopy cover. IEEE Trans. Geosci. Rem. Sens. 24: 517-526.

Evans, D.L., ed. 1995. Spaceborne Synthetic Aperture Radar: Current Status and Future Directions: A Report to the Committee on Earth Sciences, Space Studies Board, National Research Council. NASA Technical Memorandum 4679.

Hess, L., J. Melack, S. Filoso, and Y. Wang. 1995. Delineation of inundated area and vegetation along the Amazon floodplain with the SIR-C synthetic aperture radar. IEEE Trans. Geosci. Rem. Sens. (in press).

Kasischke, E., N. Christensen, E. Haney, and L. Bourgeau-Chavez. 1994. Observations on the sensitivity of ERS-1 SAR image intensity to changes in aboveground biomass in young loblolly pine forests. Int. J. Rem. Sens. 15:3-16.

Lemoine, G., G. deGrandi, and A.J. Sieber. 1994. Polarimetric contrast classification of agricultural fields using MAESTRO 1 AIRSAR data. Int. J. Rem. Sens. 15:2851-2869.

Lin, D., E. Wood, P. Troch, M. Mancini, T. Jackson. 1994. Comparisons of remotely sensed and model simulated soil moisture over a heterogeneous watershed. Rem. Sens. Environ. 48:159-171.

Melack, J., L. Hess, S. Sippel. 1994. Remote sensing of lakes and floodplains in the Amazon basin. Rem. Sens. Rev. 10:127-142.

Pope, K., J. M. Rey-Benayas, and J. Paris. 1994. Radar remote sensing of forest and wetland ecosystems in the Central American tropics. Rem. Sens. Environ. 48: 205-219.

Rignot, E. and J.B. Way. 1994. Monitoring freeze-thaw cycles along north-south Alaskan transects using ERS-1 SAR. Rem. Sens. Environ. 49: 131-137.

Rignot, E., J.B. Way, et al. 1994a. Monitoring of environmental conditions in taiga forests using ERS-1 SAR. Rem. Sens. Environ. 49:145-154.

Rignot, E., J.B. Way, C. Williams, and L. Viereck. 1994b. Radar estimates of above-ground biomass in boreal forests of interior Alaska. IEEE Trans. Geosci. Rem. Sens. 32:1117-1124.

Rignot, E., C. Williams, J.B.Way, and L. Viereck. 1994c. Mapping of forest types in Alaskan boreal forests using SAR imagery. . IEEE Trans. Geosci. Rem. Sens. 32: 1051-1059.

Williams, C. L. , L. A. Viereck et al., 1995c. Monitoring, classification, and characterization of interior Alaska forests using AIRSAR and ERS-1 SAR. Polar Record 31: 227-234.

THE SOUTHWEST COLORADO VEGETATION CLASSIFICATION PROJECT: AN EXAMPLE OF INTERAGENCY COOPERATION

Melinda L. Walker
Bureau of Land Management
National Applied Resource Sciences Center
Denver Federal Center Building 50, Denver, CO 80225
Allen E. Cook
TRW (Bureau of Land Management Operations)
200 Union Blvd., Lakewood, CO 80226
James R. Ferguson
Bureau of Land Management-Montrose District Office
2505 S. Townsend Ave., Montrose, CO 81401
Suzanne M. Noble
TRW (Bureau of Land Management Operations)
200 Union Blvd., Lakewood, CO 80226

ABSTRACT

The Southwest Colorado Vegetation Classification Project is an interagency effort to produce a general level vegetation map of the southwestern third of Colorado that is Geographic Information System (GIS) compatible to all the cooperators. Initiated and coordinated by the Montrose District Office of the Bureau of Land Management (BLM), the project involves eight federal and state agencies who all share a common need. The participants developed a hierarchial classification scheme acceptable to all and will receive both the raw satellite data and the classified data. The vegetation map will be derived from the classification of Landsat Thematic Mapper (TM) data acquired for the project area in both the spring and fall and subset by watershed boundaries. The classification employs unsupervised classification techniques coupled with field data collected by staff from all the agencies involved. The final data set will include attributes and metadata.

The paper describes the two important aspects of the project--the coordination and data processing. The project management portion includes a history of the project; the development of the Memorandum of Understanding, the project plan, and the classification hierarchy; data sharing; personnel training; funding; and project status. The data processing portion includes an overview of the project area, the image processing techniques used, the incorporation of ancillary data sets, field data collection methods, and data validation. Lastly, the paper explains the status of plans underway to extend the project into the rest of the state.

INTRODUCTION

The Bureau of Land Management's Montrose District encompasses the entire southwestern corner of the state of Colorado. Its landscape is diverse and includes valley grasslands, timbered slopes, 14-thousand-foot peaks, alpine meadows, and desert-like arid ecosystems. In late 1991, the Montrose District Office (MDO) initiated an effort to develop a comprehensive vegetation map for public lands within its boundaries. The MDO found that such a data set did not exist within state or

74

federal land management agencies and vegetation data that was available was out of date, incompatible with agency needs, or was site specific. Additionally, there was insufficient data available on non-public lands necessary to do valid ecosystem modeling and management. Compounding matters, the MDO found that development of a comprehensive land cover data set was beyond the funding and workload capacity of the office. To rectify the gap in information, and stay within existing capabilities, the MDO took the lead in initiating a cooperatively-funded project to map land cover within and immediately adjacent to their District boundaries. This project has two primary objectives: (1) to obtain a current map and GIS database of vegetation types for southwestern Colorado derived from satellite imagery data and (2) to lay the foundation for future cooperation among the various contributors based on ecosystem management activities.

This initiative was launched in fiscal year 1992 with preliminary staff work by MDO personnel that included contacts with the remote sensing staff at the, BLM's Service Center and any federal or state agencies with management activities in the area. As discussions evolved, the scope of the effort increased with the recognition of the potential benefits to be derived from interagency cooperation. Personnel from the Bureau of Reclamation, Colorado Division of Wildlife, U. S. Fish and Wildlife Service, National Biological Survey, National Park Service, Natural Resources Conservation Service, and U. S. Forest Service agreed to cooperate with the BLM on this project. The ultimate benefit to future ecosystem management initiatives is the use of a common vegetation classification as the basis for communication and coordination with other agencies.

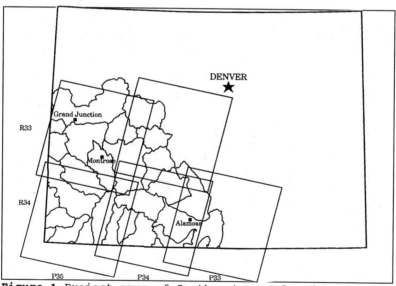

Figure 1 Project area of Southwestern Colorado, including USGS hydrologic unit watersheds and Landsat TM scene boundaries.

75

OBJECTIVES

The project has been progressing in two phases.

Phase I
The objective of this phase is to provide a vegetation data set for all of southwestern Colorado derived from Landsat TM 7-banded imagery for interagency use and development in landscape-level ecosystem management activities. The final classification is designed to incorporate individual group or agency needs, developed to the lowest common denominator of detail, and is useable on GIS by all project cooperators. A vegetation classification scheme was developed by an interagency subcommittee, incorporating several classification schemes currently in use. The scheme adopted for the project was agreed upon by all project cooperators, and is designed to "crosswalk" through a relational database with other hierarchial schemes, including the Federal Geographic Data Committee (FGDC) draft vegetation classification scheme.

Phase II
This phase is intended to add detail to portions of the vegetation data set created in Phase I using ancillary data to meet the more detailed needs of various agencies. This phase will include data archiving and relational database development to make other data sets available to all the users

METHODS

Landsat TM data represents the most efficient multispectral coverage of the study area. Data was purchased from EOSAT, using the MOU to ensure each contributor has full use of the original data for other analyses. Under the MOU terms, all project participants are entitled to possess and use a copy of the data sets purchased for this project, and data usage is not restricted to this project. The area of the project is over 7.8 million hectares, covering the southwestern portion of Colorado (Figure 1). This area includes both public and private land ownership, with public lands under the administration of multiple state and federal agencies. The TM data purchased for the project includes spring and fall data for five TM scenes, all from 1993. The original TM satellite data was rectified to the UTM projection and was then subset into manageable portions using the U. S. Geologic Survey (USGS) hydrologic unit polygon data. These subset areas from spring and fall were then merged into one image using TM bands 3, 4, and 5 from each image to constitute a six banded image for each watershed. Each watershed was then classified, using a standard non-parametric unsupervised classification algorithm, into general vegetation categories according to multispectral variability in the data. The original resolution of 25 meters is being preserved in all produced data sets, and is considered the minimum mapping unit for the project.

Field data collection for validation and accuracy assessment has and is being conducted during the 1995 and 1996 field seasons by field crews from the participation agencies. Color plots of the original TM data are provided to field personnel for field navigation and data collection. To avoid excessive expenses for field verification, field crews are asked to collect field data concurrently with their regular field activities whenever possible. A simple field form (Figure 2) has been developed to aid field data collection efforts. Whenever possible, field crews are expected to use GPS equipment to fix the sample location, and to attach a color photograph of the location to the field data form. The photographs are then scanned and become part of the database containing the field data. As a result, a unique database is being created with accurate locations that enable the data set to be used and updated in the future.

Principal project components are:

Development of an MOU: The MOU establishes the framework for cooperative data acquisition, development and sharing. This step also included establishing a Cooperative Agreement with the Colorado Division of Wildlife and writing a project plan.

Data Acquisition: Five TM scenes provide complete coverage of the project area (Figure 1). Two dates (early summer and mid-fall) for each path/row were acquired for classification purposes.

Image Registration: The TM scenes were registered to a UTM projection.

Image Processing: Subset scenes by hydrologic unit boundaries, histogram match, and mosaic scenes as necessary. Evaluate various classification techniques (i.e. band ratios and NDVI). Perform histogram matching on watersheds that include more than one scene. Run an unsupervised classification, create hard copies of the classification for field evaluation.

Field Training/Verification/Collection: Train various agency staff in field sampling techniques and digital image processing theory as required.

Data Revision/Reclassification/Recoding: Refine the classification based on field data. This is the stage the project is at as of this writing.

Database Development: Create a hierarchial classification database.

Final Products: The final vegetation classification, as well as the field data, will be provided to contributors for review and comment in digital and hard copy format.

Metadata: Metadata files will be developed and included with all data sets produced in compliance with the FGDC Metadata Content Standard.

Remote Sensing Ground Truth Field Data Form - Southwest Colorado Vegetation Classification Project

Observer:	Agency and Office:
Date:	Point ID #:
Watershed code #: Watershed Name:	Photo taken? ❑ Yes ❑ No
UTM X: Corrected X: UTM Y: Corrected Y: Zone: ❑ 12 ❑ 13 GPS Location? ❑ Yes ❑ No* Datum: ❑ NAD 27 ❑ WGS 84 ❑ NAD 83	Latitude: Longitude: Corrected Longitude: Corrected Latitude:

Elevation (MSL):
 	
Units: ❑ Feet   ❑ Meters
From: ❑ Topo ❑ GPS   ❑ Other
	Slope: ❑ Flat 0 - 5%
	         ❑ Moderate 6 - 20%
	         ❑ Moderately steep 21 - 40%
	         ❑ Steep 41 % or more

Area Size: ❑ 3 - 4.9 Acres
            ❑ 5 - 9.9 Acres
            ❑ 10 Acres or more
	Aspect: ❑ N ❑ NE
	          ❑ S ❑ NW
	          ❑ E ❑ SE
	          ❑ W ❑ SW

Land Use/Structural: ❑ Tree   ❑ Shrub
  ❑ Grass/Forb   ❑ Bare   ❑ Wetlands
  ❑ Riparian   ❑ Water   ❑ Urban
  ❑ Other
	Vegetation Distribution
	Dominant Group ❑ patchy ❑ continous
	2nd Dom. Group ❑ patchy ❑ continous

Dominant Group:
  Primary Series:
  Secondary Series:
	Group Cover: ❑ 0 - 5% ❑ 51 - 75%
	               ❑ 6 - 25% ❑ 76 - 95%
	               ❑ 26 - 50% ❑ 96 - 100%

Second Dominant Group:
  Primary Series:
  Secondary Series:
	Group Cover: ❑ 0 - 5% ❑ 51 - 75%
	               ❑ 6 - 25% ❑ 76 - 95%
	               ❑ 26 - 50% ❑ 96 - 100%

Tertiary Group:
  Primary Series:
  Secondary Series:
	Group Cover: ❑ 0 - 5% ❑ 51 - 75%
	               ❑ 6 - 25% ❑ 76 - 95%
	               ❑ 26 - 50% ❑ 96 - 100%

General Description of Area:	Comments:

BLM Service Center - Version 4 - 6/29/95   fieldform4.aw
* Use of GPS is strongly prefered. If GPS equipment is unavailable, you should select only larger areas
for sample sites. Areas of 500 meters (1640 ft.) square are only .2 inches at 1:100000 scale, therefore,
it is recommended to measure only areas larger than 500 meters. This will avoid locational errors in determining field location. Keep in mind that 1:24000 scale plots can be produced for areas you wish to
study intensively.

Figure 2 Field sampling form.

Application of Project Results

Project results and products, to be useful, must be made available to as many interested agencies/parties as possible. Spatial data products are an essential component in the construction of data sets required to do ecosystem-based management. The ability to look at a mosaic of landscape-level vegetation types in terms of their spatial distribution, integrity and abundance will facilitate resource management and help accomplish management objectives.

Deliverable products of the project are the vegetation classification data in digital format, color maps of the data at various scales, the database of field data, and a final report describing project methodology, results and accuracy assessment.

This data set is envisioned to be a dynamic entity, maintained and continually updated by the contributing agencies. The data set can be used as a reference for change detection studies, ecosystem management modeling, regulatory compliance monitoring, and interagency land management policy making.

CONCLUSION

This project has proven to be an excellent example in inter-agency cooperative efforts and coordinating individual needs to achieve a common goal. This project is producing a new data set, common to all contributors, for use in land management and ecosystem modeling activities, without the usual limitations of ownership boundaries, disparate management philosophies or data file structures.

As mentioned in the objectives, this project has been such a success that it is expanding to the east side of Colorado.

Applications of Remote Sensing on the Sandhills of the Nebraska National Forest

Virginia Emly
Nebraska National Forest
USDA Forest Service
Halsey, NE 69142

Robert Sprentall
Nebraska National Forest
USDA Forest Service
Chadron, NE 69337

Henry Lachowski, Vicky Varner, and Paul Maus
Remote Sensing Applications Center
USDA Forest Service
2222 West 2300 South, Salt Lake City, UT 84119

ABSTRACT

Ground data collection has been an important source of information about the condition and trend of range vegetation for decades. Ground data collected for rangeland analysis usually consists of collecting and/or enumerating individual plants or species populations over a specified area or along a transect. The data is then used to determine range condition and trend, seral stages, habitat types, percent cover, or forage estimates (kgs/hectare). This study explores the use of satellite and airborne remote sensing information to assist in the identification of habitat type and seral stage measurements in a grassland ecosystem. Landsat Thematic Mapper (TM), aerial videography, and Xybion multispectral imagery were acquired for an area over the Samuel R. McKelvie National Forest (a subdivision of the Nebraska National Forest) in Nebraska. The remote sensing information was correlated with ground data that showed that habitat type could be determined using Landsat TM and ancillary data. Weak correlations exist between seral stage measurements and Landsat TM imagery. The higher spatial resolution of the Xybion multispectral imagery and aerial videography showed promising application for identifying individual plant species. Seral stage correlations were difficult to evaluate since it was not possible to pinpoint the ground plot positions on the airborne imagery. However, the authors feel that technology is available to improve the location of plot location on the airborne imagery that may make this information more useful for future applications.

INTRODUCTION

The Nebraska National Forest, Rocky Mountain Experiment Station, and the Remote Sensing Applications Center (RSAC) conducted a development project to examine the use of remote sensing and geographic information systems (GIS) for use in the management of range ecosystems. The project was conducted under the Integration of Remote Sensing program overseen by the Forest Service Remote Sensing Steering Committee (RSSC). Two objectives were identified by Nebraska National Forest range conservationists and wildlife biologists as being important areas where remote sensing could assist them. The first objective was to examine how remote sensing could be used to identify and map habitat types found in the area, primarily Dry Valleys and Sands/Choppy Sands types. These habitat types are usually separated by a slope break and have slightly different indicator species. The second objective was to distinguish among the four seral stages found within each habitat type. Seral stages were determined using the methodology developed by Uresk (1991) which uses Daubenmire plots to measure the frequency and cover of three indicator species found along parallel transects within a 20 by 50 meter plot. The sources of remote sensing data acquired included Landsat Thematic Mapper (TM), airborne multispectral imagery (Xybion), and aerial videography. Eighty ground plots were collected within the same approximate time of the remote sensing data (July 1994). The ground data included Daubenmire transects for seral stage and habitat type determinations, and Robel pole readings for residual cover estimation (Uresk 1991). A global positioning system (GPS) was used in the field to identify the location of the ground plots, and was also used aboard the aircraft to navigate the predetermined flight lines for acquiring imagery over the ground plots.

Study Area

The Samuel R. McKelvie National Forest, a unit of the Nebraska National Forest, lies within the Nebraska Sandhills. The unique ecosystem of the Sandhills is due mainly to soil and geographic location. The soil (mostly Entisols with no horizon) is composed of 80% fine textured sand. Plant species from surrounding tall grass and short grass prairie ecosystems combine in the Sandhills and form a unique mixture of vegetation. Dominant grass species include sand bluestem (Andropogon hallii), little bluestem (Schizachyrium scoparium), prairie sandreed (Calamovilfa longifolia), switchgrass (Panicum virgatum), sedges (Carex spp.), Needle-and-thread (Stipa comata), Junegrass (Koeleria pyramidata), and Scribner's panicum (Panicum scribnerianum). Shrubs present in the area are chokecherry (Prunus virginiana), wild plum (Prunus americana), buckbrush (Symphoricarpos occidentalis), and rose (Rosa arkansana). The Sands/Choppy Sands habitat type is dominant on the McKelvie and consists of grassy, rounded hills that were formed and modified by wind. In between the hills of the Sands/Choppy Sands habitat type are the internodal sandy basins. The linear valleys of the Dry Valley habitat type are found between series of Sands/Choppy Sands and are influenced by closer proximity to the water table and slight soil differences. Dry Valley's are mostly in the southern portion of the study area and can range in size from under 5 acres to over 300 acres.

LITERATURE REVIEW

Numerous studies have assessed the use of remote sensing for mapping and identifying condition and trends on grasslands. The uses of photographs, both from the air and from the ground, have been used extensively to document species, productivity, and provide condition and trend estimates (Carneggie et al., 1983). Generally, large scale photographs ranging from 1:600 to 1:5,000 are used to make species and productivity estimates, and medium scale photographs ranging from 1:6,000 to 1:32,000 are used for mapping plant communities as well as condition and trend assessments. Satellite remote sensing has also been used to map and monitor grasslands. Studies have shown that good correlations exist between spectral information from Landsat TM and percentage of green vegetation in range ecosystems. However, other studies have shown that relationships between TM data and biomass and individual range species are poor (Adams, Bernier et al. 1988). Influences of senesced grass, litter, and soils can alter the spectral response and diminish the signal from the green vegetation (Huete and Jackson 1987). Combined use of aerial videography and Landsat TM was shown to be 67 percent accurate in a classification of twelve range vegetation types in the southwest US. In this study, aerial videography was used as reference data for a supervised classification (Martinez et al., 1994). Other studies have shown that visual interpretation of Landsat TM is very useful for analyzing range vegetation from satellite, particularly for assessing condition and identifying different grazing systems and their effects (Thompson, Dams et al., 1984).

METHODS AND RESULTS

Acquiring Remote Sensing and Ground Data

Remote sensing data on the McKelvie was acquired from both satellite and aircraft. Landsat TM satellite imagery was acquired over the McKelvie on July 22, 1994. This mid-summer date was close to the peak biomass period for the year. The imagery was geocoded to a UTM projection system and resampled to 25 meter pixels. The image quality was excellent and very few clouds were present. The collection of the airborne remote sensing data required flight planning. Since Dry Valleys only cover six percent of area the six flight lines were established over the southern portion of the McKelvie to maximize the opportunity for ground data collection on the highest diversity of Dry Valley seral

82

stages on the unit. Field crews, navigating with GPS, collected ground reference data along the six flightlines. The ground reference data to determine seral stage consisted of determining frequency and cover classes of three indicator species within 50 Daubenmire plots along two parallel transects. This method, developed by Uresk, 1990, uses indices and a score card system to determine ecological condition or seral stage for each ground location. Approximately, 80 ground locations distributed between the two habitat types were used in the study. The airborne imagery was acquired from a Cessna 206 airplane with a large belly hole. A configuration of cameras including a natural color Super VHS (SVHS) video, multispectral Xybion video, and 35 mm color Kodak digital camera, were mounted in the belly hole. The natural color video camera is a high-end Panasonic video camera that records imagery on SVHS cassette tape. The Xybion multispectral video has a rotating filter wheel that allows up to six different images to be captured near simultaneously in six different wavelengths ranging from visible (400 nm) to the near infrared (1000 nm). Xybion imagery is also recorded on SVHS cassette tape.

The lack of ground reference features makes collecting imagery over this area difficult without proper navigation tools. A program called GPS Touchdown was used to monitor the position of the aircraft in relation to the predetermined flightlines. This software allows a laptop computer to be used in place of a hand-held datalogger in the GPS equipment configuration. Predetermined flightlines, as well as roads and Forest Service Unit boundaries, were displayed as a background map on the monitor. The position of the aircraft is shown on the monitor as a moving, blinking cursor. When the cursor deviates from the predetermined flightline a course correction can be made. The flightlines and ground plot locations shown in Figure 5 are overlaid on the Landsat MT. The video and Xybion imagery collected from the aircraft varied in swath width from 100 feet to one quarter mile.

Analysis of Remote Sensing and Ground Data

The eighty ground plots with GPS locations were subsequently located on the TM satellite image. The analysis portion discussed in this paper is restricted to the use of six raw spectral bands of the Landsat TM data which are bands 1-5 and 7. Spectral statistics for at least ten pixels were gathered from the imagery for each GPS ground location. The standard deviation for the pixels at each location was generally under three for all bands except for band five (mid infrared) which was often between three and four. The assumption made here is that since the ground plots were located in representative areas of habitat type and seral stage the pixels surrounding the immediate plot locations were of the same type. The number of ground plots collected per seral stage within each habitat varied from 3 to 16.

Statistical analysis of the spectral information and plot information was done using discriminant analysis. Discrinimant analysis was used because the variables of habitat type and seral stage are qualitative in nature and form a continuum across the landscape rather than discrete units. In a discriminant analysis, linear combinations of predictor variables, in this case TM spectral bands, serve as the bases for classifying cases (ground plots) into groups (Norusis, 1990). The groups in this case are either habitat type or seral stages. Tables 1 and 2 show the results of the discriminant analysis for habitat type and seral stage within habitat type. The results are displayed in a matrix which show how well each of the ground plots were classified into their corresponding habitat type and seral stage.

In addition to the spectral analysis, a map of habitat type was created from digital elevation models (DEMs). A slope map was derived from the DEM and then regrouped into two classes; 0-3% and >3%. These breaks were representative of the Dry Valley and Sands/Choppy Sands habitat types respectively. The analysis of the airborne remote sensing data was limited to visual interpretation. The aerial video had GPS locations printed on each frame of imagery. This video imagery corresponded to the Xybion imagery well so that locations on each could be found by using the GPS locations on the video imagery. Despite the GPS link to the imagery, the exact locations of the ground transects were nearly impossible to find due to the lack of plot markers or visible landmarks. The GPS location printed on the imagery corresponded to the location of the aircraft and therefore did not represent the center of the video frame when the aircraft was not vertical. Because of these circumstances, the airborne imagery was not thoroughly assessed in this study. The data is being used in a subsequent study as reference data in a supervised classification of land cover types.

CONCLUSIONS AND RECOMMENDATIONS

There are many difficulties associated with correlating field variables with remote sensing data in range ecosystems. Initially plot locations were limited to sites at least a 1/4 mile apart with only one plot to be located within each seral stage polygon. (Note: seral stage polygons were developed from an earlier inventory and were used as an initial stratification for identifying ground plots for the various seral stages in this study). Also plots were only to be established in areas that had not been grazed during the current year. In some instances these limits were too restrictive so some plots were located in areas that had been grazed earlier in the spring as long as the area had sufficient regrowth. Sites were also located in pastures that were currently being grazed if cover removal was less than ten percent. Multiple sites were located within individual polygons if needed to complete the numerical goal of 10 plots per seral stage.

In conclusion, the two prominant habitat types were determined using DEM's. However interdunal and some sand sites might be initially included but could easily be eliminated by judgment or ground truthing. Other habitat types, such as riparian, as well as vegetation cover of the habitat types were mapped using TM imagery. Discriminate analysis using the TM spectral bands showed some promise in the separation of habitat types and seral stages. The results obtained could be strengthened by increasing the sample size, particularly for seral stage. The resulting matrices may be better than expected due to geographical similarities as well as similarities in vegetation cover of the plots.

Although intriguing information was present in the airborne data it was not pursued since ground targets were needed to ensure accuracy of the interpretations. The realities of current and future budgets require that no project be limited in its scope. Multifunctional goals and objectives can and should be pursued for each project. This might mean that traditional range and wildlife data collection methodologies need to be modified to accommodate sensor detection capability.

REFERENCES

Adams, B., M. B. Bernier, J.deValois, O.Dupont, and I. Sutherland, 1988. Evaluation of Landsat TM enhanced images for range applications in Alberta. Canadian Journal of Remote Sensing, Vol. 14, No. 1, pp. 4-16.

Carneggie, D. M., B. J. Schrumpf, and D.A. Mouat, 1983. Rangland Applications. Manual of Remote Sensing. Volume II. American Society of Photogrammetry, pp. 2325-2341.

Huete, A.R., and R.D. Jackson, 1987. Suitability of Spectral Indices for Evaluating Vegetation Characteristics on Arid Rangelands. Remote Sensing of the Environment. Vol. 23, pp. 213-232.

Martinez, R., L.Miller, J.Whitney, H.Lachowski, P.Maus, J.Gonzales, and J.Powell, 1994. An evaluation of the utility of remote sensing in range management. Remote Sensing and Ecosystem Management - Proceedings from the Fifth Forest Service Remote Sensing Applications Conference. pp. 136-143.

Norusis, M. J., 1990. SPSS/PC+ Advanced Statistics 4.0. pp. 1-38.

Thompson, M.D., R.V. Dams, and G.M. Beaubier, 1984. Transfer of Landsat technology for operational PFRA range management programs in southern Saskatchewan. Eighth Canadian Symposium on Remote Sensing, pp. 793-799.

Uresk, D., 1990. Using multivariate techniques to quantitatively estimate ecological stages in a mixed grass prairie. Journal of Range Management. Vol. 43, No. 3, pp.282-285.

Table 1. Matrix resulting from discriminant analysis of habitat types and TM spectral bands for the ground plots.

Classification Results for Habitat Type
Predicted Group

Actual Group	1	2	Total
Group 1 (Dry valleys)	31	4	35
Group 2 (Sands) (Choppy Sands)	10	31	41

Table 2. Matrix resulting from discriminant analysis of seral stages within habitat types and TM spectral bands for the ground plots.

Classification Results for Seral Stage

Dry Valleys
Predicted Group

Actual Group	1	2	3	4	Total
Seral Stage 1	10	2	0	0	12
Seral Stage 2	0	15	0	1	16
Seral Stage 3	2	0	1	1	4
Seral Stage 4	0	2	0	1	3

Sands/Choppy Sands
Predicted Group

Actual Group	1	2	3	4	Total
Seral Stage 1	7	0	0	1	8
Seral Stage 2	0	8	0	1	9
Seral Stage 3	1	0	8	2	11
Seral Stage 4	2	3	3	5	13

TECHNOLOGICAL BUILDING BLOCKS
FOR DERIVING TIMBERLAND SUITABILITY ANSWERS
Kim Mayeski, GIS/Remote Sensing Specialist and
Faye Krueger, Assistant Forest Planner and GIS Coordinator
USDA- Forest Service (Caribou National Forest)
250 South Fourth Avenue, Pocatello, Idaho 83201

ABSTRACT

With the advent of GIS and image processing capabilities, the process of calculating timberland suitability for Forest Planning required through the National Forest Management Act has changed. This paper compares the traditional method of determining suitability with innovative, new technology. The format of this paper begins with a description of what data is currently available, explains the process for determining information needs, and concludes with how the final product, timberland suitability, is derived. The vegetative layer is one of the most critical layers for the suitability model. The process of creating this layer using satellite imagery is described. The vegetative layer is used in several spatial analyses that can be vital in making good land management decisions. Various ways of using image processing for incorporating resource data from remote sensing systems into GIS are discussed.

INTRODUCTION

Forest plans provide broad, programmatic direction to manage the resources of the National Forest in a coordinated and integrated manner. This direction conveys the intent of governing laws, regulations, policies, and Regional guidance through Forest Plan goals, objectives, standards, and guidelines. Collectively, this guidance is developed for resource professionals to help them attain the described desired conditions anticipated on the Forest.

Direction in Forest Plans is reviewed at the District level. Districts compare the existing condition on the land with the desired conditions defined in the Plan. District personnel find opportunities to move an area toward the desired conditions and design projects to achieve the outcome. The Forest Plan becomes a commitment to the public based on their participation. Money is allocated based on projections in the Plan for goods and services. Forest Plans also describe the basis for monitoring resource conditions. Monitoring information often results in changes in management.

The planning process addresses all resources. One emphasis area in the National Forest Management Act (NFMA) relates to timber resources. Specific direction is given for determining suitable acres of timber. Suitability is defined in NFMA as "the appropriateness of applying certain resource management practices to a particular area of land, as determined by an analysis of the economic and environmental consequences and the alternative uses foregone. A unit of land may be suitable for a variety of individual or combined management practices."

IDENTIFYING LAND SUITABLE FOR TIMBER PRODUCTION

The National Forest Management Act (NFMA) directs that during the forest planning process, lands that are *not* suitable for timber production will be identified. NFMA also requires that lands not suitable for timber production be reviewed at least every 10 years. NFMA regulations outline this process in Section 219.14. The process is carried out in three stages:

Stage 1

National Forest land is tested against criteria for biologic capability, availability, or physical suitability. Lands failing to pass these tests are set aside from further consideration for timber production and classed as "not suitable." (See Land Classification List in Table 1). This paper identifies the steps taken for this stage and compares it to the process used in the initial round of planning.

During Stage 1, lands not suitable for timber production are identified by the following criteria:

- The land is not forested.
- Technology is not available to ensure timber production from the land without Irreversible resource damage to soils productivity or watershed conditions.
- There is no reasonable assurance that such lands can be adequately restocked.
- The land has been withdrawn from timber production by an Act of Congress, the Secretary of Agriculture or the Chief of the Forest Service.

Stage 2

Lands passing the tests in Stage 1 are assessed to determine the cost and benefits for a range of timber management intensities or regimes. No land is discarded at this point.

Stage 3

The lands from Stage 2 are tested against criteria representing Forest objectives, silvicultural requirements, and cost efficiency. Lands failing to pass these tests are classed as "not suitable" also.

HOW LAND NOT SUITABLE FOR TIMBER
WAS DETERMINED IN THE FIRST ROUND OF FOREST PLANNING

During the Caribou National Forest's initial planning process in 1985, non-forested land was identified by taking the vegetation types from the rangeland analysis and timber types from the 10-year timber inventory. Landtype associations were also mapped. Using clear mylar overlays, vegetation polygons were mapped. Those stands that were forested were identified by cover type and structure. Landtype capabilities were mapped on mylar and landtype groups were identified that could not sustain timber management activities. Also areas that were withdrawn from timber production were mapped on mylar, all at 1:24000 scale.

WHY GIS TODAY?

The process used in the first round of planning to identify vegetation and landtypes and to hand-draw polygons on mylar overlays was labor intensive. In many cases, different people gathered the same data across the forest, introducing their own bias into the data collection. With the advent of GIS (Geographic Information Systems) and remote sensing, data collection becomes more cost effective and efficient. GIS is an organized collection of computer hardware, software, geographic data, and personnel designed to efficiently capture, store, update, manipulate, analyze, and display all forms of geographically referenced information. Remote sensing is the art of collecting and interpreting information about an area from a remote vantage point.

When the Caribou National Forest embarked on revising their Forest Plan, different methods for developing a vegetation layer were discussed. In the past, a timber inventory was taken every 10 years and used to determine suitability and other management activities. For this new planning effort, forest managers agreed to use digital imagery to derive a consistent vegetation layer. By using GIS technology, spatial depictions are highly accurate.

STEP 1 - INFORMATION NEEDS ASSESSMENT

An information needs assessment was the first step. When assessing needs for a GIS project of this size, it is important to know what questions need to be answered through the analysis. By knowing which questions need to be answered, appropriate spatial layers can be determined and assembled. From monitoring results and other available information, the Forest needed to ascertain where suitable timber exists today and whether a change in management direction is needed. The National Forest Management Act establishes the criteria for determining suitability. One criteria is to identify non-forested lands. With GIS technology, roads, utilities, administrative sites, streams, and standing bodies of water can be buffered and eliminated from the suitable base. During the initial Forest Plan efforts, these features were not captured and were included in suitable acres.

The information needs assessment identified the following thematic layers the Caribou National Forest needed to complete a suitable timber model:

1) Vegetation Layer (forested vs non-forested)
2) Administrative Sites
3) Improved Roads
4) Utilities or Utilities Corridors
5) Private land
6) Mines
7) Standing Bodies of Water
8) Streams
9) Slopes from 0-45% and slopes from 0-65%
10) Land types that are unstable, unstockable, and have low productivity.
11) Research Natural Areas

STEP 2 - AUTOMATE DATA FOR NON-FORESTED LANDS

Each layer or theme had to be automated for input into the GIS. Automation is transferring hard copy data to a digital format that can be used in a GIS environment. There are several different ways of automating layers so they are in a usable format. Two common ways are "digitizing" and "deriving". Digitizing is probably the most common way of automating spatial data. This process converts graphic information scribed by interpreters into digital form subject to manipulation by the computer. Deriving layers from other sources is another way of obtaining necessary GIS data for an analysis. A vegetation layer was derived from Landsat Satellite Imagery, a remote sensing system that contributes data at low cost and in a form that is compatible with GIS requirements.

Lands which fall into one of the categories on the Land Classification List (Table 1) generally should not be included in the suitable timber base. Although these lands may have suitable timber, special management designation or the inability of the land to pass criteria under NFMA regulations explained earlier in this paper, eliminate any future timber harvest opportunities. The order in which these layers are discussed below follows the order of the thematic layers in the land classification list.

Vegetation Layer:

Satellite imagery from July, 1991 was purchased, and a vegetative layer was derived from the imagery, using Erdas image processing software. This layer contained the complete forest vegetation, both forested and non-forest lands. The process and guidelines for deriving a vegetative layer for use in managing national forests, published by the Nationwide Forestry Application Program, was used. The raw imagery was clipped to the forest boundary, and an unsupervised classification was completed. Landtypes that were easily recognized, like bodies of water and non-forested land, were labeled and verified. Training sites were established for classes that could not be readily identified and required further investigation and refinement. Training sites and the associated field forms were entered and used to classify areas with similar spectral signatures. Labeling of the classes was based on a visual and statistical analysis with vital assistance from Forest Service personnel who were familiar with on-the-ground conditions. For a more detailed description of deriving a vegetative layer from satellite imagery, please refer to the above mentioned publication.

Initially, "Past Harvest Units" were not identified as a required layer. However, this layer was identified as necessary during the process. If the units were cut before the date of the satellite imagery, even though saplings may have been planted, these saplings were too young to be identified as coniferous stands. They would likely have the spectral signature of a mountain brush or sagebrush type. The Forest decided to include all sales that were sold before fiscal year 1995 in the past harvest unit layer. Units were identified through fiscal year 1995 to aid in yield calculations at a later date. For the purposes of the suitable timber model, these units were digitized and added into the vegetation layer.

Administrative Sites:

Administrative sites include campgrounds, guard stations, ranger districts, warehouses or other Forest Service buildings. Location and identification of these sites were obtained from Cartographic Feature Files (CFFs). Cartographic Feature Files are digitized from Primary Base Series Maps at the Geometronics Service Center, a Forest Service mapping facility in Salt Lake City, Utah. These 1:24000 scale maps contain information on roads, trails, waterways, and administrative boundaries. Administrative sites in the CFFs are represented as point features. These points represent the site location on-the-ground, but not the actual size of the facility. It was necessary to identify the size of the site, because timber within these sites is withdrawn from harvest. From land status documentation for the Caribou National Forest, the exact acreage of each site was entered into GIS. A "look up" table was created in Arc/Info and the points were buffered to the actual acres that comprise each site.

Roads:

Arcs representing the existing transportation system were pulled from the CFFs and buffered 30 feet, 15 feet on each side of the road. Overall, road width averaged 30 feet forest-wide.

Utilities:

Utilities include powerlines, gaslines, slurry lines, and water lines. These arcs were pulled from the CFFs. The Forest determined that the average width of these disturbances was sixty feet, 30 feet on each side of the utility corridor.

Private Land:

Areas within the Forest Boundary that are privately owned were also identified. This thematic layer was pulled from the CFF's. These areas did not need to be buffered, because the areas represent the exact location and acreage of the private land.

Mining:

Open-pit mining of phosphate occurs on the Caribou National Forest. Areas were identified where mining impacts further preclude timber production. The actual lease boundary was not used, because much of the area within the boundary is not disturbed. The areas within the lease boundaries that have already been mined or disturbed were digitized. These areas represent where actual phosphate mineral is located and include the associated pit and borrow areas. Geologists identified these areas and digitized this layer for the GIS environment.

Standing Bodies of Water:

Standing bodies of water were also pulled from the CFFs. These standing bodies of water include ponds and lakes. These areas were not buffered.

Streams:

The streams, however, were buffered 30 feet for perennial streams and 15 feet for intermittent streams. Since the width of a stream on the Forest varies from streambank to streambank, 30 feet for perennial and 15 feet for intermittent streams were considered average based on specialists' recommendations. These buffered areas do not represent riparian areas. Riparian areas will be addressed in stages 2 and 3 in the Forest Planning process. (See Page 1, stages 2 and 3)

These eight layers represent the total non-forested land. (Refer to Table 1, Section 1 of the Land Classification list.) This process was more refined and more accurate than the one used during the initial round of Forest Planning. The non-forested areas can now be computed and removed from the land base; the resulting layer would be forested land.

Table 1. Land Classification List (NFMA Regulations)

Section	Type	Layer
Section 1	Non-Forest land (includes water)	Rangeland
		Administrative Sites
		Improved Roads
		Utilities
		Private land
		Mines
		Standing Bodies of Water
		Streams
Section 2	Land Withdrawn from Timber Production	Wilderness
		Research Natural Areas
Section 3	Land Not Capable of Producing Crops of Industrial Wood	Rocky or talus
		Low productivity sites
Section 4	Land Physically Unsuitable	Unstable soils
		Land types not restockable within five years
		Slopes 65% and greater

STEP 3 - AUTOMATE DATA FOR LANDS WITHDRAWN, NOT CAPABLE, OR NOT SUITED FOR TIMBER HARVEST

Land Withdrawn from Timber Production:

Other considerations were made to remove certain lands from the suitable base through legislation regarding the management of National Forest System lands. (Refer to Section 2 on the Land Classification list Table 1.) Forested land withdrawn from timber production includes wilderness areas and Research Natural Areas. Although the Caribou National Forest has two proposed wilderness areas, neither of these areas have received legislative action. Therefore, these areas are included in the suitable timber base. Research Natural Areas, on the other hand, are special management areas and are excluded from the suitable timber base. RNA administrative boundaries were pulled from the CFFs.

Land Not Capable of Producing Crops of Industrial Wood:

Areas that are rocky, talus, and have low productivity are identified as not capable of producing crops of industrial wood. (Refer to Section 3 on the Land Classification list Table 1.) The Forest's soil inventory was digitized, and contained land type attributes. Using this inventory, the land types were evaluated and a determination was made as to what land types are suitable for harvesting timber and what land types are unsuitable. Polygons that fit the description of rocky, talus, and low productive areas were excluded from the suitable base in the model.

Forested Land Physically Unsuitable:

Lands that fall into this category are describe as unstable and not restockable. These land types are unsuitable for harvesting timber.

In addition, forested land physically unsuitable was also defined as land with slopes greater than 65% or greater that 45% depending on which harvest method was proposed. The slope layer was derived using Digital Elevation Models (DEMs). DEMs are created by the U.S. Geological Survey. Using the ARC/INFO module, GRID, these DEM's were manipulated to create a layer called "Slope". Through mitigation measures and unconventional logging methods, harvest may occur on some slopes within the 45-65% range.

STEP 4 - BUILDING THE MODEL

Several different ways of running the suitability model were discussed. Consideration was given to the accuracy needed for the resulting suitable timber layer. The vegetative layer was derived using satellite imagery with 25 meter pixels in a raster format. For convenience, a decision was made to design the model in a grid environment.

To prepare the individual layers discussed above for use in the suitability model, an additional field of information was added in the vector environment in Arc/Info. A field or an item (Arc/Info term) was added to the vector coverages and then values of zero or 1 were entered into the field: an attribute of 1 indicates the land is suitable for timber harvest; an attribute of zero indicates the land is not suitable for timber harvest. For example, on the Research Natural Area layer, RNA polygons were attributed with a value of zero and areas outside the Research Natural Area was attributed with a value of 1. The Vector layers were converted into 5 meter grids.

The method used for running the first suitability model was to multiply the layers together. If one layer is unsuitable in a 5 meter grid cell, then the final suitability layer is unsuitable in that same 5 meter grid cell. If the unsuitable areas have a value of 0 and the suitable areas have a value of 1, 0 times any number will result in a determination of unsuitable. The modeling was done with Erdas Image Processing Software, using the model maker module.

The suitable timber model was tested on the Cache Range, an area on the Montpelier Ranger District. This area was chosen through a forest-wide assessment of which area might best represent all the features that are considered when determining timberland suitability. Using Arc/Info, a hard copy map of this test area was plotted with two different classes: suitable and unsuitable for harvesting timber. Each class was represented by a different color. Locational features, like roads and the Forest Boundary, were used to help the user identify real world locations.

After viewing the hard copy map of timber suitability on the Cache Range, managers were interested in finding out which layer was responsible for a particular area being classified as unsuitable. By multiplying all layers together at one time, this question was unanswered. Instead, multiplying the layers together in sequence as shown on the Land Classification List (Table 1) provided information on where the greatest change occurred between layers. (See Table 2 for results of this process.) The results in Table 2 were determined by taking the Rangeland layer times the Administrative Sites layer. The result of this multiplication was then multiplied by the roads layer and so on through the list. On Table 2 in the change column, the layer identified as causing the majority of acres to be classified as unsuitable was the Low Productive layer or Soils layer.

Table 2. Montpelier Ranger District - Cache Trial
Comparison of Unsuitable and Suitable Acres and Change
Total land base = 263,917 Acres

Land Classification		Unsuitable Acres	Suitable	Change
Non-Forested Lands	Range Land	171870	92047	0
	Administrative Sites	172070	91847	200
	Roads	172782	91135	711
	Utilities/Utilities Corridor	172803	91114	21
	Private Land	172920	90998	117
	Mines	No mines	0 change	0
	Standing Bodies of Water	172953	90964	32
	Streams	173541	90376	588
Forested Land	Forest Land	173541	90376	0
Land Withdrawn From Timber Production	RNAs	173894	90023	325
Land Not Capable of Producing Industrial Wood	Low Productivity	193341	70577	19446
Land Physically Unsuitable	Unstable/Not Restockable	203212	60705	9871
	Slope >65%	205073	58844	1861
	Slope >45%	214488	49429	9415

THINGS YOU WANT TO CONSIDER

Be Customer-Focused

The key to having a product that is usable by the customer or user is to involve them in its design and production. When the vegetation layer was derived, a representative from each Ranger District, who was familiar with on-the-ground conditions, helped classify the vegetative types. It also was vital that District representatives review suitability mapping results. Technical people can perform the automation of the layers and the modeling tasks, but if they are unfamiliar with real world conditions, they lack a key component. Employees who are familiar with real world ground conditions can fill the missing component.

Tune in to Problem-Solving

When the suitability for the test area was being reviewed, several questions arose. The first issue involved isolated stands of suitable timber averaging around ten acres in size. These isolated areas would not be economical to harvest, because of road-building costs to access the stands. Different filtering methods were tested. The best results were attained using the "eliminate" command in Arc/Info. Polygons, less than 10 acres, that were labeled suitable were eliminated. This method only eliminated polygons less than 10 acres, and did not change the shape of the other polygons.

Harvesting practices may leave narrow strips of timber between harvesting units. These narrow strips did not show up as suitable in the final layer, even though they are suitable. They did not show up as a pure signature of timber, on the raw imagery, because of their narrow width, and therefore, are classified as non-forested. The process that was used to fix this problem involved changing the existing Past Harvest Unit coverage, so it would include the stringers. A copy was made of the Past Harvest unit coverage and then the stringers added between units, using aerial photos for accuracy.

Limber Pine is a coniferous timber type that is not marketable and not suitable for timber harvesting. From the satellite imagery, limber pines have a similar spectral signature of other coniferous types and were classified suitable in the model. Limber pine only exists on the southern part of the Cache Range within the Caribou National Forest. Looking at the soils layer, limber pine showed up in areas that were identified on the soils layer as land type 101. Land type 101 is describe as being areas with high elevations, glaciated and cyroplanated slopes. Typically, this land type has steep slopes with stony soils that have low revegetation potential. During the initial identification of land types, this land type was missed as a type that is not restockable. This error was fixed by calculating land type 101 as unsuitable for harvesting timber.

SUMMARY

The technological world of GIS will make land management planning more understandable, consistent, trackable, more efficient, and more exact. It offers the ability to analyze and make decisions about how land resources can be managed. It meets needs for both site-specific and programmatic level analyses. It can serve as a tracking facility for monitoring management activities and land conditions. It can produce future scenarios for land management options. GIS links real world problems with powerful analytical tools. For these reasons, and many more, GIS is taking rudimentary planning, used in the early 1980s, into the 21st century.

GIS will be used to develop alternative management options during the Caribou National Forest's update to their current Land Resource Management Plan. It will also provide real time spatial displays and associated attributes that link to other relevant data bases for on-the-ground management activities. Change detection using GIS for monitoring will greatly improved the ability of resource professionals to adapt management practices with changing environments and ground conditions.

INTEGRATION OF REMOTE SENSING WITH VEGETATION MAPPING, INTEGRATED RESOURCE INVENTORIES, AND GIS DATABASE PROJECTS

Jeff A. Milliken, Ralph Warbington
U.S.D.A. Forest Service, Region 5 Remote Sensing Lab
1920 20th St.
Sacramento, CA 95814

ABSTRACT

The U.S.D.A. Forest Service, Region 5, Remote Sensing Lab currently utilizes remote-sensed data sets for vegetation mapping, integrated resource inventories, and GIS database projects. Primary data sets include Landsat Thematic Mapper imagery, SPOT imagery, and aerial photography. Vegetation mapping is accomplished using image classification, GIS modeling, and spectrally derived polygons. Vegetation maps are then used for post stratification of the National Permanent Plot FIA Inventory Grid as well as for area and volume expansion. Remote-sensed-derived vegetation layers are integrated into the regional GIS corporate database and used for deriving fuels attributes and Wildlife Habitat Relationship (WHR) classes, as well as for a variety of other GIS analyses. Image data sets are also used for vegetation map updates, identification of vegetation plot locations, coregistration of corporate GIS data layers, feature updates, and as image back drops in Arcview. Future plans include cooperative change detection projects with Region 5 Forest Pest Management for systematic vegetation updates to existing vegetation layers.

INTRODUCTION

This paper summarizes procedures presently being employed at the Region 5 Remote Sensing Lab integrating the use of remote sensed data, remote-sense-derived data sets, sample-based data sets, and numerous other GIS corporate data layers. Routine procedures are used for regional vegetation mapping, integrated resource inventories, and various GIS database projects. Principle types of remote sensed data currently being utilized include aerial resource photography, Landsat Thematic Mapper imagery, and SPOT panchromatic imagery.

VEGETATION MAPPING

The Remote Sensing Lab (RSL) is responsible for mapping existing vegetation at a regional scale for all National Forests in Region 5. Mapping is completed by both RSL staff and outside contractors. These mapping projects are currently on a ten year cycle (Fig. 1) with two to five year updates scheduled to coincide with regional change detection projects currently being conducted by Forest Pest Management (Levien et al. 1996). The regional vegetation maps are important for RPA reporting, Forest Plans, Timberland Suitability Class analysis[*] , landscape analyses, cumulative effects analyses, integration with resource inventories, long - term monitoring of trends and changes, and numerous other activities. Cooperative efforts to map multiple ownerships are actively being pursued. Consistent vegetation maps across all ownerships provide an important utility for ecosystem management.

---

[*] See U.S. Forest Service Region 5 (1995), Forestland and Resource Data Base, for more information.

US Forest Service Vegetation Maps

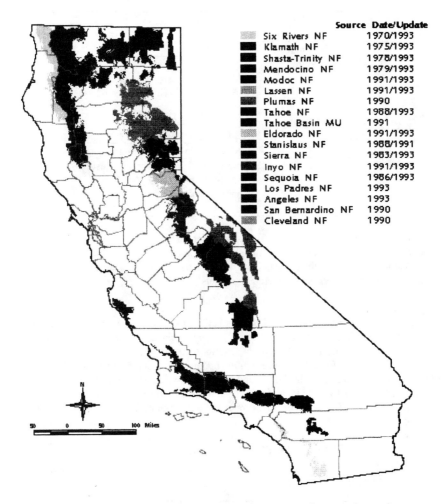

	Source	Date/Update
	Six Rivers NF	1970/1993
	Klamath NF	1975/1993
	Shasta-Trinity NF	1978/1993
	Mendocino NF	1979/1993
	Modoc NF	1991/1993
	Lassen NF	1991/1993
	Plumas NF	1990
	Tahoe NF	1988/1993
	Tahoe Basin MU	1991
	Eldorado NF	1991/1993
	Stanislaus NF	1988/1991
	Sierra NF	1983/1993
	Inyo NF	1991/1993
	Sequoia NF	1986/1993
	Los Padres NF	1993
	Angeles NF	1993
	San Bernardino NF	1990
	Cleveland NF	1990

Figure 1: Image source dates and updates for Region 5 vegetation mapping projects.

<u>Mapping Standards</u>

Consistent mapping standards and procedures employed by the RSL and designated contractors result in more consistent map products across Region 5 National Forests. Current mapping standards are shown in Table 1.

TABLE 1
MAPPING STANDARDS

ATTRIBUTES	STANDARD
Minimum Mapping Unit	One Hectare
Classification System	8-12 Lifeform Classes CALVEG Crosswalk to WHR (Mayer et al. 1988.)
Tree Canopy Closure	10% Class Breaks
Tree Size	Region 5 Class Breaks
Shrub	CALVEG (no canopy closure)
Secondary Labels	Hardwood Type, Canopy Closure, And Size In Mixed Hardwood - Conifer Stands

Mapping standards are evolving in response to recommendations being released by the Federal Geographic Data Committee, Interorganizational Resource Information Coordinating Council, Interregional Ecosystem Management Coordinating Group, and comments from customers and cooperative partners.

Mapping Procedures

Procedures developed through the RSL emphasize consistency and repeatability as well as integration with other existing data sets. Any necessary edits (primarily due to limitations in the capability of remote sensing technologies to capture all desired attributes) are tracked for use in future product updates and metadata reporting. TABLE 2 summarizes procedures used in the lab and by designated contractors. The RSL actively solicits participation by individual forests and routinely sends check plots of interim products to available forest personnel for input and quality control.

TABLE 2
MAPPING PROCEDURES

ATTRIBUTES	PROCEDURE
One Hectare Polygons	Image Segmentation
Lifeform Classes	Supervised/Unsupervised Classification
CALVEG	GIS modeling
Tree Size	Supervised/Unsupervised Classification, Li Strahler Canopy Model
Tree Canopy Closure	Li Strahler Canopy Model
Secondary Labels (mixed conifer -hardwood)	Subpixel analysis, Li Strahler Canopy Model, GIS modeling

Polygons. Polygons satisfying a 1 hectare minimum mapping unit are spectrally derived using image segmentation algorithms (Woodcock and Harward 1992; Woodcock et al.

1993), and Landsat Thematic Mapper imagery. Through this process, a base line polygon layer is produced independent of any classification process, and thus independent of any classification error. Polygons can be assigned attributes based on any ancillary data set and have the potential to be used for numerous additional applications by resource and GIS specialists.[*] Figure 2 shows an example of polygon boundaries over SPOT data.

Fig. 2: Segmentation polygons over SPOT imagery, Lassen National Forest (black area at top of figure is area of no data).

Lifeform Classification and CALVEG. Lifeform types (i.e. conifer, shrub, bare ground, hardwood, meadows, grass, water, etc.) are classified using a supervised/unsupervised approach with Landsat TM imagery (Miller et al. 1994). Segmentation polygons are assigned lifeform labels based on logic rules with an overlay of the lifeform pixel classification and segmentation polygon layer. CALVEG (U.S. Forest Service Regional Ecology Group 1981) labels are assigned to segmentation polygons through an integrated GIS modeling process. Inputs include the lifeform polygon layer, terrain data, ecological subsection boundaries, vegetation rules based on field data, and other appropriate ancillary data sets when available.[**]

---

[*] For a full description of necessary input data and the image segmentation algorithm, see Woodcock and Ryherd (1989), Ryherd and Woodcock (1990), and Woodcock and Harward (1992).
[**] For a more comprehensive description of these techniques, see Miller et al. (1994) and Woodcock et al. (1993).

Tree Size and Canopy Closure. Various methods have be used to classify tree size. These include the Li Strahler canopy model[*] , supervised classification, and unsupervised classification processes. Classifications for tree size are performed on hardwood and conifer areas which are masked out of the data after lifeform mapping and CALVEG modeling is complete. Ancillary data such as field stand data and aerial photography is routinely used to aid in tree size classification.

Canopy closure is mapped using the Li Strahler canopy model. The model requires the CALVEG type, slope, aspect, and brightness and greenness bands for each forest stand (Woodcock et al. 1990). Calibration of the model is based on collected field stand data for tree geometry, and brightness and greenness signature estimations. Canopy closure is estimated at 10% class breaks.

Secondary Labels. Secondary labels for hardwood types are generated for mixed hardwood-conifer stands. Hardwood attributes in mixed stands include CALVEG type, size, and canopy closure. Relative percentages of hardwood and conifer in mixed stands are derived from subpixel analysis and classification. Hardwood canopy closure is determined by using the relative hardwood percentage from the subpixel classification and the total canopy closure derived from the Li Strahler canopy model. Hardwood CALVEG types are modeled in GIS and size is mapped using the same size mapping procedures previously described. The addition of secondary labels in mixed hardwood-conifer stands (a relatively new procedure in our mapping program) enhances the information for these conditions and also provides for better crosswalks to other classification systems such as the Wildlife Habitat Relationship system routinely used by the California Departments of Forestry and Fire Protection and Fish and Game, as well as U.S. Forest Service, Region 5.

Vegetation Updates

SPOT panchromatic imagery is used for updating existing vegetation layers. SPOT scenes have been mosaiced at the RSL for every individual forest in Region 5 and delivered to each forest. The RSL and individual forests are updating vegetation layers by heads up digitizing on the SPOT imagery and delineating changed areas on check plots. Updates include harvested areas, fire areas, and other discernible changes in the landscape. Figure 3 shows plantation boundaries plotted on SPOT imagery. Note the power transmission corridor running through the east side of the image as well as numerous roads and small clearings.

---

[*] Refer to Strahler and Li (1981) and Li and Strahler (1985) for a complete discussion of the canopy model.

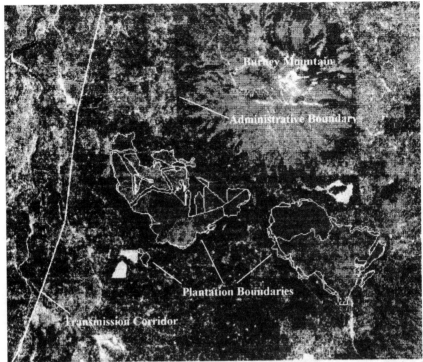

Figure 3: Plantation boundaries over SPOT panchromatic imagery, Lassen National Forest.

INTEGRATED RESOURCE INVENTORIES

Installation of the National Permanent Plot FIA Grid

The RSL administers contracts and provides regional quality control and training for the installation of the National Permanent Plot FIA Grid on Region 5 National Forests. Grid plot locations are entered in a GIS database and overlaid on SPOT panchromatic data. This aids in locating plot locations on aerial photography for use by field crews. Field crews then use the plots located on aerial photography to find plot locations in the field. Plots are surveyed in the field with GPS technology. The GIS grid plot layer is updated with the GPS controlled points after the plot is installed.[*]

Orthorectified SPOT imagery also offers utility (depending on spatial accuracy needs) in easily capturing the locations in a GIS database for any inventory points (older inventories, etc.) that may only be located on aerial photography.

Post Stratification of the National Permanent Plot FIA Grid

Strata labels based on CALVEG type, tree size, and tree canopy closure are assigned to each polygon in the vegetation database. After plot locations are updated in the GIS layer with GPS controlled locations, plot locations are overlaid on the CALVEG layer. Based

---

[*] Refer to U.S. Forest Service - Region 5 (1994) for comprehensive information on the National Permanent Plot FIA Grid.

on this overlay process, strata labels are assigned to each plot. The acreage for each vegetation strata is calculated from the GIS vegetation layer and used with the inventory data for timber volume calculations by strata (U.S. Forest Service - Region 5).

Acreage calculations, frequency of sampled types, and spatial information generated from the vegetation and grid inventory layers are also used in identifying under sampled types and areas where densification of the grid is necessary. Figure 4 shows grid plot locations on SPOT imagery.

Figure 4: National Permanent Plot FIA Grid plot locations over SPOT imagery. Square symbols represent the 3.4 mile grid, diamonds are areas where the grid has been densified.

GIS DATABASE PROJECTS

Feature Updates, Backdrops, and Registration.
SPOT panchromatic imagery is used for updating features such as roads, clearings, power corridors, and streams. Updates can be done on check plots or using heads up digitizing on screen over the SPOT data. Landsat TM and SPOT imagery are also used as image backdrops in Arcview with other GIS data layers in the Region 5 Forest Resource Database (Warbington et al. 1993).

SPOT data is also being used as a "base" layer for coregistration of other image data sets such as Landsat TM. Since many other corporate data layers are tied to the SPOT data (i.e. plantation updates), registering other image data sets to this layer for regional projects helps prevent gross misregistration problems between new remote-sense-derived vegetation layers and existing layers such as plantation boundaries. In general, we have found that terrain corrected SPOT imagery tends to be more closely registered with existing layers such as transportation, and hydrology.

Derived Data and GIS Analysis
The RSL is working with the National Forests and the California Department of Forestry and Fire Protection in developing programs to derive fuels layers from remote-sense-derived vegetation layers. These layers can be used as input to fire modeling programs. Additionally, the RSL uses vegetation layers and other corporate data layers for Land Suitability Class analysis in defining administrative land allocations. Other uses include such applications as erosion and sedimentation modeling, wildlife habitat modeling, and GAP analysis.

CHANGE DETECTION

The RSL is participating in a cooperative change detection project with Region 5 Forest Pest Management. This initial project covers areas across all ownerships in the Southern Sierra Nevada.[*] The RSL is scheduling our vegetation mapping programs such that two to five year vegetation updates will be tied to the Forest Pest Management regional change detection mapping program. In this way, remapping efforts can be optimized by focusing on areas of vegetation change.

CONCLUSIONS

As an increasing variety of land cover information becomes available in spatial (GIS) databases, the integration of various resource data, remote sensed imagery, and remote-sense-derived data sets will provide powerful analytical tools to the resource management community. Key to the successful integration of these technologies and data sets is consistent standards in metadata, data capture, data formats, inventory sampling protocols, vegetation mapping, and updating procedures, to name a few.

Region 5 is producing remote sensing and GIS products utilizing consistent standards and methods to meet the needs of different users and allow for efficient integration of data. Remote-sensed-derived and GIS databases are created at a scale compatible with that of the U.S.G.S. 7.5 minute quadrangle. This is the largest scale widely available across ownerships and provides flexibility for aggregation to smaller scales when required

---

[*] See Levien et al. (1996) for a discussion of this project and additional references.

for data integration and analysis with smaller scale data sets. Registering data sets to fine resolution imagery (i.e. SPOT data) is also compatible with 7.5 minute quadrangle and CFF feature file standards. Consistent coding standards and data formats being utilized in Region 5 also allow for data sharing and integration at the forest, region, and national level.

REFERENCES

Levien, L., C. Bell, and B.Maurizi, 1996. Large Area Change Detection in California Using Satellite Imagery: Southern Sierra Change Detection Project. Proceedings of the Sixth Biennial Forest Service Remote Sensing Conference, Aurora, CO. April 29, 1996. in press.

Miller, S., M. Byrne, H. Eng, J. Milliken, and M. Rosenberg, 1994. Northeastern California Vegetation Mapping: A Joint Agency Effort: Remote Sensing and Ecosystem Management, Proceedings of the Fifth Forest Service Remote Sensing Applications Conference, Portland, OR. April 11, 1994. pp. 115-125.

Ryherd, S.L., and C.E. Woodcock, 1990. The Use of Texture in Image Segmentation for the Definition of Forest Stand Boundaries. Proceedings: 23rd International Symposium on Remote Sensing of Environment. Bangkok, Thailand. April, 1990. pp. 18-25.

U.S. Forest Service Regional Ecology Group, 1981. CALVEG: A Classification of California Vegetation. U.S. Department of Agriculture, Forest Service, Region Five. San Francisco, CA. 168 pp.

U.S. Forest Service Region 5, 1994. Forest Inventory and Analysis User's Guide. U.S. Department of Agriculture, Forest Service, Region 5. San Francisco, CA.

U.S. Forest Service Region 5, 1995. Forestland And Resource Data Base (FRDB). U.S. Department of Agriculture, Forest Service, Region 5. San Francisco, CA.

Warbington, R. , R. Askevold, and W. Simmons, 1993. Building A Regional GIS Planning Database For The National Forests Of California. The Compiler. v.11, no.4, pp. 43-46.

Woodcock, C.E., and S.L. Ryherd, 1989. Generation of Texture Images Using Adaptive Windows. Technical Papers, ASPRS/ACSM Annual Convention. April 2, Baltimore, MD. 2:11-22.

Woodcock, C.E., and J. Harward, 1992. Nested-Hierarchical Scene Models and Image Segmentation. International Journal of Remote Sensing. 13(16):3167-3187.

Woodcock, C.E. and J. Collins, V. Jakabhazy, and S. Macomber, 1993. Forest Vegetation Mapping Methods Designed for Region 5 of the U.S. Forest Service. Unpublished Technical Manual.

FOREST BIOMETRICS FROM SPACE

Timothy B. Hill
Lead Remote Sensing / GIS Analyst
Geographic Resource Solutions
1125 16th Street, Suite 213
Arcata, CA 95521

ABSTRACT

Geographic Resource Solutions (GRS) recently completed mapping the
Applegate River watershed in southern Oregon for existing vegetation. This was
a cooperative effort between the USDA Forest Service, and the Bureau of Land
Management. GRS used Landsat TM satellite imagery, Digital Elevation Models,
measured field data, GIS, and GPS. The final database estimated polygon
attributes using continuous variables including canopy closure, average tree size,
species composition, trees per acre, and variance for tree size and canopy
closure. This paper describes the various methodologies used in the project, that
include: field data collection, image processing techniques for removing the
effects of topography, hybrid supervised and unsupervised image classification
techniques, ecological rule-based pixel aggregation, and quantitative accuracy
assessment.

INTRODUCTION

Biodiversity, Watershed Management, and Ecosystem Management are a few
of the terms being bandied about the natural resources community. These "buzz"
words and phrases exemplify the changing values society places on the forest
ecosystem. As resource professionals, we must address these changing values in
our decision making processes. In addition to these values, management decisions
must also be based on the most current and accurate information available. A
Geographic Information System (GIS) is a useful tool which may provide this type
of information to decision makers. Organizations, both public and private, use
GIS to help resolve complex current and future natural resource management
issues. Remotely sensed data have become increasingly popular for providing
information for GIS analysis. Digital imagery may be used to produce
information describing the characteristics of a forest ecosystem that is both
current and accurate. However, satellite image processing can only produce
consistent, accurate, and hence reliable maps when used with methodologies that
account for the tremendous variability found in forested environments. Several
recent attempts at mapping forested ecosystems have met with limited success.
These attempts have used methodologies which separate a forested ecosystem
into separate components (i.e., canopy closure, species type, or tree size) and
then attempt to recombine the components during final polygon formation.
These popular methodologies ignore the interrelatedness of these characteristics
within an ecosystem. These methods further inhibit the potential of image
processing for vegetation inventory by describing the separate components as a
series of classes or groups. While useful for generalizing the data, these

artificial classes rarely occur in nature (Congalton, 1991.) The characteristics of any ecosystem are an interrelated gradient. This fact must be accounted for when using satellite image processing techniques to produce reliable and accurate vegetation inventories.

Figure 1. The Applegate River watershed.

The Applegate River watershed was distinguished as an Adaptive Management Area (AMA) in President Clinton's Forest Plan. This watershed, located in southern Oregon and northern California (Figure 1.), is managed by the Bureau of Land Management (BLM), USDA Forest Service (FS) and private land owners, each managing roughly one third of the watershed. Resource professionals identified existing vegetation as the single most important GIS layer needed to complete their analysis. The existing GIS vegetation layers varied in quality and extent, lacked consistency in stand typing, and included no information on private lands. This paper describes a project Geographic Resource Solutions (GRS) recently completed that was a cooperative effort between the BLM and FS to map the Applegate River watershed. The goal of the project was to create a GIS theme of vegetation which was flexible and accurate. The information would be suitable for use by all resource managers (i.e., timber, silviculture, wildlife, watershed, fire, and others). The project incorporated techniques and methodologies developed by GRS during previous remote sensing projects.

METHODS

In October of 1994 GRS began work on mapping the Applegate River watershed. Landsat TM data, acquired August 29, 1993, was obtained from the BLM office in Portland, Oregon. Existing GIS layers for the watershed were incorporated into the project GIS. These layers consisted of: Public Land Survey (PLS), ownership, hydrography, and transportation. A digital elevation model (DEM) of the project area was also incorporated into the project s GIS database.

Classification Scheme

FS and BLM cooperators choose to use plant community descriptions from the Management of Wildlife and Fish Habitats in Forests of Western Oregon and Washington (Anonymous, 1985) for cover type attributes. GRS described canopy closure, average tree size, trees per acre; and percent hardwood, conifer, shrub, and grass as continuous estimates rather than the traditional approach which utilized categorical classes or groupings. Continuous estimates of vegetation characteristics were used because the data could be later adapted to any series of classes depending on the needs of the user. This project also included variance estimates for canopy closure, and average tree size estimates. GRS summarized

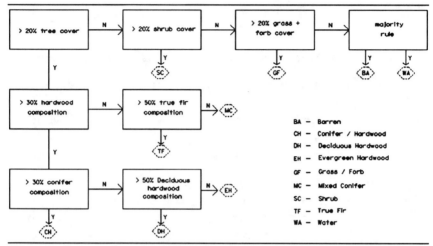

Figure 2. Cover typing scheme flow chart.

the discrete estimates of average tree size and canopy closure using nine size classes and eight canopy closure classes at the request of the BLM and FS officials. Figure 2 describes the cover typing scheme and decision process.

Field Data Collection

Vegetation mapping projects that rely on digital image classification require a mechanism for identifying vegetation characteristics on the ground (i.e., "ground truth"). Popular methods utilize photo interpretation (PI) as a source of "ground truth." PI is well suited for general land cover classification schemes, however its application in a detailed classification scheme (one which uses canopy closure, average tree size, and species groupings) will produce maps of questionable reliability. Research has found photo interpretation to be highly subjective. This research also concluded "photo interpretation may not be as accurate as many people have believed." (Biging, etal., 1991). The main reason why PI has been frequently used for "ground truth" is the cost of acquiring actual ground data is too expensive (Congalton and Green, 1993.) Field data collection costs can be the most expensive portion of a mapping project. However, even greater costs can be incurred if the maps produced are inaccurate and unusable. Decisions based on the information contained in an inaccurate map can result in costs not limited to money including habitat loss and job loss. If a mapping project is started with errors in the "ground truth" these errors are compounded through the various stages of the mapping process. The goal of the Applegate project was a detailed vegetation inventory. With this goal in mind, GRS started the project using the most reliable and accurate ground truth available: measured field data.

Total Cover Summary:

Size Class: Species:	0-4"	5-8"	9-12"	13-16"	17-20"	21-25"	26-31"	32-47"	48"+	Tree Cover	Non-Tree Cover	Total Cover
Douglas-fir	1.0%	2.5%	1.0%	2.0%	6.0%	8.0%	12.0%	25.0%		57.5%		57.5%
sugar pine									3.0%	3.0%		3.0%
white fir	2.0%	4.0%	2.0%	3.0%						11.0%		11.0%
hardwood			1.0%							1.0%		1.0%
madrone		8.5%	7.0%	3.0%						18.5%		18.5%
misc shrub											1.0%	1.0%
forb/herbac											3.0%	3.0%
duff/debris											5.0%	5.0%
Total Cover	3.0%	15.0%	11.0%	8.0%	6.0%	8.0%	12.0%	25.0%	3.0%	91.0%	9.0%	100.0%

Tree Cover Summary:

Size Class: Species:	0-4"	5- 8"	9-12"	13-16"	17-20"	21-25"	26-31"	32-47"	48"+	All Sizes
Douglas-fir	1.1%	2.7%	1.1%	2.2%	6.6%	8.8%	13.2%	27.5%		63.2%
sugar pine									3.3%	3.3%
white fir	2.2%	4.4%	2.2%	3.3%						12.1%
hardwood			1.1%							1.1%
madrone		9.3%	7.7%	3.3%						20.3%
Total Tree Cover	3.3%	16.5%	12.1%	8.8%	6.6%	8.8%	13.2%	27.5%	3.3%	100.0%

Quadratic Mean DBH (by Cover) Summary:

Size Class: Species:	0-4"	5- 8"	9-12"	13-16"	17-20"	21-25"	26-31"	32-47"	48"+	All Sizes
Douglas-fir	4.1"	6.5"	12.0"	15.5"	18.4"	23.8"	29.3"	37.6"		30.3"
sugar pine									48.0"	48.0"
white fir	3.6"	7.5"	9.0"	15.0"						9.9"
hardwood			12.0"							12.0"
madrone		7.3"	10.7"	13.3"						9.8"
QMean DBH	3.7"	7.2"	10.6"	14.5"	18.4"	23.8"	29.3"	37.6"	48.0"	26.3"
QMean DBH - Con	3.7"	7.1"	10.1"	15.2"	18.4"	23.8"	29.3"	37.6"	48.0"	29.2"
QMean DBH - Hwd		7.3"	10.8"	13.3"						9.9"

Trees Per Acre Summary:

Size Class: Species:	0-4"	5- 8"	9-12"	13-16"	17-20"	21-25"	26-31"	32-47"	48"+	All Sizes
Douglas-fir	5.5	10.6	1.0	1.0	4.4	3.6	3.6	7.0		36.8
sugar pine									0.6	0.6
white fir	18.2	9.6	4.3	1.0						33.2
hardwood			1.0							1.0
madrone		35.3	18.9	6.6						60.8
All Trees	23.7	55.5	25.1	8.6	4.4	3.6	3.6	7.0	0.6	132.2
All Trees - Con	23.7	20.3	5.3	2.0	4.4	3.6	3.6	7.0	0.6	70.5
All Trees - Hwd		35.3	19.8	6.6						61.7

Quadratic Mean Crown Size Summary:

Size Class: Species:	0-4"	5- 8"	9-12"	13-16"	17-20"	21-25"	26-31"	32-47"	48"+	All Sizes
Douglas-fir	10.0ft	11.9ft	24.0ft	35.4ft	29.1ft	36.9ft	43.7ft	51.2ft		43.3ft
sugar pine									55.0ft	55.0ft
white fir	8.1ft	15.3ft	16.0ft	40.0ft						24.1ft
hardwood			24.0ft							24.0ft
madrone		12.6ft	15.8ft	17.7ft						14.8ft
QMean CD	8.8ft	13.3ft	17.6ft	32.1ft	29.1ft	36.9ft	43.7ft	51.2ft	55.0ft	37.5ft
QMean CD - Con	8.8ft	14.1ft	19.0ft	38.2ft	29.1ft	36.9ft	43.7ft	51.2ft	55.0ft	41.5ft
QMean CD - Hwd		12.6ft	17.0ft	17.7ft						15.4ft

Table 1. An example of summarized transect data used for field data collection.

GRS methodologies for creating a GIS database of detailed vegetation information require measured field data. GRS vegetation sampling technique used a line-point transect method. For this project, transects were 1188 feet in length, with points spaced 12 feet apart. A total of 100 points were sampled along the transect. At each point along a transect a vertical sighting was taken using the GRS_densitometer (vertical sighting device). This device projected a vertical line of sight, with cross hairs to identify the vertical point. If the vertical point intercepted a tree crown, field personnel recorded the following information: species, canopy position, diameter at breast height (DBH) to the nearest inch, crown diameter to the nearest foot, shrub cover, herbaceous cover, and non-vegetative cover. Table 1 illustrates a summary of the information collected by the line-point transect method. Transects were oriented like a triangle for training data. This type of orientation enabled field personnel to install transects in small stands. Transects were elongated for accuracy assessment to facilitate sampling as much of a polygon as possible. Each transect was assigned a unique sequential number representing its order of placement. Field data were recorded using software loaded on a hand held computer. GPS data were collected as reference points (RP) for each transect's location. Distance and azimuth from the RP to the start of the transect were measured and stored in the header of the transect data files. Upon completing a transect, the field crew described the sampled stand. Forest Service personnel differentially corrected the GPS data. The transect data, stand descriptions, and GPS RP locations were incorporated into the project GIS. The field crew used the GIS layers displayed on the satellite imagery, RP locations, and distance and azimuth measurements to place the transect locations in the GIS. The transect data were summarized and loaded as attributes for each transect. In addition to field data collected for this project, additional field data were used from a previous GRS project with the California Department of Forestry and Fire Protection. These transects were collected during 1992 and were immediately adjacent to the Applegate River watershed.

Image Processing

The Applegate River watershed contains rugged terrain typical of western Oregon. These steep slope angles and orientation combine with the solar angle and azimuth to cause a tremendous amount of reflectance variation in satellite imagery. Differential illumination can be a significant source of classification error in areas of high relief. Upon receiving the satellite data from the BLM, the imagery was topographically normalized using the Backwards Radiance Correction Transformation (BRCT) based on a non-Lambertian assumption and a Minnaert constant (Colby, 1991). This technique uses estimates of slope and aspect from the DEM, and sun angle and azimuth parameters during image acquisition to correct for differential illumination caused by terrain. The project area was initially divided into three eco-regions. These regions were developed to help resolve confusion between vegetation types that were spectrally similar but had different vegetation properties. FS personnel familiar with the project area assisted in developing the eco-regions. After the initial data sets were reviewed by BLM, FS, and GRS representatives, four eco-regions were redefined as follows: below 3500 foot elevation, low elevation; 3500- 5500 on west to south east aspects and 3500-6200 foot elevation on south to south west aspects, mid elevation; above the mid-elevation, high elevation; and serpentine and peridotite soils. Each eco-region was treated as a separate classification. Supervised training sets were developed for each eco-region. Transect locations from the GIS were used as seed points for generating supervised spectral statistics. Spectral statistics were loaded into a database table and associated with each respective transect. Variability and normality were analyzed for each supervised training area. Training area boundaries were modified as needed. Results from Jeffries-Matusita (J-M) divergence analysis were also loaded into a database table to facilitate analyzing vegetation and spectral differences between transects. Table 2 shows the results of J-M analysis with respect to vegetation. This process was iterative, depending on the results from each analysis.

CONFUSION SUMMARY FOR TRANSECT#:		10						
	MC	Douglas-fir	91.0%	26.3	3448	NE	C	
TRANSECT#	VEG TYPE	PR SPECIES	DENSITY	QM DBH	ELEVATION	ASPECT	SLOPE CLASS	JM DISTANCE
552	MC	Douglas-fir	77.0%	30.7	4005	E	S	1.12050
31	MC	Douglas-fir	99.0%	30.3	3776	SE	S	1.14560
572	MC	Douglas-fir	85.0%	51.7	4639	NE	M	1.27500
16	MC	Douglas-fir	80.0%	25.1	5101	N	S	1.29330

Table 2. An example of J-M divergence analysis with respect to the class' vegetation characteristics.

Once "clean" supervised training sets were completed for each eco-region, final classifications and unsupervised techniques were utilized. All the following methodologies were implemented using batch processing within each eco-region. An initial maximum likelihood (ML) classification was performed with a 90% probability threshold. The resultant unclassified areas in the class map were then used as a mask for developing unsupervised clusters. An ISODATA algorithm was used to develop 60-100 spectral classes from an initial 255 clusters. Two ML classifications were then run at a probability threshold of 95%: one using the supervised statistics, and one using the unsupervised statistics. A spatial overlay was performed between the supervised and unsupervised class maps. The unsupervised classes were used to augment the unclassified areas in the supervised class map. Three products were produced from this overlay process. The first was the merged class map. The second was a report of the overlay indicating the supervised classes corresponding to each unsupervised class (which pixels share the same spatial location). This report was later used in unsupervised class labeling. The third product was a mask of areas which remain unclassified in the merged class map. These areas, generally edge types or some anomaly, typically represented less than two percent of the area. Another ML classification was run on these unclassified areas with a probability threshold of 100%, and the resultant class map was merged creating a final merged class map. After this hybrid supervised/unsupervised classification was completed, there were no unclassified pixels in the class map.

Unsupervised Class Labeling

The unsupervised class labeling algorithm (Fox and Brown, 1992.) used was similar to stratified sampling for forest inventory. Weighted vegetation characteristics were calculated for each unsupervised class based upon the supervised classes which share the same spatial location. Each unsupervised class vegetation characteristics were then checked for integrity. This process was also used to validate the supervised classes. Table 3 is an example of the report

Unsupervised Class	Supervised Class	Pixel Count
2104	0	1670
2104	3	655
2104	30	3866
2104	35	372
2104	48	140
2104	518	437
2104	531	980

Table 3. An example of the results for an unsupervised class from the GIS overlay between supervised and unsupervised class maps.

produced from the GIS overlay process. Upon completion of the unsupervised labeling procedure, all eco-regions were merged to form a final pixel map of the entire project area. In this map, every grid cell contained an estimate of the vegetation characteristics similar to those illustrated in Table 1.

Polygon Formation

A major obstacle in many mapping projects that rely on image classification is how to develop a vector database where all stands - polygons - meet or exceed the minimum mapping unit (mmu). The solutions to the problem of how to build an accurate and reliable database with an mmu far above that represented by a single pixel are not well publicized. Conventional methods for eliminating heterogeneous pixel data utilize various pixel smoothing algorithms. Such techniques as modal, mean and/or majority filtering are abundant in commercial image processing software packages. However, these techniques are mathematical solutions. In the case of vegetation mapping, a mathematical solution applied to an ecological problem will result in a poor quality map.

This mapping project had a mmu of 5 acres. There were 445 unique classes represented in the final classified pixel map. When these 445 classes were displayed there was a "salt and pepper" effect. This heterogeneity is typical of most results of image classification. GRS used a ecological rules-based polygon formation routine (Stumpf, 1993) to produce polygons of similar vegetation characteristics. This process compared the vegetation characteristics of a subject pixel (or group of pixels) to all adjacent pixels (or groups of pixels). Pixels, or groups, were merged with their most similar neighbor. Similarity was estimated by evaluating vegetation characteristics such as: percent canopy closure, average tree diameter, species composition, percent hardwood composition, percent conifer composition, and trees per acre. Table 4 illustrates the aggregation process and similarity calculations. There were nine iterations in the aggregation process, each with progressively larger mmu s and different similarity thresholds up to the desired mmu of 5 acres. Final polygon attributes were summarized from all the original pixel data within the polygon boundaries. Each polygon had a vegetation summary similar to Table 1. A sample database record is shown in Table 5. Aggregating the pixel map for the project and producing an attributed vector GIS database was accomplished in 26 hours of

Stand ID #	Veg Type	Predominant Species	Tree Cover	Percent Conifer	Ave. QMD	Trees Per/Acre	Similarity Index
5534	MC	Douglas-fir	75%	70%	24"	187	
5533	MC	Douglas-fir	95%	80%	16"	250	
diff =	0	0	5.0	1.5	4.0	0.6	= 11.1
5534	MC	Douglas-fir	75%	70%	24"	187	
5532	CH	Madrone	89%	35%	14"	232	
diff =	10	12	3.5	3.5	5.0	0.5	= 34.5
5534	MC	Douglas-fir	75%	70%	24"	187	
5545	GF	Annual Grass	0%	0%	0"	0	
diff =	25	25	18.8	10.5	12	1.9	= 93.2
5534	MC	Douglas-fir	75%	70%	24"	187	
5541	DH	White Oak	35%	0%	10"	104	
diff =	14	15	10.0	10.5	7.0	0.8	= 57.3

Similarity was based on the combined difference in vegetation characteristics.
In this example, the subject stand 5534 was merged with stand 5533.

MC = Mixed Conifer
CH = Conifer Hardwood
GF = Grass/Forb
DH = Deciduous Hardwood

Table 4. An example of the determination of the most similar adjacent stand based on many vegetation characteristics.

processing. Errors in preliminary data sets were identified by FS and BLM officials. Since there was no human intervention in the pixel aggregation process, these errors were usually systematic and easy to track down. Since GRS methodologies were highly automated, GRS analysts were able to validate and correct the errors rapidly.

Accuracy Assessment

Accuracy assessment is the process of comparing map data to some assumed 100% correct reference data. Though this appears to be a straight forward task, many methodologies for this vital step in the mapping process are severely flawed. The reason is that assumed correct reference data have traditionally consisted of existing maps, PI, and/or ocular estimates. These types of reference data test for agreement between the map and reference data, not map accuracy. Existing maps may or may not have estimates of reliability. PI and/or ocular estimates, while useful for general land cover information, are subjective and have questionable reliability when estimating percent canopy closure, average tree diameter, species composition, percent hardwood composition, percent

114

conifer composition, and stems per acre. GRS used both measured field data for forest characteristics and ocular estimates only for non-forest cover types (i.e., shrub, grass, barren, and water). In addition, FS personnel used PI to check only cover types in the non-forest sampled polygons.

GRS utilized a stratified random sampling (SRS) scheme with replacement. This scheme provides information on all map categories regardless of the amount of area consumed by any one stratum. Research has shown SRS to be well suited for accuracy assessment of maps derived from remotely sensed data (Congalton, 1991). The first step in the accuracy assessment was generating a GIS theme of sample points. Random UTM coordinate pairs were generated throughout the project area. These coordinate pairs were used as sample points. Each sample point was assigned a unique sequential number representing its order of placement. A spatial overlay in the GIS associated the map strata with each sample point. Polygons were selected for sampling based on their sample number, within each stratum. Since cost was the primary factor in determining sample size, GRS and FS agreed that 20 samples per stratum would suffice and a collapsed series of classes would have to be used. GRS used the following classes for canopy closure estimates: 0-19%, 20-39%, 40-59%, 60-79%, and >=80%; and non-forest, 0-4", 5-12", 13-20", 21-32", and >32" for average tree size estimates.

GRS collected field data for accuracy assessment using the same technique as was used during training data phase of the project. Transect data were recorded on hand held computers with the same software as was used during training data collection. Transect summaries were processed using the same program that summarized training data and final polygon attributes. These steps insured consistent reference data. Field personnel were supplied with maps which had the sample point, polygon boundaries (without labels), transportation, hydrography, and PLS. Transects, installed within sampled polygon boundaries, were orientated to facilitate sampling as much of the polygon as possible. GPS positions were used to check the actual transect placement relative to the desired location. Transect summaries were loaded as attributes for their respective samples. The summaries served as reference data for generating error matrices. Correspondence between map and reference data will be determined by confidence intervals established for the discrete estimates of canopy closure and average DBH. Unfortunately, at press time these procedures were not yet finalized. However, they will be presented during the symposium. GRS used a more conservative estimate of correspondence between map and reference data by using a sliding class width (Hill, 1993). GRS used this technique to account

COLUMN	VALUE
pri_key	36506
veg_type	MC
closure_class	8
pct_closure	80.3
pct_conifer	77.3
pct_hdwood	22.7
size_class	5
qmdbh	17.6
qmdbhcon	19.2
qmdbhhwd	10.6
pix_ct	497
grid_val	29232
class_status	10
acreage	76.7
pr_species	Douglas-fir
pred_sp_pct	44.2
cv1	7.7
cv2	13.2
cv3	14.8
cv4	13.8
cv5	12.0
cv6	10.6
cv7	3.9
cv8	3.3
cv9	0.9
cv_shr	5.8
cv_hrb	3.0
cv_bar	10.9
cv_oth	0.0
tpa_tot	297.5
tpa_con	186.7
tpa_hwd	110.8
tpa1	171.9
tpa2	49.2
tpa3	29.9
tpa4	21.8
tpa5	10.7
tpa6	10.1
tpa7	1.9
tpa8	1.2
tpa9	0.9
tpa_class	3

Table 5.   An example database record.

for those cases in which class estimates between the map and reference data did not correspond, but the discrete estimates were within a specified range. GRS used a range of five inches for average DBH estimates, and ten percent cover for canopy closure estimates. The ranges corresponded with the original classes used at the start of the project. Class width match determination for cover type was based on the dominant cover. This addressed the differences found in mixed types such as Mixed Conifer and Conifer Hardwood and cover types such as True Fir and Evergreen Hardwood respectively. Table 6 illustrates the situation described above.

	stand ID	VEG TYPE	Dominant Species	Closure Class	Closure	Size Class	Average DBH
reference	271	MC	white fir	2	59%	4	20.6"
map	12254	TF	white fir	3	61%	5	22.4"

Table 6. An example of two estimates of a polygon's vegetation characteristics. The two estimates have very similar characteristics, but the do not have the same class values.

RESULTS

Error matrices were developed for each major forest characteristic: canopy closure, average DBH, and cover type. Each error matrix contained both "producer's" and "user's" accuracy measures by class stratum, an overall percent correct figure, an overall percent correct figure weighted by the area consumed by individual stratum, and a Kappa coefficient. The canopy closure error matrix is presented in Table 7. The overall percent correct was 86%, percent correct

MAP DATA	REFERENCE DATA	NON-TREE 0-20%	SPARSE 20-40%	OPEN 40-60%	MODERATE 60-80%	DENSE 80% +	TOTAL	PERCENT CORRECT	ACRES	CORRECT ACRES
	NON-TREE	20					20	100.0%	67,677	67,677
	SPARSE		1	1	3		5	20.0%	77,290	15,458
	OPEN			3	2		5	60.0%	100,610	60,366
	MODERATE			1	11	2	14	78.6%	156,790	123,192
	DENSE					21	21	100.0%	192,845	192,845
	TOTAL	20	1	5	14	25	65		595,212	459,538
	PERCENT CORRECT	100.0%	100.0%	60.0%	78.6%	84.0%		86.2%	527,535	391,861

TOTAL PERCENT CORRECT ACRES 74.3%

Kappa 0.4574
Var(Kappa) 0.0032

Table 7. Canopy Closure Error Matrix.

weight by area was 74%, and a Kappa of .46. While the overall figures are good, some individual stratum have poor results (i.e. 20% correct for the SPARSE class). Table 8 presents the error matrix for average DBH. The overall percent correct was 92%, percent correct weight by area was 88%, and a Kappa of .51. The results for average tree size were excellent. Table 9 shows the error matrix for cover type. The overall percent correct was 88%, percent correct weight by area was 85%, and a Kappa of .86. The cost of the Applegate River watershed mapping project are shown in Table 10.

		REFERENCE DATA									
		0 non-forest	1 0-5"	2 5-13"	3 13-21"	4 21-32"	5 +32"	TOTAL	PERCENT CORRECT	ACRES	CORRECT ACRES
M A P D A T A	0	20						20	100.0%	67,677	67,677
	1		1					1	100.0%	299	299
	2			13	1			14	92.9%	163,788	152,089
	3	1		1	13			15	86.7%	263,881	228,697
	4				1	5	1	7	71.4%	92,047	65,748
	5						2	2	100.0%		7,470	7,470
	TOTALS	21	1	14	14	6	3	59		595,162	521,979
	PERCENT	95.2%	100.0%	92.9%	92.9%	83.3%	66.7%		91.5%		

TOTAL PERCENT CORRECT ACRES — 87.7%

Kappa — 0.5126
Var(Kappa) — 0.0023

Table 8. Average Tree Size Error Matrix.

	REFERENCE DATA												
	BA	CH	DH	EH	GF	MC	SC	TF	WA	TOTAL	PERCENT CORRECT	ACRES	CORRECT ACRES
BA	2									2	100%	3,871	3,871
CH		5		1						6	83%	230,481	192,068
DH		2	4		1					7	57%	32,331	18,475
EH		3		2						5	40%	33,410	13,364
GF					20					20	100%	41,035	41,035
MC						14	1			15	93%	193,733	180,818
SC							16	1		17	94%	21,429	20,169
TF							1	14		15	93%	21,429	20,001
WA	1								4	5	80%	1,391	1,113
TOTAL	3	10	4	3	21	14	18	15	4	92		579,111	490,913
PERCENT	67%	50%	100%	67%	95%	100%	89%	93%	100%		88%		

TOTAL PERCENT CORRECT ACRES — 85%

Kappa — 0.8589
Var(Kappa) — 0.0015

Table 9. Cover Type Error Matrix.

DISCUSSION AND CONCLUSION

Although the accuracy information presented indicate high levels of accuracy compared to similar projects, the results should be viewed with some skepticism because the sample sizes in most stratum are extremely low. The reason for the small sample sizes was the lack of funds available for such field data collection. However, given the high quality reference data, the results indicate that the goal of the project was met - producing an accurate, flexible, wall to wall vegetation inventory for the Applegate River watershed. More samples are needed in each map stratum to facilitate an adequate accuracy assessment and remove any skepticism. This is essential in any remote sensing project. The cost information presented in this paper are from GRS invoices for the project.

Applegate Cost Summary	
ITEM	COST
Project Administration	$4,254
Field Data - Training Phase	$24,067
Field Data - Accuracy Phase	$24,875
Image Training	$10,505
Image Processing	$3,396
Image Classification	$5,285
Pixel Aggregation	$21,073
Data Conversion	$2,566
Total	$96,020
Cost Per Acre	$0.16
Cost Per Hectare	$0.40

Table 10. The Applegate River watershed project costs by task.

There were additional costs incurred by FS and BLM personnel involved with the project. Over half of the money spent was used to collect measured field data. While field data collection was a significant component of the project cost, it provided the opportunity to develop very specific estimates of cover, tree size, and species composition. User needs will dictate the necessity of developing specific or generalized categorical estimates. GRS has developed and applied a successful methodology for developing a detailed vegetation inventory by utilizing image processing techniques with measured field data.

REFERENCES

Anonymous, 1985. Management of Wildlife and Fish Habitats in Forests of Western Oregon ad Washington. U.S. Department of Agriculture. Forest Service. Pacific Northwest Region. Publication No.: R6-F&WL-192-1985

Biging, G., R. Congalton and E. Murphy, 1991. A Comparison of Photo interpretation and Ground Measurements of Forest Structure. In: Proc of the 57th Annual Meeting of American Society of Photogrammetry and Remote Sensing, Baltimore, MD (3):6-15.

Colby, J., 1991. Topographic Normalization in Rugged Terrain. Photogramm. Eng. Remote Sens. 57(5): 531-537

Congalton, R., 1991. A Review of Assessing the Accuracy of Classification of Remotely Sensed Data. Remote Sens. Environ.(37):35-46

Congalton, R., and K. Green, 1993. A practical Look at the Sources of confusion in Error Matrix Generation. Photogramm. Eng. Remote Sens. 59(5):641-644

Fox ,L. and G. Brown, 1992. Digital Classification of Thematic Mapper Imagery for Recognition of Wildlife Habitat Characteristics. In: Proc. 1992 ASPRS/ACSM Convention, American Society of Photogrammetry and Remote Sensing, Bethesda, MD. (4):251-260

Hill, T., 1993. Taking the "" Out of "Ground Truth": Objective Accuracy Assessment. In: Proc. 1993 Pecora 12, A Symposium on Land Information from Spaced Based Systems, Sioux Falls, SD 389-96.

Stumpf, K., 1993. From Pixels to Polygons: The Rule-Based Aggregation of Satellite Image Classification Data Using Ecological Principles. In:Proc. Seventh Annual Symposium on GIS in Forestry, Environment, and Natural Resources. Vancouver B.C. Canada. (2):939-945

THE IMPLEMENTATION OF GIS TO ASSESS THE ROLE OF INSECTS AND PATHOGENS IN FOREST SUCCESSION

Lowell G. Lewis
GIS/Image Processing Specialist
Management Assistance Corporation of America
Forest Health Technology Enterprise Team
USDA Forest Service
3825 E. Mulberry Ft. Collins, CO 80524
Phone: (970) 498-1749

Lawrence E. Stipe
Entomologist
Forest Health Protection
USDA Forest Service
PO Box 7769 Missoula, MT 59807
Phone: (406) 329-3289

ABSTRACT

The Forest Health Protection staff in the Forest Service's Northern Region are advancing the principles of ecosystem management by developing a groundbreaking methodology to analyze the role of insects and pathogens in forest succession. The use of GIS technology is critical to this Technology Development Project due to its ability to link and accurately manage the spatial and attribute data from two distinct time periods. The two sets of stand maps from subcompartment data were co-registered and 'unioned' together in order that the two attribute data sets could be linked and compared. Additional attributes were derived through a series of queries of ground survey data from the Forest Service's relational database. The methodology employed should now enable the Forest Service to predict future forest succession trends that more accurately reflect the complex and interrelated roles of insects and pathogens.

INTRODUCTION

Project Description

Forest planning methodology has been evolving toward an emphasis on *successional modeling* as a primary prediction tool in broad-scale analyses. The aim of successional modeling is to use vegetation *succession* trends and rates to predict changes in the composition of a stand.

Among the factors involved in forest succession are the impacts of insects and pathogens (diseases). Insects and pathogens are the primary and most predictable agents of change in most natural forests, recycling more biomass than is typically consumed by fires in even the most fire-intensive ecosystems (Hagle and Williams 1995). In 1991, Forest Health Protection (FHP) staff from the Northern Region (Region 1) of the USDA Forest Service began to examine these considerations by identifying specific insects and pathogens in the Region which affect forest succession, categorizing and defining their specific actions, and deriving *successional functions* related to them.

Recognizing the national potential for this initiative, FHP's Washington Office provided special project funding for a two-year Technology Development Project (TDP) to develop prototypes and provide guidance for future projects of a similar nature. The intent of this TDP was to develop methods for defining the spatial and temporal patterns of insect and pathogen actions and describing how these actions can be used to predict future patterns of forest succession. In order to be successful, the project staff had to develop completely new analytical processes to assess the transitions in *successional stage* caused by a variety of insect and pathogen agents.

This TDP has been given high national priority due to its utilization of the principles of *ecosystem management* and its conceptual design that the processes developed will be able to be implemented at a variety of locations and scales: Region-wide by other Forest Service Regions; on an individual forest using 100%-sampled forest data; on a single watershed; or even on a subwatershed.

Note: This paper uses many terms and concepts related to an evolving multi-use philosophy of resource management known as "ecosystem management." Such terms and other terms specific to this project are indicated in the text with italics, and definitions are provided in a Glossary section at the end of this paper.

Project Area

The study area for this TDP was the Forest Service's Northern Region. This area of the Northern Rocky Mountains falls within the Dry Domain and the Temperate Steppe Regime Mountains Division (Bailey et al. 1994), is in close proximity to the Canadian border, and is characterized by glaciated mountains, alpine meadows, and coniferous forests. The Northern Region contains thirteen National Forests, most of which are located north of the Salmon River in Idaho and west of the Continental Divide in Montana (see Figure 1). Data from over 8000 stands (see Figures 2 and 3) within 1133 randomly sampled subcompartments in twelve of the Region 1 forests were used in this analysis. The methods and procedures for this TDP were initially developed at the *land unit scale*, incorporating queries of attributes from these randomly-sampled subcompartments.

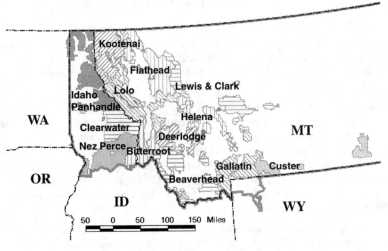

Figure 1: National Forests in Region 1.

Forest Name	Forest Area (sq. mi.)	Number of 1935-era Subcompts.	Number of Stands	Number of 1975-era Subcompts.	Number of Stands
Beaverhead	3,482	0	0	72	548
Bitterroot	2,614	51	313	51	678
Clearwater	2,915	27	138	27	358
Custer	2,030	0	0	36	181
Deerlodge	2,140	35	174	59	598
Flathead	4,093	68	261	68	657
Gallatin	3,391	0	0	60	372
Helena	1,822	26	135	70	462
Idaho Panhandle	4,343	76	450	108	769
Kootenai	4,041	75	298	75	577
Lolo	4,019	60	247	60	665
Nez Perce	3,512	7	62	22	255
TOTAL	38,402	425	2,078	708	6,120

Figure 2: Project forest data.

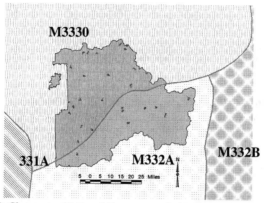

Figure 3: Clearwater National Forest, sampled subcompartments and surrounding ecoregion section ecocodes.

Why GIS Was Critical to this Project

There were three critical reasons for using GIS in this TDP:

1. **Greater accuracy in registering sample data locations.** Stand simulation models developed by the Forest Service and others have, by not fully considering the effects of insects and pathogens, provided an idealized view of future forest conditions in that they frequently overestimated future tree growth (Hagle and Williams 1995). To examine with greater accuracy what changes actually take place over time, sampled forest stand data from two distinct time periods must first be linked to the same exact piece of ground for a rigorous comparison of forest conditions. GIS can meet this need: once the stand boundaries for both datasets have been digitized and co-registered, and the attributes attached for both time periods, those attributes may be compared within a GIS by unioning the two spatial data sets together. All attributes from both periods may then be accessed by querying any of the 'child' polygons formed from the union of the two sets of 'parent' stand coverages. By accurately maintaining the data sets from both time periods, the GIS supports the validation and calibration of prior models and database queries, as well as the creation and validation of new queries specific to this project.

2. **Assignment of attributes.** The earlier of the two data sets (described later) did not carry the attribute of *habitat type*: at the time of the earlier survey, the concept of habitat type had not yet been introduced, whereas habitat type had been assigned to all stands during the latter survey. As habitat type, reflecting predominant conditions in the physical environment, changes very slowly, this attribute could reasonably be assigned to stands of the earlier survey from the more recent survey via the same GIS union process.

3. **Area weighting.** Some of the various combinations of habitat type, cover type, and successional stage had to be summarized for each forest or *ecoregion section* by weighting their values by area for each class. This type of calculation is efficiently and accurately performed by a GIS. Since area is automatically calculated for each polygon as soon as polygon topology is established, the total number of hectares assigned to each class is easily derived and that value used in calculating area-weighted statistics.

123

METHODS

The key step in determining insect and pathogen roles in forest succession was to determine successional stages for sampled stands at two different time periods and then compare the actual succession between the two periods. (Obviously, to isolate insect and pathogen functions, only stands that were not subject to other major agents—clearcutting, fire, etc.—would be compared.) Steps in this process included establishing a pilot site, acquiring data from two distinct time periods for the same stands, digitizing and overlaying the data from these two time periods, comparing stand conditions for attributes of interest, and determining potential successional functions for each combination of insect or pathogen agents and habitat groups. The final product is a transitional matrix, which defines probabilities for pathways of change in the landscape. The complex processes for deriving successional functions and the transitional matrix are beyond the scope of this paper.

Establishing a Pilot Site

A pilot site was chosen in order to discover and perfect all techniques prior to processing data from the remaining sites. The Nez Perce National Forest was selected as the pilot forest for the project. Although, because of limited 1930s photo coverage, it provided the smallest of the twelve forest samples to ultimately be analyzed, the Nez Perce promised to be the first National Forest to have both data sets available. Because of the project's time constraints, the Nez Perce was the logical choice.

Obtaining the Maps

Project area base maps showing stand-level attribute information for the two time-periods were required to determine the successional stage for each stand within the selected subcompartments. Data for this project were collected in the 1930s (hereafter referred to as "1935-era") and in the 1970s (hereafter referred to as "1975-era").

Stand type maps produced by the Forest Survey group were used to record the 1935-era stand attributes for most of the locations subsequently mapped during the 1975 era. These early maps (Figure 4) were hand-drafted township-size (36 square mile) maps at a scale of 1:31,680 (two inches per mile). Forest type information was color-coded and placed on the original base map, and an alphanumeric coding system was developed to describe the main stand characteristics (see Figure 4). This map series was produced for all Forest Service ownership lands in Region 1 except for most areas east of the Continental Divide: thus, the absence of 1935-era data from the Beaverhead, Custer, and Gallatin National Forests (as noted in Figure 2).

As is common in GIS work, data acquisition was an arduous aspect of this project. The only remaining complete set of 1935-era maps was originally stored at the Regional Office Headquarters in Missoula, Montana. Sometime during the late 1960s, however, an employee at the Regional Office decided to dispose of these maps as outdated. Responding to a tip from a friend on the janitorial crew at the Federal Building, Dale Johnson (Archivist at the University of Montana's Mansfield Library) found the many-volume map set in a dumpster, destined for the local landfill and oblivion. With permission from Regional Office staff, the maps were recovered and are now kept at the University library.

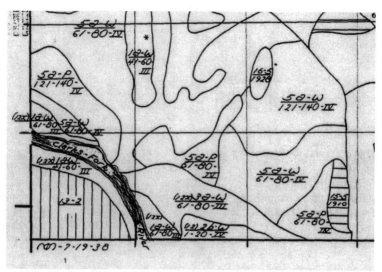

Figure 4: Portion of a 1935-era township map. Each square represents 1 square mile.

Unfortunately for the pilot project, maps were recovered for all but the Nez Perce and Clearwater National Forests. Data for these areas had to be recreated from aerial photographs taken during the 1930s (stored at the Regional Office) using photointerpretation techniques of the era. Moreover, because photographs were available for only 7 of the 22 subcompartments of the Nez Perce, the pilot project could be run only on that limited sample.

Collecting the 1975-era base maps at each National Forest headquarters office turned out to be another major task. Many individuals involved in the survey had either moved to other offices or were retired, and the important historic value of these maps was not reflected in their treatment. Many calls and site visits were required to locate the maps (maps for the Lewis and Clark National Forest were never found).

The base maps used for the 1975-era were found to have been drawn atop the standard 7.5-minute topographic series at a scale of 1:24,000. These subcompartment maps (see Figure 5, below) usually showed little else but the outer-most boundary and identification number. Mylar map overlays had been used to delineate stand boundaries, habitat types, transect lines, etc. For many maps, line quality and readability were poor: lines and numbers had to be retraced and the overlays rescaled to fit the base map prior to copying the maps and overlays as digitizing source material.

Digitizing the Maps

Assigning stand identifiers. To identify each stand on the hundreds of maps now ready for digitizing, a modification of the nested-hierarchical scheme used in the timber stand database was chosen. In the 1975-era stand-numbering convention, codes for forest, district, compartment, subcompartment, and stand were combined in a unique 10-digit identifier (see the breakdown, below). Not all stand polygons in the 1975-era survey subcompartments had been assigned a stand number: where stands were unlabelled, they were assigned a stand number of 999.

125

Stands on the 1935-era maps did not contain stand numbers, and so were assigned an alphabetical character in the final place of an eight-character alphanumeric identifier (see the breakdown, below). As a result, all 1935-era and 1975-era stands within the same subcompartment could be related through the same first seven digits in their ID.

	1975-era Stand-ID	1935-era Stand ID
Stand identifier format:	(05 1 16 21 012)	(05 1 16 21 E)
Forest code	digits 1 and 2	digits 1 and 2
District code	digit 3	digit 3
Compartment #	digits 4 and 5	digits 4 and 5
Subcompartment #	digits 6 and 7	digits 6 and 7
Stand #	digits 8, 9, and 10	character 8

To help manage the numerous seven-digit subcompartment ID numbers, these numbers were exported to a PC database file. To this file were added township, range, and section data for each subcompartment. Self-adhesive tags were created from these numbers for each source map and a duplicate tag set was used to mark the map check print after digitizing. The township, range, and section data were used to aid the digitizing contactor in locating reference and tic-mark locations on the numerous map sheets.

Digitizing subcompartment boundaries and stand polygons. The first digitizing step was to produce a separate layer, by forest, of the 1975-era survey subcompartment boundaries. Using these 1975 subcompartment boundaries as a backdrop (template), the stand boundaries for both eras were digitized up to or just beyond the extent of each subcompartment. The digitizers used additional registration aids, such as Public Land Survey System corners and Digital Line Graphs (DLG) of hydrography, to properly position and register the 1935-era maps. Stand boundaries were then clipped (snapped) to the extent of the subcompartment boundaries. This process produced the following three layers:

- 1975 subcompartment boundary polygons (no internal stand lines)
- 1975 stand polygons
- 1935 stand polygons

Establishing the 1935-era attribute database. All attribute data needed for the 1975-era maps were available in the Regional timber stand database in ORACLE format and required no conversion. However, for the 1935-era map set, the stand attribute codes had to be entered by hand as a comma-delimited field in the data table. Two examples of the stand attribute codes and their corresponding comma-delimited descriptor fields are:

$$\frac{3-P}{200-III} \quad 3, P, 200, III; \qquad \frac{5a-W}{121-140\ IV} \quad 5a, W, 121-140, IV$$

All attribute data were entered exactly as found on the hand-drawn map. Any code interpretations and parsing were done subsequently by the analysis staff.

All registration and digitizing tasks were done by Geodata Services, a local contractor from Missoula, Montana, using PAMAP, a non-topological GIS software package. From PAMAP, the spatial subcompartment data were saved in DLG file format for importing into ARC/INFO by the analysis staff. The contractor also provided letter-size plots of each subcompartment: one for the 1975-era stands and one for the 1935-era stands (see Figures 6 and 7 for examples): these copies were used to check for digitizing and data entry errors. 1935-era attribute data associated with the PAMAP polygons were also stored in dBase files, and were later linked to individual stand polygons through common identifier-tags. Having a local contractor close to the Regional office and the University library not only made it convenient to transfer both map documents and data files, but it also made quality control (QC) checks much easier.

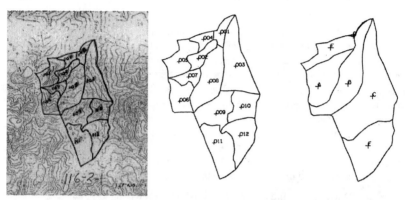

Figures 5, 6, and 7: Sample of 1975-era subcompartment map and digitizing check print, and 1935-era digitizing check print.

Hardware/Software Selection

At the time that this TDP was in its planning stages, so was the Forest Service's Project 615, a service-wide acquisition and implementation of GIS and office automation technologies. By the completion of the TDP, GIS tools were becoming available throughout the Forest Service for future projects such as this one. It was project staff's early intention to utilize Project 615 components for the TDP: in large part, they succeeded. The key hardware and software components selected in Project 615 for GIS applications (and throughout this project) include IBM RS6000 workstations running ORACLE7, ARC/INFO 7, and ArcView2.

GIS Processing Issues

During the design and pilot phase of this TDP, several key GIS processing issues had to be resolved.

Projection. Seven of the twelve forests analyzed on this project were digitized in UTM Zone 11. For uniformity of display, the remaining five forests were re-projected from Zone 12 to Zone 11. Although the re-projection of Zone 12 forests did cause a slight relative increase in polygon areas (Custer National Forest had the greatest average area increase at 1.0%), this did not affect the comparison of attributes and measurement of change over time for identical plots of ground (area being constant for both the 1935-era and 1975-era digitized maps).

Ease of use. ArcView2 was selected as a key component for this project based on its ease of use by non-GIS professionals, and its ability to easily do 'virtual joins' between ARC/INFO coverages and ORACLE or dBase tables. (ArcView2 incorporates a less-expensive and more user-friendly subset of ARC/INFO that allows users to display, query, plot, create charts, and perform limited statistical and analytical procedures to ARC/INFO coverages.) Although these tables can be stored in a remote location and edited only by permitted owners, any edits will be automatically reflected the next time that ArcView2 is initialized or re-accesses the table. This assures that current data was always used in the analyses.

Data structure. It was originally felt that the raster data structure would be superior to the vector data structure due to its more efficient data storage and more sophisticated analytical capabilities. Furthermore, since cell resolution is selected once during project design, the time-consuming process of dissolving polygon 'slivers' becomes a moot point with raster analysis. However, within ArcView2, no raster attributes are currently available for either viewing or analysis. At the time of this study, ArcView2 treated a raster as a 'dumb' image without attributes, similar in nature to a scanned image. ArcView2 had been deemed an essential component to the accessibility and analysis of the ARC/INFO coverages within this project. Although the ability to analyze raster attributes may be incorporated in later ArcView2 versions, it was primarily this present inability which determined that raster processing be dropped from further consideration.

AML scripts. Due to the nature of this project as a tool for technology transfer, it was determined early on by project staff that Arc Macro Language (AML) scripts would be utilized for the more complicated and redundant ARC/INFO processing steps. Although the actual time savings gained by using the three most sophisticated AML scripts described later was not quantified, it was substantial. The advantages become more apparent when considering the potential for exporting this TDP to other regions or ecosystems. These three AMLs were written to be sufficiently flexible that different variable names could be inserted as parameters to the AML command line. In addition, the application of automated procedures establishes a uniformity of process that cannot be duplicated by the manually-implemented methods that were initially employed on trial runs of the first few forests.

Other parameters. During the initial design of this project and the processing of its pilot forest, three key project parameters were established:

1. All stands which, for any reason, contain no data (non-forested land, rock outcroppings, or water bodies) were deleted and ignored in further analysis.

2. For field surveys, general stand delineation guidelines establish two hectares (five acres) as the minimum stand size. Accordingly, in order to minimize processing time for very small polygons, the project staff selected a minimum polygon resolution of 2 hectares for stands from both time periods.

   Therefore, each post-union polygon of less than 2 hectares in size was treated in one of two ways: if it had neighbors of the same habitat type group, it was merged into the neighbor with which it shared the longest common boundary arc (Figure 8); if there were no neighbors of the same habitat type group, it was retained. (Groupings of habitat types were used as the distinguishing attribute since the retention of individual habitat types would have resulted in an unmanageable number of classes during the analysis.)

128

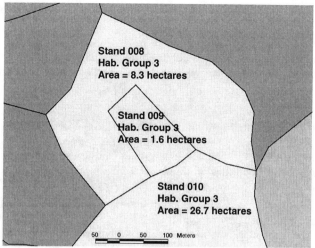

Figure 8: Candidate polygons for merging. Small stand 009 would merge into stand 008, with which it shares the longer common boundary arc.

3. Occasionally, separate stands within the same subcompartment had the same stand-ID. A naming convention had to be formulated which acknowledged the ID duplication, yet allowed the correct piece of ground to be referenced. The solution was to add an additional digit to the duplicated 1935 stand-ID (two stands identified as "1462312A" would become 1462312A1 and 1462312A2) or to index by one the leading numeral of the final three digits of the duplicated 1975 stand-ID (three stands identified as "1462312001" would become 1462312001, 1462312101, and 1462312201).

GIS Processing Sequence

As the spatial and attribute digitizing for each forest was completed in Missoula, the appropriate DLG, dBase, and ASCII data were sent to the Forest Service's Forest Health Technology Enterprise Team (FHTET) offices in Ft. Collins, Colorado, for processing. The shipment was accompanied by previously-QC'd 1935-era and 1975-era check prints of all stands in each subcompartment. All data was stored in forest-specific ARC/INFO workspaces on FHTET's IBM RS6000 workstation. The DLG files were imported as ARC/INFO vector coverages, cleaned, and dangling nodes removed. All polygons were checked for labelling and node errors, and visually verified against the check prints supplied by the digitizers where necessary.

The ORACLE data, including 1975-era successional stage attributes, were also transferred to Ft. Collins by the Missoula FHP staff and made available to ARC/INFO. (In future applications of similar projects, all ORACLE stand data will likely be stored on RS6000 workstations at each forest, and will be made available to multiple locations simultaneously via a Forest Service-wide network.)

Following the importing, editing, cleaning, and re-projection of the forests' coverages in ARC/INFO, the general ORACLE, ARC/INFO, and ArcView2 processes used for a typical forest were:

1. The 1935-era attributes that originated in comma-delimited format were imported into ORACLE tables.

2. The stand tables from both eras were joined to the 1935-era and 1975-era polygons using SQL*Net (the required ORACLE networking library) in ArcView2 via the unique tag-ID record within both the coverage and the appropriate attribute file.

3. The 1975-era ORACLE attribute table was also exported as an INFO file for subsequent use in identifying stands which contain no data.

4. The 1935-era and 1975-era coverages were unioned in ARC/INFO and the first two AMLs were run. These AMLs compared each post-union polygon to its related INFO file, identified those polygons which have no 1975-era data, flagged them, and dissolve them into ARC/INFO's 'universe polygon'—in effect, deleting them—so that they were ignored in further analysis.

5. The third AML was then run on the output of the second AML to find any remaining unioned polygon that was smaller than 2 hectares—a 'sliver'—and merge it with the neighbor of the same habitat type group with which it shared the longest common boundary arc, if such a neighbor existed.

6. The output coverage from the three AMLs (consisting of all unioned polygons that met all prior criteria for existence) was added to ArcView2 as the final 'theme', and was joined to both the 1935-era and 1975-era attributes.

7. An export ASCII file containing key attributes was created in ArcView2, combined with other attributes within ORACLE and a query was run to derive the 1935-era successional stage of each polygon.

8. The table of successional stages and other essential derived attributes was made available to ArcView2 and was joined to all unioned coverage polygons. Tables were prepared within ORACLE and transmitted to other project staff for final statistical analysis and derivation of the transitional matrix.

RESULTS

Two sets of analyses from the Nez Perce National Forest illustrate the effectiveness of using GIS to manage spatial and attribute data for a project such as this; other groups have now taken advantage of data made available through this project; and documentation of this project will aid future similar projects.

Accurately managing the data. The results of the analysis completed during this TDP provided the opportunity to make a comparison between the use of conventional stand-based model predictions and actual site conditions. The Forest Vegetation Simulator (FVS), a widely-used, stand-based growth-and-yield model which incorporates the "best successional guesses" of forest ecologists, was modelled without the addition of root disease, dwarf mistletoe, or Douglas-fir beetle: all of which showed a high potential for presence in the Nez Perce 1935-era data set. Only 41 percent of the sites underwent the FVS-predicted transition by the 1975 era. The remaining 59 percent converted to other cover types, controlled to varying degrees by the actions of multiple insects and pathogens. The result could be seen in greatly reduced total cubic volume production from that predicted by FVS. Values derived from the analysis of accurately managed data sets such as these provide the Forest Service's best hope of more accurately predicting future vegetation conditions.

Querying and selecting the data. A typical result of mountain pine beetle (MPB) outbreaks is the killing of lodgepole pine (lp) cover type and increasing the stands' rate to climax by converting the cover type to subalpine fir. The analysts found thirteen of the 272 Nez Perce post-union polygons which had undergone a change in cover types from lodgepole pine to subalpine fir from the 1935 era to the 1975 era. The appropriate polygons were isolated in ArcView2 by making logical queries against the cover types for the two time periods (see Figure 9). Upon selecting those polygons, they also found that the mean value of the MPB_lp functional index from the 1935-era data set was 1.38, with a standard deviation of 0.33. When running summary statistics within ArcView2 on the remaining 259 polygons from the Nez Perce, they found the mean MPB_lp index value was 0.81, with a standard deviation of 0.59. This comparison of central tendencies from the two data sets suggests that, where the classic cover type transition associated with mountain pine beetle attack had occurred, the index value which indicates likelihood of agent occurrence was much higher than in the remainder of the data set. The accurate management of data and the ease with which analysts could logically select spatial features based on attribute values allowed them to more effectively visualize cause-and-effect relationships and to incorporate such findings into successional models.

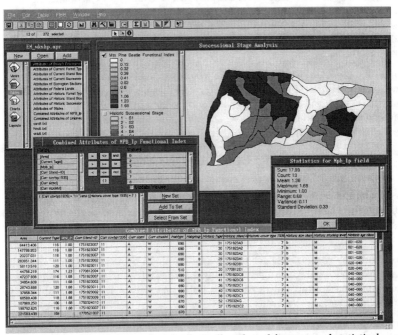

Figure 9: ArcView2 screen showing logically-selected features and statistical results.

Access to spatial and attribute data. While the TDP was still underway, the project staff received, and was able to fulfill, five requests for specific tabular results of the analysis or maps of subcompartments that met specific criteria. The Northern Region was being introduced to the advantages of having GIS analytical and mapping capabilities, and was able to see these capabilities applied to specific ongoing projects.

Documentation of the process. In addition to this paper, three other forms of documentation have been prepared regarding the GIS-related aspects of this project. They are available to those wishing to pursue a similar project in other Forest Service regions or ecosystems. They are:

- A more detailed outline of the steps employed in ARC/INFO, ORACLE, and ArcView2.

- Complete text and documentation on the three AML scripts which act on post-union polygons to delete those without 1975-era data and merge slivers with neighbors of the same habitat type group.

- Step-by-step documentation of all ARC and INFO steps on a keystroke level, as captured in ARC/INFO 'watch' files.

CONCLUSION

Prior forest plans have tended to communicate to the public that, barring direct management actions, forest composition remains unchanged. Forest ecologists are finding that, by incorporating insect and pathogen activities, they can dramatically increase their accuracy in predicting future vegetation conditions. Processes from this project can be employed to generate more meaningful forest plans, to analyze alternative actions, and to more accurately communicate those alternatives to the various publics served by the Forest Service.

To fulfill this goal, projects of this nature require the ability to manipulate enormous amounts of data. The ability of GIS to accurately manage spatial and attribute data, query the data set and derive additional information, map the results in a human-consumable form, and generate reports and tabular data is essential, and could not be done otherwise with the same accuracy or efficiency.

RECOMMENDATIONS

In order to enjoy the expertise of a local digitizing contractor familiar with Forest Service procedures and the 1935-era data, project staff had to convert PAMAP spatial files and dBase attribute tables to ARC/INFO format prior to data preparation and analysis. Where possible, digitizing should be accomplished in the same GIS data format as that used for analysis.

Environmental Systems Research Institute, the developer of ARC/INFO software, is currently reported to be investigating giving ArcView2 the ability to assign attributes to their raster data structure (GRID) module. If that materializes, it may be useful to revisit the results of our raster vs. vector analysis.

The GIS-related processes developed in the course of this TDP have proven successful in application, and could be adapted wholesale for use on other Regions, ecosystems, and watersheds. To further verify the general applicability of these processes, they should also be tested at other scales, with other vegetative types, and other agents of concern. This groundbreaking study has added yet another valuable tool to address the complex demands of ecosystem management, harnessing a new technology for use in the diverse lands under the care and custody of the Forest Service.

GLOSSARY

ecoregion section The smallest unit of the nested hierarchical ecoregion delineations (Bailey et al. 1994), including domains, divisions, provinces, and sections. Seventeen sections are required to encompass the thirteen national forests in the Northern Region.

ecosystem management Ecosystem management is the use of an ecological approach to achieve multiple-use management of national forests and grasslands by blending the needs of people and environmental values in such a way that national forests and grasslands represent diverse, healthy, productive, and sustainable ecosystems. It is a principle developed in the National Hierarchical Framework of Ecological Units widely embraced by the Forest Service under the leadership of its Chief, Jack Ward Thomas.

habitat type Habitat type is defined as "an aggregate of all land areas potentially capable of producing similar plant communities" (Pfister et al. 1977, quoted in Perry 1994, p. 87). The exact habitat type is typified by the expected climax vegetation based on climate, soil type, slope, and aspect.

land unit scale A land unit scale is the largest of the four hierarchical scales used for planning and analysis within small ecological units, corresponds to map scales ranging from 1:250,000 to larger than 1:24,000, and generally utilizes polygons ranging from thousands of acres to less than ten acres (see the National Hierarchical Framework of Ecological Units, Table 3).

succession The progressive change in plant communities toward a hypothetical dynamic equilibrium (climax) state.

successional functions The results of pathogen and insect activities that lead to changes in successional stage or alter the course or timing of vegetative succession.

successional modeling Computer modeling of probable changes in succession based on present and projected factors.

successional stage A point in the biotic succession of a site characterized by the dominance of a single type of vegetation.

REFERENCES

Bailey, Robert, et al. 1994. Ecoregions and Subregions of the United States. USDA Forest Service.

Hagle, Sue, and Williams, Steve. 1995. A Methodology for Assessing the Role of Insects and Pathogens in Forest Succession. In Analysis in Support of Ecosystem Management: Analysis Workshop III. USDA Forest Service.

Perry, David. 1994. Forest Ecosystems. Johns Hopkins University Press, Baltimore.

Pfister, R.D., Kovalchik, B., Arno, S., and Presby, R. 1977. Forest habitat types of Montana. USDA Forest Service Technical Report INT-34. Intermountain Research Station, Ogden, Utah.

National Hierarchical Framework of Ecological Units. 1993. USDA Forest Service, Washington D.C.

ACKNOWLEDGEMENTS

The authors would like to extend their thanks to the following persons, without whom this project could not have been successfully completed:

Dale Johnson, Archivist, University of Montana, for his rescue of the 1935-era maps.

Chuck Siefke, Computer Specialist, MACA, for his expertise in the manipulation of ORACLE data.

Steve Williams, Project Leader, USDA Forest Service, for his review of the manuscript and comments on contents.

Sue Hagle, Plant Pathologist, USDA Forest Service, for elucidating the mechanisms of insect and pathogen modeling.

Mike Badar, Application Specialist, ESRI, for writing the three principal AMLs.

Ken and Robin Wall, GeoData Services, for their persistence in accurately digitizing from less-than-clean data sources.

Mark Riffe, Technology Transfer Specialist, MACA, for his editorial acumen.

Jeanine Paschke, Imaging Systems Specialist, MACA, for assistance in preparing figures for this paper.

SPATIALLY EXPLICIT ECOLOGICAL INVENTORIES
FOR ECOSYSTEM MANAGEMENT PLANNING
USING GRADIENT MODELING AND REMOTE SENSING

Robert E. Keane, Cecilia H. McNicoll,
Kirsten M. Schmidt and Janice L. Garner

USDA Forest Service
Intermountain Research Station,
Intermountain Fire Sciences Laboratory
P.O. Box 8089
Missoula, MT 59807
Phone: 406-329-4846, FAX: 406-329-4877

ABSTRACT

Planning future ecosystem management projects will greatly depend on comprehensive, spatially-explicit ecological inventories. This paper presents a study-in-progress that will develop a mechanistic, process-based sampling, analysis and mapping system that efficiently creates spatial data layers of ecosystem attributes across multiple spatial and temporal scales using a combination of gradient modeling and remote sensing. Gradient models describe the response of ecological variables, such as vegetation structure and composition, along a range of environmental conditions. Remote sensing and image processing are used to map general successional and ecological characteristics that are treated as gradients in the model. Linkage of gradient models with remote sensing will allow an economical, detailed, and comprehensive classification of ecosystem features into thematic data layers portable to Geographic Information Systems (GIS) that describe the ecological properties of a landscape. Over 950 plots were established on the Nez Perce and Kootenai National Forests of Idaho and Montana to quantify important environmental gradients, describe ecosystem characteristics, and define ground-truth and training areas. Collected data are now being analyzed and the computer models integrating gradients with image processing are now being written.

INTRODUCTION

Successful ecosystem management and planning projects will require accurate, spatially-defined ecological data describing ecosystem characteristics and ecosystem processes in a landscape context (see Jensen and Bourgeron 1993). Moreover, the quantification of these ecosystem properties is greatly reliant on efficient, economical, credible, and ecologically-based land system inventories. Mapped ecosystem attributes are critical inputs to many ecosystem planning efforts because they can be used by a wide variety of analysis and simulation tools that evaluate trends, compare management alternatives, and ensure conservation of ecosystem components and processes.

Remote sensing coupled with ecological inventory and analysis offers an alternative to extensive systematic sampling for describing spatial attributes across an entire landscape (Gosz 1993, Jensen 1986). However, the spectral, spatial, and temporal resolution of affordable remotely sensed imagery products are sometimes inappropriate for describing some important ecosystem characteristics such as fuel loadings, biomass amounts, and fire regime. The process of classifying remotely-sensed imagery is more an art than a science, and it is not always repeatable. Achieving high imagery classification accuracy often results in either an extensive and costly ground-truth sampling effort or a compromise in the usefulness of the classification to natural resource management. New image processing technologies allow the use of biophysical and plant autecological information to improve classification accuracy. However, this requires an extensive knowledge of the relationships between site and vegetation conditions across the classified landscape. Interestingly, this is exactly the type of ecological information used in traditional models of vegetation structure and composition known as gradient models (Kessell 1979).

What is needed then is a sampling, analysis and mapping system that efficiently quantifies important ecosystem attributes across multiple spatial and temporal scales. This paper describes a research project that is developing such a system using gradient modeling and remote sensing. The project's objective is to develop cost-effective methods to rapidly generate spatially-explicit inventories of ecosystem characteristics at a landscape-level. The system would economically prepare thematic data layers portable to Geographic Information Systems (GIS) that comprehensively describe ecological properties of a landscape in a spatial domain (USA CERL 1990). The system would use environmental, biological and spectral data to quantify important ecological gradients affecting ecosystem processes and conditions. These gradients would then be used to generate maps of ecosystem characteristics at various scales. The generated maps can be used in ecosystem management planning or "real-time" situations such as input for fire behavior predictions.

This study will be accomplished in four phases. First, a field sampling methodology was designed and implemented to obtain comprehensive inventories of ecosystem attributes at various spatial and temporal scales. Second, a mechanistic gradient model will be developed from the sampled field data and, also from data generated from simulation models, to quantify ecosystem properties across a landscape. Third, image processing of remotely-sensed data products will be used to classify the landscape into general ecosystem categories that are represented in the gradient model. Lastly, the gradient model and image classification procedure will be linked to create a comprehensive mapping system for generating spatially-explicit ecological inventories called the **Landscape Ecosystem Inventory System (LEIS)**.

STUDY AREAS

The Upper Kootenai Study Area (UKSA) on the Kootenai National Forest in northwestern Montana (Figure 1) and Lower Salmon River Study Area (LSRSA) on the Nez Perce National Forest in central Idaho (Figure 2) are two large (5,900 and 6,000 km^2, respectively) and diverse regional landscapes selected for this study. These watersheds are bounded by the Hydrologic Unit Code delineation at the 4th code level (Seaber et al. 1987). They were selected because they are topographically, geologically, and vegetationally different, but they are ecologically representative of surrounding land areas. In addition, there is an estimated 10 to 20 percent overlap in environmental gradients across the two study areas which will allow expansion of ecological gradients to areas outside the study watersheds. A extensive Geographic Information System (GIS) for each study area includes many spatial data layers georeferenced in UTM coordinates stored in both raster and vector formats.

136

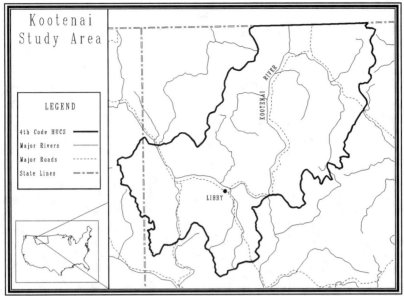

Figure 1 -- *The Upper Kootenai Study Area (UKSA), Montana, USA.*

ECOLOGICAL INVENTORY DESIGN AND IMPLEMENTATION

A comprehensive, hierarchically-nested sampling design was developed to inventory important ecosystem characteristics on the landscape using the USDA Forest Service Northern Region's ECODATA system as a starting point (Jensen et al. 1993, Hann et al. 1988, Keane et al. 1990). Data collection emphasizes the sampling of ecosystem processes, thereby ensuring adequate coverage of ecosystem attributes. Moreover, ecosystem process sampling provides the context in which to interpret ecosystem conditions and dynamics. This ecological inventory was designed for three purposes: (1) obtain ecological information along process gradients to construct the gradient model, (2) delineate ground-truth and training areas for image classification, and (3) quantify initialization and parameterization data for ecological simulation models.

The primary sampling unit was a circular ECODATA macroplot approximately 0.04 ha in area. Stratification of potential plot locations was based on the frequency of ecosystem processes across several spatial and temporal scales within the study areas as discussed in the next sections.

Regional Stratification

Both study areas were divided into units called **landscapes** based on watershed delineations done at the sixth-level Hydrologic Unit Codes (HUC) (Seaber et al. 1987) (UKSA landscape divisions shown in Figure 3). Approximately 20 percent of these watersheds were selected for sampling based on accessibility, geographical distribution, and presence of ecosystem processes (see labeled landscapes in Figure 3). Accessibility and geographic location was assessed from road and trail maps and GIS data layers. Distributions of regional ecosystem processes were assessed from coarse-scale climate, geomorphology, and hydrology GIS data layers developed for the Interior Columbia River Basin (ICRB) Scientific Assessment (Keane et al. 1996b). Climate was represented

137

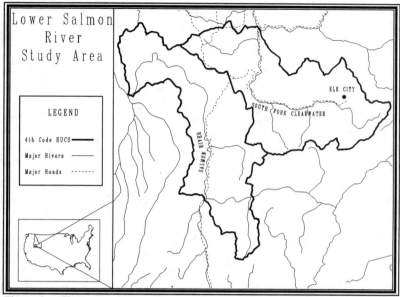

Figure 2 -- *The Lower Salmon River Study Area (LSRSA), Idaho, USA*

by precipitation and temperature maps (2 km² resolution) simulated from extensive weather data gathered for 1989 weather year (Thornton et al. 1996). Geomorphological processes were mapped using regional delineations of subsections (Bailey 1995) as created from the ICRB effort (Nesser et al. 1996). Hydrology was implicitly accounted for in the landscape stratification since landscapes are delineated using hydrologic boundaries.

Landscape-Level Stratification
Hierarchically nested under landscapes are **landscape polygons (LANDPOLYs)** that define areas having uniform biophysical, vegetation, and disturbance conditions. Not all LANDPOLYs within a landscape were delineated and sampled in this effort. Instead, only LANDPOLYs representing important environmental, vegetational and biophysical gradients were sampled. Landscape polygons are similar to stands except LANDPOLY's define a homogeneous portion of a stand. Each sampled LANDPOLY was delineated on maps in the field and later digitized into a GIS. Potential vegetation type (Pfister et al. 1977, Cooper et al. 1992), cover type, and structural stage classifications provided the primary criteria for the selection of LANDPOLYs to sample. Secondary criteria included community type, soil, aspect, slope, slope position and drainage. Aerial photos and maps were used to detect major changes in these criteria along the environmental gradients.

Stand-Level Sampling
The finest sampling unit, the **macroplot,** was established in each delineated LANDPOLY. Ecological conditions within a macroplot represented the ecological conditions of the entire landscape polygon (Mueller-Dombois and Ellenburg 1974). Gradient transects (termed GradSects) are straight-line routes traversing the greatest amount of vegetational and environmental variability within a selected landscape, and they were used to establish LANDPOLYs and their associated macroplots on a landscape.

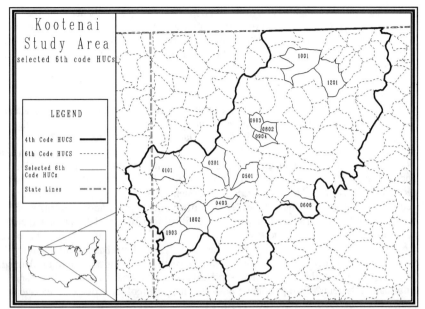

Figure 3 -- *Distribution of 6th code HUCs in the Upper Kootenai Study Area. Numbered 6th code HUCs are the* **landscapes** *selected for this study.*

<u>Ecosystem Inventory Plots.</u> It took over 3 hours to measure the many ecological variables on a macroplot. Biophysical parameters evaluated at each macroplot included elevation, aspect, slope, soil characteristics, and potential vegetation type. Diameters, heights, ages, growth rates, and crown dimensions were estimated for all trees within a macroplot boundary. The forest floor was described from measurements of downed woody fuel loadings, and fuel, duff and litter depths. Percent canopy cover and average height were estimated for all vascular and non-vascular (mosses and lichens) species. Ecophysiological measurements included leaf area index (LAI), leaf longevity by tree species, and specific leaf area. The disturbance history of insects, diseases, grazing, harvesting, and fire was also evaluated for each macroplot.

<u>Temporal Process Sampling.</u> Several permanent macroplots were established in four representative LANDPOLYs on four landscapes. Many primary ecosystem processes were measured at various time intervals on these plots to quantify simulation model parameters and understand the temporal dynamics in process classifications. Litterfall, soil and needle respiration, photosynthesis, and transpiration are among the more important variables measured at hourly, daily and monthly timesteps for this study.

<u>Field Sampling Results.</u> Over 950 macroplots were established by 8 crews of 2 people over the 20 landscapes selected in both sampling areas during the 1995 field season. These data have been entered into database structures and have been scanned for errors.

GRADIENT ANALYSIS AND MODELING

A gradient model will be constructed to predict distributions of ecosystem characteristics across the landscape based on distributions of sampled environmental gradients in study areas. This project will derive environmental gradients using a mechanistic approach that quantifies the processes that control ecosystem properties rather than inferring processes from surrogate environmental measurements such as elevation (Gauch 1982, Kessell 1979). These mechanistic gradients will be computed from the field data and simulation modeling.

Background
Gradient modeling has seen limited use in natural resource management (see Kessell 1979) because gradient analysis of ecosystem data requires detailed knowledge of complex mathematical analysis tools (Gauch 1986, Kessell 1979, ter Braak 1987) which often discourages many people. In addition, most easily-measured environmental gradients are often the secondary or indirect factors influencing vegetation distribution and abundances. Primary ecological factors are often inferred from the these surrogate, secondary gradients. However, recent ecosystems research has identified a suite of ecosystem processes-oriented variables that often dictate vegetation dynamics across several scales (see Waring and Schlesinger 1985). Simulation models can now quantify these processes in a spatial and temporal domain (Running and Hunt 1993). This paves the way for a new generation of gradient modeling that will mechanistically quantify the important environmental gradients influencing vegetation. Quantification of mechanistic gradients from a combination of field sampling, computer simulation, and Geographical Information Systems (GIS) provides a powerful mapping tool for ecosystem management projects (Greer 1994, USA CERL 1990).

Advantages of using gradient modeling over other classification schemes are numerous. First, gradients are often scale independent, flexible and portable. Gradients are similar in lands outside the sampled areas thereby expanding the utility of the ecological data. Many gradients are static and do not change over time (e.g., topography) so replicated sampling is not necessary. Computed gradients can be modified and refined as additional land areas are sampled and more environmental variables are measured.

Gradient Analysis
Quantification of the important gradients within each study area will be accomplished directly from the field data and from simulation efforts. Gradients will be directly related to ecosystem characteristics using multivariate statistical analysis (i.e., regression, constrained ordination, logistic regression, discriminant analysis) and higher-order mathematical analysis. Gradient analysis of independent, continuous ecosystem process variables, such as net primary productivity and elevation, will also use multivariate statistical and mathematical methods such as regression, discriminant analysis and generalized linear modeling. All variables measured in the field will be used as independent variables in the direct gradient analysis, including any synthesis and summary of the field data. Other variables computed or simulated from field data will also be incorporated in the gradient analysis. These variables include plant biomass, organic matter weight and primary productivity. Ordination, detrended correspondence analysis and canonical correspondence analysis are additional methods of indirectly investigating gradients that control ecosystem characteristics using vegetation composition and structure (Gauch 1986, Kessell 1979).

Mechanistic Gradients
The development of mechanistic gradients will be accomplished using simulation techniques. Field data will be used as input to various ecosystem process models to compute the hypothesized causal mechanisms that dictate presence and abundance of

ecosystem components. These models will be parameterized using the sampled field data and the detailed ecosystem process measurements obtained from the permanent macroplots where process measurements were taken over time. Possible simulated gradients for investigation include evapotranspiration, net primary productivity and nitrogen content. Various ecosystem models such as BIOME-BGC (Running and Hunt 1993), MTCLIM (Hungerford et al. 1989), CENTURY and Fire-BGC (Keane et al. 1996a) will be used to quantify these important processes.

Many parameters will need to be determined prior to simulation modeling. Weather data will be collected for all years on record for each weather station within and near each sampled macroplot. Species and stand ecophysiological parameters will be quantified from a thorough literature search (Keane et al. 1996a) and from measured and historical field data. Soils and geological information will be obtained from digitized soils maps and surveys. The temporal measurements of ecophysiological variables will also be used to test and validate simulation model results.

The Gradient Model
Results of the gradient analysis will be efficiently stored in digital format. However, the number and complexity of gradients and ecosystem characteristics in this study preclude the storage of gradient data using conventional means. Complex mathematical techniques and data structures will be developed to efficiently store gradient relationships in easily accessible databases. The gradient model will be programmed in the C++ language on a UNIX workstation. It will be intimately linked with GIS software to provide the full "glory" of spatial ecosystem analysis.

REMOTE SENSING AND IMAGE CLASSIFICATION

Most remotely-sensed, image classification projects concerned with natural resource inventories identify vegetation communities or land cover types on the landscape. These community type classifications are then passed to land management for application to current projects (see Greer 1994). Land management will typically assign resource-oriented attributes to each land cover class to map ecosystem characteristics on the landscape for planning (Greer 1994). As a result, errors in the spectral classification are compounded with errors resulting from attribute assignments to yield maps that do not always portray a true spatial representation of ecological components.

This study will relax the number of land cover categories in the image classification to improve accuracy and applicability. Instead of predicting 20 to 50 specific cover categories, this effort will only predict 5 to 10 classes. These classes will be general and describe broad biome-level differences (e.g., forest, rangeland, water). These broad categories will be represented as successional and land-use gradients in the gradient model.

Image Acquisition
Commercial and noncommercial satellite imagery products have been obtained for both study areas. SPOT, TM and MSS scenes will be assessed along with other imagery products identified through a product search. Imagery products NOT available to land management agencies because of high cost, high data storage requirements, or limited license distribution will not be used. We will also use the same imagery products from two or more time periods to integrate temporal changes in spectral signatures in the imagery classification.

This remote sensing portion of the study will be accomplished under two separate efforts. First, ecosystem characteristics will be classified using common imagery products to evaluate optimal spectral and spatial resolution for the mapping of these characteristics for ecosystem management. A second effort will evaluate the utility of high spectral and spatial resolution imagery in ground-truthing efforts.

LANDSCAPE ECOLOGICAL INVENTORY SYSTEM

There are major limitations to using only remote sensing or only Gradient Modeling to construct maps of ecological characteristics. Remote sensing relies on the reflective properties of a stand to consistently predict locations of other similar stands (Jensen 1986). Unfortunately, light reflection is not always an accurate or consistent variable to use in the spatial prediction of biotic elements. Many physical factors limit its predictive ability including shadow, atmospheric distortion, composite pixels, sensor inadequacies, scattering, and sensor resolution. It is common for two very different vegetation communities to have the same spectral signature. Next, the mathematics and statistics used to "train" spectral distributions to predict vegetation characteristics are complex and may be limiting.

Predicting existing conditions is a major limitation of the gradient modeling approach. Quantification of successional pathways and the factors that control successional trajectories is costly and enigmatic using gradient modeling, especially considering the detail needed for ecosystem management projects. Moreover, existing conditions mapped using a gradient model are often dictated by the often coarse spatial resolution of mapped gradients. Abrupt changes in the biota at smaller scales are difficult to quantify using coarsely mapped gradients. For example, riparian communities and fine scale successional changes are difficult to map using a gradient approach because most GIS layers do not have the detail to identify the environmental factors that regulate these communities.

The integration of gradient modeling with remote sensing is a logical direction for future, successful ecosystem management map-making projects. The ability of remote sensing to detect subtle changes in landscape composition, coupled with the ability of gradient modeling to predict geographic distributions of biotic communities will enable land managers to quickly construct ecological maps of project areas for land management planning. Gradient modeling also allows the spatial description of important processes on the landscape which are very important in trend analysis. This study will develop an automated system that will create maps of ecosystem characteristics for any area using a combination of remotely sensed digital data and gradient analysis.

A system of gradient data structures will be developed and stored on a main computer (Figure 4). These data structures will be designed so they can be modified as additional areas are sampled. The data structures can then be used to make GIS spatial data layers for any area encompassed by the gradients rather than only the effective sampling area. Remotely sensed data will be used to generate spatial representations of those gradients that change over time (i.e., successional status can be assessed from new imagery, Figure 4). This entire system will be implemented on a UNIX workstation and automated so that users can submit the required input layers and quickly obtain maps of ecosystem characteristics, hopefully within a few days.

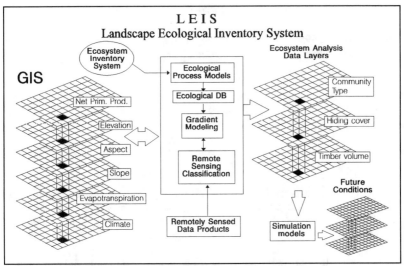

Figure 4 -- *Diagram of the Landscape Ecosystem Inventory System.*

POTENTIAL APPLICATIONS

Future applications of spatial data generated from LEIS are numerous and diverse. General ecology uses include the quantification of landscape structure and composition. Fractal dimension, patchiness, fragmentation, corridors and other landscape attributes can be assessed and used in wildlife analyses. Fire, insect and disease disturbance regimes can be directly classified on a landscape using mapped gradients. Productivity is explicitly modeled for GIS layers. Input parameters to many ecosystem simulation models can be mapped with this system allowing the projection of future landscape dynamics as a consequence of management policies (Figure 4). Fuel model and fuel loading maps can be generated for fire behavior simulations for planning and "real-time" applications. Fuel loading maps can be used to predict fire effects such as smoke and tree mortality. Stand attributes influencing animal species distributions, such as snag density, hiding cover and thermal cover, can be delineated spatially using the LEIS system. Mapping of threatened and endangered species habitats is easily accomplished by explicitly defining the gradients important for their distribution. Timber volume can be coarsely mapped to provide information as to harvest schedules.

ACKNOWLEDGMENTS

We thank Wendel Hann, Northern Region USDA Forest Service; Michele Wasienko-Holland, Lolo National Forest, USDA Forest Service; Dan Leavell, Kootenai National Forest, USDA Forest Service; Pat Green, Nez Perce National Forest, USDA Forest Service; Colin Hardy, Bob Burgan, James Menakis, Don Long, Scott Mincemeyer, and Todd Carlson, Intermountain Fire Sciences Lab, USDA Forest Service; Joseph White and Peter Thornton, University of Montana, Missoula, MT; and Kathy Schon-Rollins, Brian Paulson, Myron Holland and John Pierce, Missoula, MT for their valuable help during this effort.

REFERENCES

Bailey, Robert G. 1995. Description of the ecoregions of the United States. Washington, DC: Misc. Publ. No. 1391. U.S. Department of Agriculture Forest Service. 108 p.

Cooper, S.V., K.E. Neiman, and D.W. Roberts. 1991. Forest habitat types of northern Idaho: A second approximation. USDA Forest Service General Technical Report INT-236. 143 pages.

Gauch, H. 1982. Multivariate analysis in community ecology. Cornell University Ithaca, New York: Cambridge Studies and Ecology. 298p.

Greer, Jerry D (Editor). 1994. Remote Sensing and Ecosystem Management. Proceedings of the 5th Forest Service remote sensing applications conference. American Society of Photogrammetry and Remote Sensing. Bethesda, MD. 377 pages.

Gosz, James R. 1993. Ecotone hierarchies. Ecological Applications 3(3):369-376.

Hann, W.J., M.E. Jensen, and R.E. Keane. 1988. Chapter 4: Ecosystem management handbook -- ECODATA methods and field forms. USDA Forest Service Northern Region Handbook. On file at Northern Region, Missoula, Montana.

Hungerford, R.D.; Nemani, R.R.; Running S.W.; Coughlan, J.C. 1989. MTCLIM: A mountain microclimate simulation model. Research Paper INT-414. Ogden, UT: U.S. Department of Agriculture, Forest Service, Intermountain Research Station. 52 p.

Jensen, J.R. 1986. Introductory digital image processing. Prentiss-Hall, Englewood Cliffs, New Jersey. 379 pp.

Jensen, M.E. and P.S. Bourgeron (editors). 1993. Eastside Forest Ecosystem Health Assessment, Volume II, Ecosystem Management: Principles and Applications. USDA Forest Service, National Forest System Information Report. 344 pages.

Jensen, M.E., W.Hann, R.E. Keane, J.Caratti, and P.S. Bourgeron. 1993. ECODATA--A multiresource database and analysis system for ecosystem description and evaluation. In: M.E. Jensen and P.S. Bourgeron editors, Eastside Forest Ecosystem Health Assessment, Volume II, Ecosystem Management: Principles and Applications. USDA Forest Service, National Forest System Information Report. pg 249-265.

Keane, R.E., M.E. Jensen, and W.J. Hann. 1990. ECODATA and ECOPAC - analytical tools for integrated resource management. The Compiler 23:11-24.

Keane, Robert E., Penelope Morgan and Stephen W. Running. 1996a. Fire-BGC -- a mechanistic ecological process model for simulating fire succession on coniferous forest landscapes. USDA Forest Service Research Paper INT-484. 122 pages.

Keane, Robert E., J. Jones, and W. Hann (Compilers). 1996b. Broadscale landscape dynamics of the Interior Columbia River Basin. USDA Forest Service General Technical Report PNW-GTR-(in preparation).

Kessell, Stephen R. 1979. Gradient modeling: resource and fire Management. Springer Verlag, New York. 432 pages

Mueller-Dombois, D., and H. Ellenburg. 1974. Aims and methods of vegetation ecology. New York, NY: John Wiley and Sons. 547 p.

Pfister, Robert D.; Kovalchik, Bernard L.; Arno, Stephen F.; Presby, Richard C. 1977. Forest habitat types of Montana. Gen. Tech. Rep. INT-34. Ogden, UT: U.S. Department of Agriculture, Forest Service, Intermountain Forest and Range Experiment Station. 174 p.

Running, S.W.; E. Raymond Hunt. 1993. Generalization of a forest ecosystem process model for other biomes, BIOME-BGC, and an application for global scale models. In: Scaling Physiological Processes: Leaf to Globe. Academic Press, Inc. Pages 141-157.

Seaber, P.R., P. Kapinos, and G.L. Knapp. 1987. Hydrologic unit maps. US Geological Survey Water-Supply Paper 2294. 63 pages.

ter Braak, C.J.F. 1987. Unimodal models to relate species to environment. Agricultural Mathematics Group, Box 100, NL-6700 AC Wageningen, The Netherlands. 13 pages.

Thornton, P.E, S.W. Running, and M.A. White. 1996. Generating surfaces of daily meteorological variables over large regions of complex terrain. Journal of Hydrology (In press).

Waring, Richard H. and William H. Schlesinger. 1985. Forest ecosystems: concepts and management. Academic Press Inc. New York, New York. 340 pages.

USA CERL. 1990. GRASS 4.0 Reference Manual. United States Army Corps of Engineers, Construction Engineering Research Laboratory, Champaign, Illinois. 208 p.

145

IMAGERY USED IN SUPPORT OF GIS ON THE STIKINE AREA OF THE TONGASS NATIONAL FOREST

James R. Schramek
Stikine Area, Tongass National Forest
P. O. Box 309
Petersburg, Alaska 99833

ABSTRACT

The Stikine Area has 8 years experience in ARC/INFO GIS, 3 years experience with airborne video imagery and 4 years experience with GPS data collection. We've also some experience with satellite imagery. We scanned prints of existing orthophoto imagery creating our own georeferenced, digital orthophotos. We've moved from the DG to IBM UNIX processing for our GIS and have both 615 and very similar non-615 systems on the Area using ARC/INFO software. We've developed user-friendly GUI tools accessing both our orthophotos and satellite imagery for use with GIS data. We are working in cooperation with other Areas of the Tongass and Pacific Meridian for multispectral analysis of LANDSAT TM imagery in creation of an "existing vegetation" inventory. We've been operating the Region's Trimble GPS base station for nearly 4 years and have done a variety of processing to bring GPS data into GIS. We've taken airborne video both from fixed wing and rotary aircraft and we've made use of imagery both as raw video and as processed, georeferenced imagery in addition to creating referenced vectors from the imagery. Our remoteness has probably been the main reason that we seem to have developed quite a variety of our own capabilities here at one Supervisor's office. This experience on a full spectrum of GIS and remote sensing capabilities offers a useful perspective to share some 'do's and don'ts" and to offer some suggestions for Agency emphasis in the future.

INTRODUCTION

The Stikine Area encompasses the central 4 million acres of the 17 million acre Tongass National Forest. It has one of the three Supervisor's offices on the forest with the office located in Petersburg. There is a Ranger District office in Petersburg and another in Wrangell. The Tongass is made up of coastal mountain ranges with deeply incised coastal fiords and over 1,000 islands. There are over 11,000 miles of shoreline, and much of the terrain is steep and heavily vegetated. Spring and summer are cool and wet. Torrential rains are common in the fall. Winters are usually near freezing, and vary between snow and rain. The lands are sparsely populated. The remote islands and rugged terrain can make traveling hazardous as well as difficult. Many areas are accessible only by boat or aircraft, often in marginal weather conditions. Although Southeast Alaskan skies can be spectacularly clear and blue, they are more frequently dark and gray. Many clear days also have heavy morning ground fog that often extends well past midday. Few areas on earth can be more challenging to schedule the taking of imagery whether by satellite or aircraft.

The author participated in efforts to acquire GIS capability for the Stikine Area starting in 1982. A comprehensive study of resource information needs and implementation plan for the Tongass National Forest was completed on contract in 1985 (ESRI 1985). Modified plans were developed and the necessary technical approval documentation for a Tongass-wide procurement was finished in late 1986. ARC/INFO was installed on the Area's Data General system in June of 1987. The first year and a half included mostly an organized panic to complete automation of existing resource inventory to provide a database for revising the Tongass Land Management Plan (TLMP). Focus broadened in 1989 to include most project work. Initial enthusiasm for GIS was gradually tempered by the reality that our computer systems were marginal at best to support user-friendly graphical interfaces and most processing in anything close to "real time". One of the early "casualties" was the notion that most resource specialists would routinely make use of GIS in their project work. Quite a lot of work was accomplished using GIS but at pretty high levels of effort and mainly by GIS specialists writing macros that were processed in batch mode.

Four IBM 58H workstations were leased by the Tongass just over a year ago, one for the TLMP Revision Team and one for each supervisor's office. The ranger districts on the Stikine were each equipped with Project 615 contract workstations several months later. All GIS processing on the Area now occurs on UNIX workstations with ARC/INFO Rev. 7.03. Migration into the workstation environment has required a learning curve and some operational adjustments but it also has already provided a slight increase in access to terminals, a doubling in available GIS commands, a ten fold increase in data storage, more than a thousand times faster processing and the ability to develop powerful and easy to use tools for resource specialists. Many of the most exciting new accomplishments use remotely sensed imagery. The vision of resource specialists being routine users of GIS is becoming reality on the Stikine Area.

(Product names are used solely for clarification and do not constitute endorsement by either the author or the USDA Forest Service)

GIS DATA CAPTURE

Imagery was a primary source for nearly half the GIS layers but was only of use in hard copy form either for interpretation from resource photos or as orthophotos used for the manuscripting base. Resource specialists worked with photos to create manuscripts or overlays to orthophoto prints for input to GIS. The remaining layers were automated from maps.

Imagery, especially in the form of traditional 1/15,840 scaled, 9x9 inch resource photography has been an integral part of field work on the Area for many decades. Orthophotography initially became available following a State of Alaska sponsored project to capture high altitude imagery in 1979. The CIR imagery was converted into orthophoto prints at 1/31,680 with flight line orientation and varying print sizes. Unrectified enlargements of the original images (/160,000 enlarged to 1/31,680) were used for landform, soils and vegetation mapping efforts along with some stream channel inventories. Imagery for quarter quad centered (1/4th of 15 minute quads) orthophoto production was obtained for the Stikine Area in 1985, with 1/31,680 referenced prints available the following year. All this imagery was used either for interpretation, for use as overlays to other photo prints, or for capture of features to transfer onto maps. Mylar overlay manuscripts of the 1/31,680 quad centered orthophotography formed the basis for soils, streams and watershed layers that were automated for the Stikine Area, with all the remaining layers generated mainly from vectors drawn on maps through a variety of means. Most of the data automated for TLMP was of "project level" or nearly so in the "paper and photo print paradigm". They were definitely the best available information.

Several of these layers automated for TLMP each contained versions of the same features. Consequently, there may be 4 or 5 similar but not identical vectors representing the same feature but from different layers. For TLMP, these layers were sampled with an overlay process where the attributes of polygons were collected onto the center points of twenty acre hexagons. This points database preserved a statistically reliable representation of the diversity and detail found in the polygon layers more effectively than would have been feasible with the limitations of polygon overlay processing over such a large area (Reed 1991). The points database also effectively overcame the problems of automation slivers for use in revising TLMP. Project level work however requires a scale where resolving sliver detail is sometimes significant. There are also, of course, different vectors in our layers that represent each inventory professional's estimate of where to draw the class boundary where there really isn't a distinct feature to map. The challenge on projects is to efficiently update these layers with new information and corrections that reconcile feature locations between layers.

AIRBORNE VIDEO

The Remote Sensing Applications Center (RSAC, formerly named NFAP), a part of the Washington Office of the USDA Forest Service, provided equipment and technical expertise to acquire airborne video imagery of active floodplains of the Bradfield River on the Wrangell Ranger District, near the southern end of the Stikine Area (Bobbe 1994). Real-time GPS was used on those flights to obtain georeferencing for screen captures. The plane used was flown with a more pronounced "nose up" attitude than other aircraft used previously with that equipment. The GPS coordinates were not centered on the image and made it difficult to register captures accurately. This turned out to be a fortunate problem because simple scans of an orthophoto print ended up being used as a more accurate backup for referencing. The process of using scanned orthophoto prints became the lower cost basis for capture referencing with the airborne video system purchased by the Area. The scanned orthophotos are also especially useful in GIS on the new workstations.

The Area's video camera system follows closely on the recommendations from the Forest Pest Management, Methods Application Group (Myhre 1992). Processing is accomplished on an IBM compatible 486/66 PC with memory expanded to 64 mb of RAM and with two 1.2 gb hard drives. It has an HP Scanjet IIc flatbed scanner and a read/write optical drive with 600 mb storage on each removable disk. Video captures and most other image processing is accomplished using TNT MIPS from MicroImages Corporation. This video system has proven adequate for acquiring vectors that more accurately reference new harvest unit and road locations than using photo or map locations for proposed harvest units and roads. Features on the captured video images compare favorably with positions where these new features have been located with GPS. The system was used to scan prints of existing 1985 orthophotos in 16 bit TIFF format. Reductions in the original 1/31680 orthophoto prints to 1/48000 were ordered through the Geometronics Service Center in Salt Lake City. These reductions fit the flatbed scanner and were scanned at various trial densities. A density of 400 dpi (dots per inch) seemed to provide the best compromise between resolution and file size. Greater densities included more detail but also picked up the inherent graininess of the orthophoto. After referencing to Alaska Stateplane Zone 1, resolution of the scanned orthos is just slightly over 10 feet per pixel. These "pseudo-digital" orthophotos, although very useful, are a low cost, temporary solution to obtaining real DOQ's from new imagery (Winterberger 1996). New orthophotos developed from new imagery will utilize the additional geodetic control points that were surveyed on the Tongass the past two years. The new orthophotos will then provide a basis for accurate referencing of GIS features. Capabilities of the system to scan, manipulate and make plots of other imagery were not anticipated in justifying the procurement, but have been in significant demand for a variety of purposes on the Area.

Use of video has been limited on the Area by lower budgets, the processing time required to cover large areas with sufficient resolution and a desire for higher resolution imagery. The system contains a very small camera which is used for navigation when employed in the full. This camera has also been used separately video taken at low elevation from a helicopter to document details along stream channels and stream channel buffers. These tapes were used simply as video and not processed for capture and georeferencing of images. The Area recently acquired a Kodak DCS 420 digital color camera to meet the needs for better resolution. This camera has been added to a locally fabricated mount for the belly port on a DeHavilland Beaver and now can be deployed with both the video cameras. The digital camera produces images of higher resolution that do not require processing to capture from video. The video tapes however, can be run during the entire flight, offering virtually unlimited imaging while the Area's digital camera is currently limited to the available drive storage in the camera backs on board. Video tapes are also of interest to our resource specialists and have even been the subject of FOIA's from the public.

GLOBAL POSITIONING SYSTEM

Interest in using GPS became The Stikine Supervisor's office building in Petersburg is almost centrally located on the Tongass and has a suitably clear "sky plot" for use as a GPS base station. The site was selected as the site for a single base station for the Tongass in 1993. Trimble Pathfinder Professional field units and a Trimble Community Base Station were obtained under Forest Service-wide contracts. The Area has supported the data collected daily by the base station, making several weeks of correction files available for routine access by Forest Service offices through the RIS (information transfer) capabilities of the DG computer system. Daily correction files are also kept on tape and have been made available to other users on request.

On the Area, GPS has been tested for automating proposed harvest units, existing clear-cut harvest units not already available on imagery, classification and mapping of small stream channels not already inventoried in addition to inventory of roads and transportation structures. The "nesting" features of the software are used to create "data dictionaries" (both definition and collection format routine) for collecting both road and stream data. The collection was designed to use the more accurate positions of points to describe line locations. Attributes of stream reaches or roads are entered while obtaining positions. Trimble's Pfinder software is used for post processing differential corrections. A line file and it's nested points file that contain the point and line locations and the associated attributes for each are output from the PC. Macros were written for ARC/INFO operating Rev. 5.01 to build line coverages with attributes from these files. These macros reliably produced correct coverages as long as the files were in the proper format and there were no gaps in the position data. The idea was to provide a process for field specialists to transfer their information into GIS without the need for attributing and editing, which took considerable time and a GIS specialist on the DG. The dense, wet canopies and steep terrain foiled attempts to obtain positions under stands and along small, incised stream channels approximately 20% of the time when satellites should have been obtainable. This was particularly frustrating as it prevented collecting attribute information in a format that would fit the GIS macros. The Pathfinder units also proved to be too heavy for packing over the rough terrain as a routine practice. Edit capabilities in GIS on the UNIX workstations offer easier solutions to migrating the GPS data into GIS.

GPS is currently in use from vehicles for updating the entire inventory of roads, bridges, culverts, turnouts and other transportation structures on the Stikine Area. It is also sometimes used for finding landmarks to reference traverses and for collection of point sample plot data. Newer equipment and upgraded software would likely produce better results however, additional investments in GPS technology will likely require proof of more reliable performance under canopies before GPS use on the Area becomes routine.

VEGETATION MAPPING

In March, 1994 the Tongass National Forest contracted with Pacific Meridian Resources for developing a pilot vegetation classification from LANDSAT Thematic Mapper data. An ARC/INFO polygon GIS layer and raster GIS layers for overstory species, crown cover, and tree size/structure were produced for a 35,000 acre portion of Etolin Island on the Wrangell Ranger District in the southern part of the Stikine Area (Pacific Meridian 1994). The project included accuracy assessments and showed promise for extending the methods to map the entire Tongass. The Wrangell Ranger District is currently using the products along with other GIS and inventory information in analysis of prospective timber harvest options on Etolin Island. Pacific Meridian was contracted again in 1995 to develop mapping for over a million acres on the Chatham Area north of Sitka. This site showed considerably greater variability and presented greater difficulty for classification than the relatively homogeneous, northern end of Etolin Island that was piloted the previous year (Pacific Meridian 1995). Additional accuracy assessments are planned this year with assistance from RSAC prior to completing inventory of the remaining Tongass acreage. LANDSTAT Thematic Mapper files have also been very popular as landscape scale images for plots and viewing on screen.

CHANGES IN GIS DATA CAPTURE

Careful examination of new GIS layers automated for analysis of proposed timber harvest and road options clearly showed that the majority of feature location errors occurred in the transfer from 9x9, 1/15840 field photos to clear mylar manuscripts over orthophotos. Check plots of digitized and edited ARC/INFO coverages matched orthophoto manuscripts almost exactly, but the manuscripts sometimes contained considerable differences from the features drawn on field photos. These errors were primarily responsible for purchase of Global Positioning System (GPS) technology on the Area. Costs and difficulties encountered for GPS use, especially under canopies, led to acquisition of Mono Stereo Digitizing Software (MDSD), a product of Carto Instruments Incorporated. MDSD contains a suite of digitizing and photogrametric utilities (This package has been used routinely for digitizing orthographically corrected vectors from lines drawn on 9x9 resource photography. The process requires a significant learning curve and a well planned editing sequence in GIS but it produces accurate digitizing in approximately the same time as formerly devoted to making mylar overlays to orthophotos for digitizing. Most new GIS data automation on the Stikine Area is now accomplished either through GPS, MDSD or the capabilities for use of image backdrops for on screen digitizing within ARC/INFO. This on screen capability is particularly efficient for automating any features visible on the backdrop imagery and also quite useful for editing features automated from either GPS or MDSD. Success with early efforts at command based editing rekindled enthusiasm to create user-friendly and efficient applications for making GIS routinely useful to non-GIS experts.

GIS APPLICATIONS

Imagery formed the basis for the first of several applications that were designed to meet the needs of individual specialists. These tools were quite simple to use and only intended to do a limited task. The first was a complete overhaul to an earlier digitizer tablet menu which was one of the very few friendly tools that actually worked efficiently on the DG. It was used to automate relocation points for radio-telemetered goshawks. Hundreds of relocation points were collected as marked points on 1/15840 scaled, 9x9 inch resource photography during aerial relocations. These points were then transferred, digitized and attributed over 1/15840 enlargements to orthophoto prints over a digitizing tablet. One of many adjustments to the UNIX platform included difficulties in getting digitizing boards to work properly. Finding an alternative using the newly scanned and georeferenced orthophotography was an obvious solution. The menu process developed was simple enough to learn that the technician who did the processing was able to learn how to use it and successfully process 6 months of collections in one week. This was at least 5 times faster than the method used over the digitizing board. One of the advantages stemmed from not needing to find the appropriate orthophoto and then mount it on the board. Sometimes relocation points fall on the boundaries between the quarter quad sheets or even near the corners. Hours were lost in sorting through photos and remounting orthophotos on the digitizer because a "missed" point that was thought to be on a particular ortho was really on one previously mounted. The GOSHAWK menu simply finds and uses whatever orthophoto is needed from the nearly 200 available for the Stikine Area, with a simple click of the mouse. The operator doesn't need to know which orthos are needed, what they are called, or where they are stored on the system. This caused a bit of local interest from selected resource specialists who wanted similar applications. These didn't get fully completed before it became apparent that creating a more generalized application would be more efficient and quite a bit more useful.

IDTEDIT was developed to meet the needs of an interdisciplinary team (IDT) in the early phases of their project analysis. It's a menu driven application that operates in ARCEDIT and provides on screen viewing, digitizing and editing capabilities. It's considerably more complex and powerful than it's GOSHAWK predecessor and a bit harder to learn. However, it has the advantage that team members can help teach each other while doing their work. This leads to team members who have a better understanding of other specialists data in addition to their own. This is of great benefit because the data can be less friendly to learn than the GIS software, which is usually far better documented! Viewing and attribute editing are simple enough that persons with no formal GIS training have used the package successfully after less than an hour of one on one instruction. Most of the feature editing capabilities are not meaningful to someone who lacks a basic understanding of GIS, editing GIS data, and familiarity with the layer to be edited. The menus however, provide a variety of help and a good basis for learning. ARCEDIT terminology is purposely used to provide consistency with ARC/INFO help utilities. IDTEDIT has multiple menus. They are designed so they can be all open at the same time. This allows for very flexible and fast editing that is especially efficient for those who already have some ARCEDIT expertise. The application is designed to copy library layers for project based editing and ease the process of keeping corporate data current. Edited layers are forwarded to the database administrator for checking and update back into the library.

Nearly everyone who sees IDTEDIT or tries it, asks the same question, and usually very near their first minute of experience. "Can I get a plot of that?", was heard so often that the FASMAP application simply had to be developed. This application currently produces plots at a user-specified paper size, scale and for areas based on either "center of interest" or "corner to corner" extent. It is intended to produce standard plots based on identified needs. Several options are being considered, including a feature to approximately replicate screens from IDTEDIT. Currently, the options that produce features similar to those found on a quad map in either color or gray shade are operational. Products are similar to those often seen after ordinary quad sheets are transformed through scissors and rubber cement exercises, but with up to date information and options to add additional information. These are VERY easy to use menus that have GIS novices producing maps within a few minutes and half a dozen mouse clicks. These maps by no means are intended to resolve all the plotting needs for GIS. They just put very commonly needed working maps into simple routines that allow GIS specialists to work on analyses, models, printer separates files, developing more tools, and maintaining the database! FASMAP is much quicker and easier to use than ARCVIEW, but it isn't anywhere near as flexible for a wide variety of uses. It's simply more efficient to use FASMAP for what it does and go to ARCVIEW or other means in ARCPLOT when more is needed. There is a virtual explosion in possibilities for tools that increase productivity! There are a great many needs that can be met, especially if we can learn from each other more efficiently and keep application development close to real users doing real production work. So many possibilities and such an enormous variety of powerful new capabilities are now available that the biggest challenge is how to share progress and build off what has been learned.

OBSERVATIONS AND INSIGHTS

We've been striving toward the goal of putting GIS capabilities routinely in the hands of our resource specialists for almost a decade. Most of our specialists work on IDT's at least part of the time. Most of our pressing needs are focused on ecosystems and require understanding that spans many disciplines. Yet, much of the professional, organizational and interpersonal relationships and cultures remain "within a function". This, unfortunately, can result in conflict, especially in times of tight budgets and competing issues. Imagery and interdisciplinary applications can provide a key element to overcome "turf battles". Provide everyone something that is obviously easier and better and they're far more willing to change and work together. Attempt to give them something that isn't really better, expect them to switch, and they will work together, just to gang up on you.

Accuracy of positions though expensive, appears to be the cheapest route to precision between layers of information. Accurate coordinates aren't of particular value as long as all the features in the database share locations that replicate their true positions relative to each other on the ground. The merging of information from a mixture of sources can require successive adjustments of all the data if there is insufficient base information accuracy to match common new coordinate capture methods (GPS). Accurate referencing of the database allows newly collected data that meets accuracy standards to be used with confidence. Not having accurate base data coordinates requires an edit session whenever new data is collected to ensure that the road is on the correct side of the stream etc. New orthophotos from new imagery for the Tongass planned for acquisition this summer will have sufficient accuracy to warrant general updating of GIS feature positions. Until then, the scanned orthos provide only an interim solution to create precision between where needed for projects and GPS data, likely with more accurate coordinate locations will continue to need adjustments to match the database locations for analysis purposes.

Standards for data and meta-data are important. Trouble is they are often difficult to find, let alone follow. Applications (menu tool packages) appear to offer a key to efficiently standardizing. Standards for data provide for development of more efficient applications that can be widely used. Applications also can be developed to standardize, or encourage the standardizing of data. Rather than simply produce documents that outline standards in great detail, perhaps even untested standards, application tools can be developed that *make the right thing to do also the easiest thing to do.* Project 615 technology can provide the basis for sharing expertise for application development in addition to supporting the processing of the applications. The synergism of ideas between disciplines and the serendipity that occurs when real, on-the-ground problems are addressed with teamwork usually provides the most progress. Developing standard tools that are available to everyone without delaying needed work and stifling on-the-ground creativity is the challenge.

Managers are currently under great pressure to cut costs. Forest Service budgets are very compartmentalized and funding is focused along resource functional or disciplinary lines. However, much of the information developed by one discipline may be critical to efficient work by other disciplines. For example, timber cruise foresters who traverse planned unit boundaries may decide to increase their productivity by not taking the time to obtain reference coordinates for their traverses. This could save as much as 15% of their field costs, allowing them to do more traverses and not adversely affecting their primary goal of determining a traverse acreage for the unit. Not having those coordinates however, deprives the other users of unit boundary information. This is likely to require them to expend several times greater sums from their budgets to recover the loss and jeopardize future work of timber cruise foresters due to inability to credibly monitor and evaluate management practices. The interdisciplinary availability of and need for spatial resource information which is characteristic of GIS and image processing requires well informed management decisions that cross functional boundaries. Maintaining complex budget options with wise choices will be difficult until there is wider familiarity with and use of GIS and image processing tools.

Effective training is necessary to efficient use of GIS. Experience on the Stikine Area has indicated that only 25% or fewer of those attending intensive generic GIS training retained the ability to use GIS after only a few months. Training efforts aimed at individuals and providing specific tools to perform work they need, enjoy a much higher rate of success. Experience gained increases benefits from subsequent generic training courses.

Progress with spatial information, imagery and GIS is now so fundamental that we are likely to see major shifts in customary and traditional map products that have been "standard bearers" for decades. Quad maps and similar "fixed" rectangular format maps may all but disappear from agency use in the next 3 to 5 years and may even phase out of public use, being replaced with maps of custom paper size and extent yet retaining all the richness in map detail while being produced on demand, directly from databases containing cooperatively managed feature and attribute files that are current within days or months instead of being updated once a decade. The challenge will be to make the transitions while retaining and evolving the skills of the work force.

ACKNOWLEDGMENTS

The author wishes to thank Ken Winterberger for his collaborations on orthophotography, Ron Hall for his dedication and skill for refining orthophoto scanning techniques, Steve Alarid for his collaboration on IDTEDIT and his programming of FASMAP and Felicia Acrea, Rob Aiken, Jim Cariello, and Jim Brainard for their patience and ideas for testing and improving IDTEDIT.

REFERENCES

Bobbe, T., Reed, D. and Schramek, J. 1993. Georeferenced Airborne Video Imagery: Natural Resource Applications on the Tongass. Journal of Forestry, 91 (8): pp. 34-37.

ESRI. 1985. Final Design Report and Implementation Plan, Tongass National Forest (Ketchikan and Stikine Areas), Natural Resource Management Information Study. prepared for: USDA Forest Service, Alaska Region, Contract No. 53-0109-4-00161.

Myhre, D., Russell, B., and Sumpter, C. 1992, Airborne Video System User's Guide, USDA Forest Service, Forest Pest Management Methods Application Group, Fort Collins, Colorado.

Pacific Meridian Resources. 1994. Tongass National Forest, North Etolin Island Vegetation Mapping Project, Final Report.

Pacific Meridian Resources. 1995. Tongass National Forest, Vegetation Mapping Project, Final Report.

Reed, E. Using a Distributed GIS Database in Forest Planning, in GIS Applications in Natural Resources, GIS World Inc. 1991, pp. 163-166.

Winterberger, K. 1996. Using "Pseudo" Digital Orthophotography or the Good, the Bad and the Ugly, in preparation for the Sixth Biennial USDA Forest Service Remote Sensing Applications Conference, Aurora, Colorado.

DEVELOPMENT OF A GEOGRAPHIC INFORMATION SYSTEM

FOR THE CHUGACH NATIONAL FOREST

by
Carl J. Markon, Hughes STX Corporation,
USGS/EROS Alaska Field Office,
4230 University Drive, Anchorage, AK 99508-4664
and
Bruce E. Williams, Chugach National Forest,
U.S. Forest Service,
3301 C Street, Anchorage, AK 99503

ABSTRACT

The Chugach National Forest is the second largest national forest in the United States covering over 3.8 million hectares of land and offshore waters, of which 0.6 million hectares are current inholdings. More than 1 million hectares of offshore waters are within the forest boundary.

In 1987, a geographic information system (GIS) was developed by the U.S. Forest Service with assistance from the U.S. Geological Survey to produce an environmental impact statement for the Big Islands management area within the forest. In 1990, the development of the GIS was expanded to include the Forest Plan Revision.

The Chugach National Forest GIS currently involves the use of both vector and raster data. The vector data layers are derived from information digitized from hard-copy maps. The raster data sets are derived from satellite images, rasterized vector data layers, and digital elevation models. Applications of the database to Chugach National Forest planning and management activities consist of forest wide management, timber management, wildlife management, and recreational planning.

The vector component of the database consists of biophysical, landcover, land status, and historical and current use layers. Examples include soils, timber type, streams, anadromous fish habitat, roads, recreation sites, ownership, archaeology, and management area boundaries. There are 35 data layers, 20 of which have been completed; the remaining 15 will be finalized in the next 6 months.

Image classification was used to produce the integrated land cover map. Data sources included Landsat thematic mapper, Landsat multispectral scanner, and SPOT multispectral scanner images. The different data types were necessary to provide near-complete, cloud-free coverage of the forest. Dates of the images range from August 1977 to August 1991. Parts of the 20 different satellite scenes were used and each satellite scene was analyzed independently. Image classification involved the use of standardized isodata and Baysian classifiers. Ancillary data such as elevation, slope, and aspect as well as rasterized vector components were used to stratify the preliminary classes of each satellite scene into one of 25 land cover classes. After the application of ancillary data, the preliminary classifications were merged to produce one map of the entire project area covering approximately 3.8 million hectares including offshore water.

INTRODUCTION

The Afognak Forest and Fish Culture Reserve was created by an executive order dated

December 24, 1892, and encompassed the 0.16 million hectare Afognak Island. In 1907, more than 2 million hectares of land in southcentral Alaska were added to the reserve and the name was changed to the Chugach National Forest (CNF). At that time, the city of Anchorage and the current Kenai National Wildlife Refuge were also included but were later removed reducing the forest to 0.89 million hectares. In 1980, the Alaska National Interest Lands Conservation Act (U.S. Congress, Public Law 96-487) increased the acreage to its present size of 2.3 million hectares (excluding off shore waters).

To properly manage the forest, the U.S. Forest Service requires the development of an automated resource database. This database requires the collection, management, analysis and output of cartographic, remotely sensed, and geographic or resource spatial data in a timely and cyclical manner. The current and ultimate use for this database will be land management planning, management and ecosystem analysis, forest plan implementation, and project analysis.

PROJECT AREA

The CNF encompasses approximately 3.8 million hectares of land and water in southcentral Alaska (Figure 1), extending roughly across 2° of latitude and 6° of longitude. It is bordered on the north by the ice capped Chugach Mountains and the south by the Gulf of Alaska and is contained in the Kenai-Chugach Mountains physiographic division (Wahrhaftig 1965). The forest is described by Gallant and others (1995) as being in the Pacific Coastal Mountains ecoregion with steep rugged mountains and hills, isolated

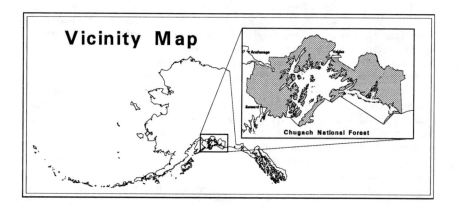

Figure 1. Location of Chugach National Forest in southcentral Alaska.

masses of permafrost and annual precipitation from 200 to 700 cm/yr. McNab and Avers (1994) have a similar description, however, they use the terms Pacific Gulf Coastal Forest and Meadow Province. Much of the forest is snow and ice covered, or open water. Low and scrub vegetation is common on the upper slopes, with conifer forests on lower slopes. Soils generally are humic cryothods on mountain foot slopes or moraines or terric cryohemists in poorly drained organic soils on gentle to steep slopes (Reiger and others, 1979).

METHODS

Data Base Design

All data were georeferenced to a Universal Transverse Mercator (UTM) projection with the origin in zone 6. Final raster database resolution was 30 m by 30 m for each picture element, or pixel; vector database resolution varied, depending on data source. All vector data and a minimum of raster data were stored online in the USFS geographic information system (GIS).

Vector Data

The vector data components consist of 35 layers representing biophysical, land cover, land status, historic, and current use components. Many of the data layers were obtained by digitizing hand drawn information overlaid on, or derived from one or more of the following products: U.S. Geological Survey topographic maps, orthophotoquads, or aerial photography. Dates of the different data sources vary from the early 1950's to the present. Vector data were accessed via a hierarchical structure in which each component represents a 'theme' and each theme has a subdirectory containing forest wide coverage. Themes are listed in Table 1.

Raster Data The primary raster data layer consisted of multispectral satellite data derived landcover and digital elevation data. A schematic diagram of the development of the digital land cover map is shown in Figure 2.   Data from three different satellite sensors were used to obtain complete and near cloud-free coverage of the forest: Landsat multispectral scanner (MSS), Landsat thematic mapper (TM) and SPOT multispectral scanner. Table 2 lists satellite scene aquisition information. The remainder of the report covers the development of the raster data.

PREPROCESSING

Each satellite scene was reviewed for quality and cloud cover and then georeferenced to UTM zone 6. Control points were selected from each Landsat scene, located on USGS 1:63,360-scale topographic maps, and used to define a second-order least squares polynomial transformation relating UTM zone 6 coordinates to the line and sample of each scene. Criteria for selecting control points required that each point be recognizable on both the topographic maps and the satellite scene. The mean residual errors associated with the second-order transformation indicated a registration accuracy within 0.50 to 1.25 pixels depending on data source.   In addition to the satellite data, digital elevation data also were obtained. These data were acquired from two sources. Digital elevation data for each 1:250,000-scale quadrangle were acquired from the Defense Mapping Agency (DMA). The DMA generated the data by digitizing 200-foot contour lines from USGS 1:250,000-scale topographic maps and converting them to rectangular grid values. Elevation data also were acquired from the U.S. Forest Service, which generated the data by digitizing 100-foot contour intervals from selected 1:63,360 scale USGS topographic maps. All data were resampled to a 30 meter pixel and merged to form one complete dataset of the entire project area. Slope and aspect were also generated from the digital elevation data (Markon, 1994).

Table 1. Themes used in the vector portion of the database.

Data Theme	Theme Description	Type
AN_AREAS	Analysis Areas	POLY
ARCH	Archeological/Cultural Resource Sites	POINT & POLY
ARCH_SURV	Archaeological Survey Status	POINT, LINE, & POLY
ELEVATION	Digital Elevation Model (DEM)	GRID
ASPECT	Terrain Aspect from DEM	GRID
BOUNDARY	Forest, District, and Management Area Boundaries	LINE & POLY
CONTOURS	Contours Generated from DEM	LINE
COVTYP	Cover Type	POLY
EAGLES	Eagle Nest Sites	POINT
ECOSECTIONS	Ecosections and Sub-Sections	POLY
EMU	Ecological Mapping Units	POLY
EVC	Existing Visual Condition	POLY
GEOLOGY	Geologic Map	LINE & POLY
INSECTS	Insect Infestations	POLY
LAKES	Lake Shore Polygons	POLY
LANDFORM	Landform Types	POLY
LANDNET	Township, Range, & Section	LINE & POLY
LANDSTAT	Ownership Boundaries	POLY
LANDSTAT_LN	Landstatus Lines (Powerlines, Pipelines, etc.)	LINE
MINERALS	Mineralized Zones from USDI Bureau of Mines	POLY
MINES	Shafts, Tunnels, Tailings, or Prospects	POINT
NWI	USF&WS National Wetlands Inventory	LINE & POLY
OTHER_LIN	Miscellaneous Lines (Airstrips)	LINE
OTHER_PTS	Miscellaneous Points (Fire Occurrences)	POINT
PRECIP	Mean Annual Precipitation	LINE
QUADS	15-Minute USGS Quad Index	POLY
REC_SITES	Recreation Point Features	POINT
ROADS	Road System	LINE
ROS	Recreation Opportunity Spectrum	POLY
SHORE	Saltwater Shoreline Polygons	LINE & POLY
SLOPE	Terrain Slope Grids (DEM)	GRID
SNOW_500K	Mean Annual Maximum Snowpack	LINE
SNOW_MOD	Snow Depth for Wildlife Model	POLY
SOILS	Soil Survey for Kenai Road Corridor	POLY
SPEC_ADM	RNA's, Experimental Forests, Wilderness Study Areas	POLY
STREAMS	Freshwater Streams	LINE
SUBSIST	Subsistence Survey Data	POLY
TIMTYP	Timber Types	POLY
TRAILS	Recreation and Other Forest Trails	LINE
VQO	Visual Quality Objectives	POLY
WATERSHED	Watershed Polygons	POLY
WILD_HAB	Wildlife Habitat Areas	POLY
WILD_SHU	ADF&G Game Management Units	LINE & POLY

Table 2. Satellite scene identificaiton and acquisition date.

IMAGE	Path/Row	Date	Image	Path/Row	Date
(SPOT) Kenai	453-225	08/14/90	TM 6817	068/017	08/01/85
	452-226		TM 6717	067/017	07/28/86
(SPOT) Seward	454-226	08/14/90	TM 6517	065/017	07/27/85
	454-227		TM 6718	067/018	08/13/86
(SPOT) Perry Isl.	457-225	06/11/89	TM 6618	066/018	07/24/87
	457-226		MSS 6818	068/018	09/05/83
(SPOT) Valdez	458-225	06/11/89	MSS 7117	071/017	07/10/78
(SPOT) Hawkins Isl.	461-226	07/03/89	MSS 7018	070/018	08/12/84
(SPOT) Martin River	465-225	08/15/90	MSS 7118	071/018	08/20/77
(SPOT) Bering River	465-226	07/20/90			
(SPOT) Kayak Isl.	465-227	08/14/90			

DEVELOPMENT OF SPECTRAL STATISTICS

Training statistics used to generate the preliminary classification on each individual scene were derived using an isodata algorithm (Swain and Davis,1978; Fleming, 1988) on a systematic sample (every 10th line and sample). Each spectral cluster was defined in terms of descriptive statistics (means, standard deviations, covariance matrices between bands) for all bands. Redundant or overlapping statistics were removed or combined to form one statistical data set that provided an independent estimate of the spectral characteristics of each scene being analyzed.

CLASSIFICATION AND SPECTRAL CLASS LABELING

Final statistical files were applied to the satellite scene being analyzed using a maximum likelihood classification algorithm (Fleming, 1988). The algorithm uses a complex mathematical decision rule for evaluating each pixel in the satellite scene. The pixel brightness value was compared to the descriptive statistic values obtained for each spectral cluster. Each pixel was assigned to one of the spectral clusters developed, based on the probability that the pixel belonged to the cluster. The result was a classification where each pixel fell into a cluster that represented a land cover class.

The spectrally classified training blocks were evaluated and each cluster was assigned a vegetation or land cover class name based on interpreted color-infrared aerial photographs and field data descriptions (when available). Spectral class inconsistencies within and between geographical areas were noted for possible refinement after classification. Frequently, classes contained more than one vegetation type due to similar spectral response from (1) similar vegetation types, (2) the effects of shadow and water on the overall reflectance of the vegetation, and (3) variation of the vegetation due to elevation, slope, or aspect changes.

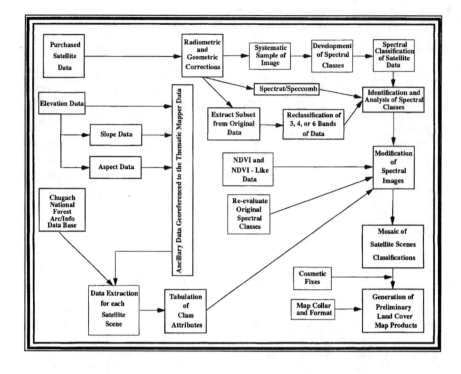

Figure 2. Process flow for development of landcover data layer.

POST-CLASSIFICATION REFINEMENT

Preliminary classification accuracies were often low, due to similar spectral responses of different land cover types, variable terrain, and moisture presence. To increase the accuracy of the classification, ancillary data and other post-classification techniques were applied to separate classes that were known to be misclassified. For example, (1) small areas from the original scene were extracted and reclassified, (2) normalized difference vegetation index-like data (Markon and others, 1995) were applied, (3) spectral stratification and recombination (see SPECSTRAT and SPECCOMB, LAS Users Guide, 1992) were applied, (4) slope and aspect were used to separate shadow from water, (5) elevation was used to separate shrub riparian types from alpine types, (6) ecological strata were used to separate spectrally similar but physiognomically distinct vegetation classes, and (7) landform was used to separate some conifer types from shadowed areas.

Accuracy Assessment

An accuracy assessment has not been performed on a number of the data layers, including the satellite derived land cover layer. Before any analyses are performed using these data, the user should contact Chugach National Forest GIS personnel. Any comments regarding the usefulness of the products is based on personal observations and comments from coworkers and are strictly qualitative.

RESULTS

Land Cover Classification

The land cover classification for the CNF resulted in 6 major classes: forest, scrub, herbaceous, water, barren and other. Within these classes are 39 subclasses (Table 3). and follows, in part, vegetation classes by Viereck and others (1992). The most prominent class identified on the CNF was clear water (25.69 percent), followed by ice/snow/clouds (21.44 percent), closed conifer (8.66 percent), turbid water (8.10 percent) and tall closed shrub (6.49 percent). Table 4 gives acreage estimates for the entire national forest.

Table 3. Classes and subclasses used for the Chugach National Forest satellite derived map.

CLASS DESCRIPTION			**CLASS VALUE**
Forest	Needleleaf	Closed	1
		Open	2
		Woodland	3
	Broadleaf	Closed	4
		Open	5
	Mixed	Closed	7
		Open	8
Scrub	Dwarf Tree		11
	Tall Shrub	Closed	13
		Open	14
	Low Shrub	Closed	15
		Open	16
Herb	Graminoid/Forb	Dry/Mesic	17
		Wet	19
	Bryoid	Moss	23
		Lichen	24
	Aquatic	Fresh	25
		Brackish	26
Water	Clear		28
	Turbid		29
Barren	Bedrock or Unconsolidated		33
	Sand/Mud		35
Other	Ice/Snow/Clouds		36
	Shadow		38
	Sparsely Vegetated		39

Tabel 4. Area summary for each class value in the preliminary landcover map.

Class #	Pixel Count	Hectares	Acres	%
1	3736481	336283	830965	8.66%
2	692648	62338	154040	1.61%
3	566536	50988	125993	1.31%
4	721813	64963	160526	1.67%
5	110234	9921	24515	0.26%
7	33886	3050	7536	0.08%
8	14117	1271	3140	0.03%
11	153647	13828	34170	0.36%
13	2798769	251889	622425	6.49%
14	364643	32818	81094	0.85%
15	1508979	135808	335586	3.50%
16	784443	70600	174454	1.82%
17	1901893	171170	422967	4.41%
19	653255	58793	145279	1.51%
23	22956	2066	5105	0.05%
24	138378	12454	30774	0.32%
25	42483	3823	9448	0.10%
26	9459	851	2104	0.02%
28	11080166	997215	2464145	25.69%
29	3492172	314295	776633	8.10%
33	2446255	220163	544029	5.67%
35	534330	48090	118831	1.24%
36	9244760	832028	2055965	21.44%
38	1601781	144160	356224	3.71%
39	468990	42209	104300	1.09%
Total	43123074	3881077	9590246	100.00%

CONCLUSIONS

The U.S. Forest Service (U.S.F.S.) is required to fulfill a multitude of land management needs, for both public (U.S. citizens) and private (industry, native corporations) entities. In order to accomplish this mission on the Chugach National Forest, the U.S.F.S. requires a multitude of information covering all aspects of land cover and land use, both historically and in the present. The Chugach National Forest GIS that cureently being completed will be (and is currently being) used to help fulfill its mission as a land steward. The GIS will provide a means by which multiple data layers can be analyzed to meet the needs of various resource managers. For example, the database is currently being used in a series of environmental assessments for timber salvage in stands affected by bark beetles. Also, being in a GIS configuration, the database will not become static, but will be continuously updated as new information becomes available.

REFERENCES

Galant, A.L., Binnian,E.F., Omernik, J.M., and Shasby, M.B. 1995, Ecoregions of Alaska: U.S. Geological Survey Professional Paper 1567.

Fleming, M.D. 1988, An integrated approach for automated cover type mapping of large inaccessible areas in Alaska: Photogrammetric Engineering and Remote Sensing, Vol. 54, pp.357-362.

LAS Users Guide, 1992, Overviews, Vol. 3: U.S. Geological Survey, EROS Data Center, Sioux Falls, S.D.

Markon, C.J. 1994, Development of a digital land cover data base for the Selawik National Wildlife Refuge: U.S. Geological Survey Open File Report 94-627.

Markon, C.J., Fleming, M.D., and Binnian, E.F. 1995, Caracteristics of vegetation phenology over the Alaskan landscape using AVHRR time-series data: Polar Record, Vol. 31, pp. 179-190.

McNab, W.H., and Avers, P.E. 1994, Ecological subregions of the United States: Section Descriptions: U.S. Department of Agriculture, Forest Service, WO-WSA-5, Washington, D.C.

Reiger, S., Schoephorster, D.B., and Furbush, C.E., 1979, Exploratory soil survey of Alaska: U.S. Department of Agriculture, Soil Conservation Service, 213 p.

Swain, P.H., and Davis, S.M. 1978, Remote sensing: The quantitative approach: McGraw-Hill, New York.

Viereck, L.A., Dyrness, C.T., Batten, A.R., Wenzlick, K.J. 1992, The Alaska vegetation classification: U.S. Department of Agriculture, U.S. Forest Service, Pacific and Northwest Forest Experiment Station, PNW-GTR-286, Portland, Oregon.

LARGE AREA CHANGE DETECTION
USING SATELLITE IMAGERY:
Southern Sierra Change Detection Project

Lisa M. Levien
USFS Region 5, Forest Pest Management
Cynthia L. Bell
California Department of Forestry and Fire Protection
Barbara A. Maurizi
Pacific Meridian Resources
1920 20th Street
Sacramento, CA 95814

I. Abstract

The U.S.D.A. Forest Service, Region 5 Forest Pest Management (FPM) staff and the California Department of Forestry and Fire Protection, Forest Pest Management (CDF-FPM) and Fire and Resources Assessment Program (FRAP) are conducting a large area change detection project in the southern Sierra Nevada. Landsat Thematic Mapper imagery from 1990 and 1995 are being used to assess changes over 16.7 million acres. The project has been divided into two Phases. Phase I includes developing a landscape-level change map identifying a continuum of change classes. Phase II includes determining the causes of those changes and quantifying them by using ancillary information and field work. A multi-temporal Kauth-Thomas transformation is being used to determine changes in brightness, greeness and wetness. Previous studies have indicated that forest changes are highly correlated with the multi-temporal Kauth-Thomas transformation. Specific causes of landscape change will be determined across all ownerships in the project area using a multistage sampling program.

II. Introduction

Detecting vegetation change and determining its causes has become increasingly important to natural resource managment organizations, local planning organizations, individuals and others concerned with the degradation of the environment. Changes in vegetation result in changes in wildlife habitat, fire conditions, aesthetic value, ambient air quality, etc. which in turn impact management decisions and policy. To detect change over relatively small areas, multi-temporal aerial photos are an effective tool; however, satellite imagery has proven to be more effective over large regions.

Satellite imagery and image processing for monitoring changes in landcover has been well documented over the past 20 years. The US Forest Service FPM and CDF FPM in cooperation with Boston University (BU), Department of Geography for the past 5 years have conducted change detection studies to determine conifer mortality using satellite imagery (Macomber, S. and Woodcock, C.E., 1994, and Collins, J.B. and Woodcock, C.E., 1994). Various methods have been investigated indicating a strong ability to monitor forest canopy changes.

164

The objective of this project is to implement the methodology developed in cooperation with BU to create a long-term monitoring program which will identify trends in forest health, provide interim map updates, assess changes in vegetation extent and composition, and maintain a regional approach to change detection. The approach employed is a multi-temporal Kauth-Thomas transformation using Landsat Thematic Mapper (TM) imagery from 1990 and 1995. The project area is in the southern Sierra Nevada's extending from the Stanislaus National Forest south to the Sequoia National Forest, east to the California-Nevada border and west to the foothills, encompassing 16.7 million acres. Figure 1 shows the study area and ownership.

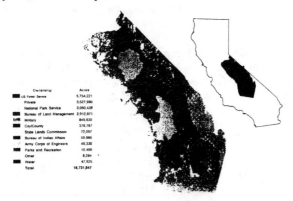

Figure 1. Project area and ownership

II. Background

Previous studies conducted cooperatively with USFS Region 5 FPM and the Ecosystem Conservation staff (EC), CDF-FPM and BU assessed the utility of using remote sensing for monitoring and mapping conifer mortality. In 1991 an initial study was undertaken to evaluate the potential for measuring and mapping conifer loss due to drought related effects using Landsat TM imagery (Macomber, S.A., and Woodcock, C.E., 1994). The approach, an extension of a vegetation mapping project, used a geometric canopy reflectance model to determine canopy closure and tree size (Woodcock, C.E., et.al., 1994). The assumption was that changes in canopy closure could be mapped and related to changes due to mortality. Intensive field work was required to collect enough data on the ground to correlate it with the canopy model. This investigation was conducted on the Lake Tahoe Basin Management Unit (LTBMU) in the Sierra Nevada's

In a later study from 1994-1995, FPM and CDF-FPM entered into another cooperative agreement with BU to develop a more simplified methodology for implementing change detection over large areas using remote sensing to monitor mortality. Various methods were investigated and a multi-temporal Kauth-Thomas transformation was employed (Collins, J.B. and Woodcock, C.E., in press). Results were promising and required much less field work than the previous study. Field work has been a limiting factor in many remote sensing studies, but this approach simplifies the amount and type of field data necessary for implementation. The approach developed to map forest mortality levels showed much promise for application in other situations. This is critical as these methods have been employed in this current project.

IV. Methods

The methods are discussed in three parts: database preparation, image processing, and analysis.

Database Preparation

Erdas Imagine, ESRI Arc/Info and Image Processing Workbench (IPW) (Frew, 1990) were employed for this project. Most of the processing was done on a Sun SPARC 2 workstation donated for this project by California Department of Forestry Fire and Resource Assessment Program. A Sun SPARC 10 was used for some of the more CPU intensive processes.

Six Landsat Thematic Mapper (TM) scenes covering the project area for the early date (1990) had been acquired previously by CDF. The same six scenes were acquired for 1995, as close to the 1990 dates as possible, taking clouds into consideration. The scenes used for this project are detailed in Table 1 below.

Path	Row	Acquisition Date	Processing	Resampling
43	33	8/13/89	Auto C.P.P.	CC
43	33	9/15/95	Terrain	NN
43	34	8/16/90	Manual C.P.P.	CC
43	33	9/15/95	Terrain	NN
42	34	8/25/90	Auto C.P.P.	CC
42	34	9/8/95	Terrain	NN
41	34			CC
41	34	9/17/95	Terrain	NN
42	35	8/25/90	Manual C.P.P.	CC
42	35	9/8/95	Terrain	NN
41	35	9/3/90	Manual C.P.P.	CC
41	35	9/17/95	Terrain	NN

CC = cubic convolution
NN = nearest neighbor
Table 1: Landsat TM Scenes Processed

One of the most important steps in any change detection project is assuring that multidate imagery from the same Path and Row are registered to each other, to ideally, within one pixel. This will ensure that changes due to misregistration do not appear on the final change map. The 1990 TM data was used as the base year, and to ensure that these scenes were properly registered to the UTM coordinate system, we compared them to SPOT panchromatic imagery. The 1995 images were then visually compared on-screen to the 1990 imagery, matching road intersections and other invariant, prominent features. All scenes registered well except for 1990 TM Path 41 Row 34. To improve registration, we registered this scene to the 1995 TM Path 41 Row 34 by collecting over 100 ground control points. A third order transformation resulted in an overall root-mean square error (RMS) of 1.8. The transformed 1990 Path 41 Row 34 registered well to the 1995 imagery except in areas of high elevation. Overlap from adjacent images enabled us to reduce the area of 1990 Path 41 Row 34 needed for analysis to a minimum.

Differences in atmospheric conditions, illumination conditions, and sensor calibration between different dates cause differences in pixel values of unchanged features. Since change detection is ultimately based on interdate differences in pixel level data, radiometric corrections must be used. Collins and Woodcock (in press) concluded that regression-based radiometric correction patterned after Schott, et.al., 1988 was sufficient for change detection and is relatively easy to perform. The technique adjusts pixel values in one image so that invariant features contain the same spectral values in the other image. First homogeneous areas of invariant features such as bare rock outcrops (bright) and water (dark) areas were photointerpreted and selected from the 1990 and 1995 images. Coincident clusters of pixels were then selected from both images. Spectral value means for each of the six TM bands (bands 1-5 and 7) for the clusters of pixels for both years were exported to ASCII text files and used as input for the regression calculation. Table 2 shows the results. The slope and intercept values from the regression were applied to the 1990 image values according to the equation below to derive the normalized (radiometrically corrected) values. The normalized 1990 images were used in all subsequent processes.

$$1990_{(normalized)} = (slope)1990_{(raw)} + intercept$$

Image Path/Row	Band	Intercept	Slope	r_squared	Image Path/Row	Band	Intercept	Slope	r_squared
TM 43/33	1	1.0316	0.6305	0.9836	TM 41/34	1	7.7182	0.6843	0.9525
	2	-2.7859	0.6987	0.9929		2	2.0348	0.7309	0.9746
	3	-2.4212	0.6278	0.9924		3	1.3483	0.7907	0.9922
	4	-2.9069	0.7426	0.9926		4	0.0901	0.8201	0.9936
	5	-2.1560	0.7517	0.9892		5	-4.8519	0.8950	0.9935
	7	-1.8133	0.7430	0.9869		7	-2.4631	0.8930	0.9966
TM 43/34	1	-22.2246	0.8567	0.9886	TM 42/35	1	-8.6940	0.7945	0.9668
	2	-11.1569	0.9017	0.9852		2	-6.3145	0.8825	0.9755
	3	-9.4471	0.7745	0.9870		3	-5.6283	0.7625	0.9866
	4	-7.4774	0.8630	0.9889		4	-5.8865	0.8753	0.9978
	5	-4.0760	0.7986	0.9889		5	-1.8808	0.7918	0.9937
	7	-3.1313	0.7925	0.9870		7	-1.5727	0.7816	0.9975
TM 42/34	1	-6.8068	0.7746	0.9918	TM 41/35	1	8.1614	0.7897	0.9897
	2	-5.8239	0.8421	0.9898		2	-5.0345	0.8396	0.9904
	3	-4.8230	0.7352	0.9900		3	-4.6801	0.7436	0.9909
	4	-4.6202	0.8612	0.9922		4	-4.1703	0.8822	0.9976
	5	-5.5954	0.8545	0.9931		5	-3.6942	0.8520	0.9938
	7	-3.3803	0.8429	0.9918		7	-3.0933	0.8424	0.9912

Table 2: Regression Results for Normalizing 1990 Images to Radiometrically Match the 1995 Images

Hardware, software and time constraints made it necessary to divide the scene pairs into smaller processing areas. To maintain the integrity of ecologically similar areas during the segmentation process, boundaries from the USDA Ecological Units of California Subections, 1994, and planning watershed coverages developed by CDF-FRAP (CALWATER 1.2) were used for clipping the imagery into smaller processing areas. Figure 2 shows the project area shaded by scene boundaries overlaid with the California subsections. Within each scene, windows were created so processing areas could be extracted along subsection boundaries. This was important so that a processing area had no more than 1.8 million pixels, a maximum number we have found to be the limit for portions of our processing. When it was necessary to break up a subsection, boundaries from the CALWATER coverage was used. The final result was 54 processing areas within 11 windows.

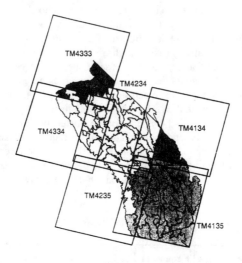

Figure 2. Project area, scene boundaries and subsections

<u>Image Processing</u>

The image processing portion of this project involved image segmentation, multitemporal Kauth-Thomas transformation (MKT), and classification.

Image segmentation is a process which creates regions (polygons) from imagery based on the spectral similarity of pixels at a local level. It is a multipass algorithm utilizing several parameters defining the threshhold of similarity between neighboring pixels in a region, the minimum size for a final region, the maximum size of a region that can be merged, as well as other parameters which come into play during processing. Band selection is based on the type of discrimination desired. Regions define forest stands somewhat like manual photo interpretation, with boundaries that reflect the underlying forest structure, vegetation and landcover composition and topography. In order to improve stand definition by image segmentation beyond simple spectral measurements a spatial component, texture, is added. We use segmentation to reduce noise in the change data by analyzing polygons as opposed to pixels.

Previous studies indicated that an image composed of three bands, TM bands 3, 4 and the texture band generated from band 4, result in satisfactory polygons (Ryherd, S.L., and Woodcock C.E., 1990). For this study, the 1995 imagery was used for generating the region maps. A minimum region size of 2.5 acres and a maximum of 100 acres was specified for this project. On the average regions are between 15 and 50 acres.

The Kauth-Thomas transformation is a well-established linear transformation applied to TM data (Kauth & Thomas, 1976). Information from all six TM bands (1-5 and 7) can be reduced to three components termed brightness (B), greenness (G) and wetness (W). These three components exhibit a high correlation to their labels and thus have intuitive meaning. The coefficients used for the transformation are stable for the TM sensor and independent of the season and illumination conditions at the time of image acquistion so they can be applied to all TM scenes from the same sensors.

For this study, a modification of the Kauth-Thomas transformation was applied to the multi-temporal 12 band image (TM bands 1-5 and 7 for the each date) to compute the change in brightness, greenness and wetness (B, G, W) values between the two dates (Collins and Woodcock, 1996). The signs of the coefficients were reversed for the bands of the 1990 image to obtain the difference B, G, and W in one pass of the data. Table 3 lists the coefficients used.

		Band Coefficients					
	Feature	TM1	TM2	TM3	TM4	TM5	TM7
Applied to	Brightness	0.2147	0.1975	0.3354	0.3949	0.3593	0.1317
1995	Greenness	-0.2017	-0.1724	-0.3848	0.5116	0.0589	-0.1275
Data	Wetness	0.1067	0.1395	0.2318	0.2408	-0.5029	-0.3233
Applied to	-Brightness	-0.2147	-0.1975	-0.3354	-0.3949	-0.3593	-0.1317
1990	-Greenness	0.2017	0.1724	0.3848	-0.5116	-0.0589	0.1275
Data	-Wetness	-0.1067	-0.1395	-0.2318	-0.2408	0.5029	0.3233

Table 3. Multitemporal Kauth-Thomas Transformation Coefficients

This transformation was performed using IPW, a public domain image processing package (Frew, J.E., 1990). To reduce "noise" and minimize edge effects and misregistration at the individual pixel level between the two dates of imagery, the region maps were applied to the pixel level delta brightness, greenness and wetness images. The output from this step is a new image with the mean values of brightness, greenness or wetness for the pixels within each region. These region mean change images for each processing area were mosaiked to create 3 seamless bands and imported into ERDAS Imagine for classification.

An unsupervised classification was performed on the three-band brightness, greeness and wetness change image as a data reduction technique to aid us in interpretation. Processing scene 2 resulted in a 50 class image. Scenes 3 and 6 were first divided into ecologically similar sections and then each section was classified separately. This provided better representation of classes within an entire scene.

Change Analysis

Spectral plots were developed by plotting the classes on an x,y axis in terms of their mean band brightness, greenness, and wetness values of the classified scene. Greeness vs. wetness and brightness vs. wetness were the most useful band combination to separate classes from one another. For example, an area that experienced extreme vegetation loss, such as a burn would appear low on the greenness axis, high on the brightness axis and low on the wetness axis. Image appearance, photo interpretation, vegetation and topographic maps also aided with interpretation. Our objective was not to identify the causes of change, although that was sometimes apparent, but to identify levels of change. A change continuum identifying seven catgories of change was created.

1	Large decrease in vegetation
2	moderate decrease in vegetation
3	small decrease in vegetation
4	Little or no change
5	small increase in vegetation
6	moderate increase in vegetation
7	large increase in vegetation

Two additional classes were also identified.

8	non-veg change
9	shadow/wet

The non-vegetated change class resulted from changes in reservoir and lake water levels. 1990 was a drought year and 1995 was a wet year resulting in dramatic changes in water levels. Other non-vegetated changes included rock to snow. The shadow class occurred primarily in coniferous forests on north facing slopes.

V. Discussion

Analysis of scene 2 (TM 4334) has been carried the furthest and will be the focus of discussing the results. Figure 3 shows the location of scene 2 in relation to the project area.

Figure 3. Location of scene 2

Areas of various silvicultural activities and fire areas that occurred after appear clearly on the change image as large and moderate decreases in vegetation depending on the amount of regeneration that has occurred. Areas that were burned or logged even earlier show increases in vegetation (Figure 4).

170

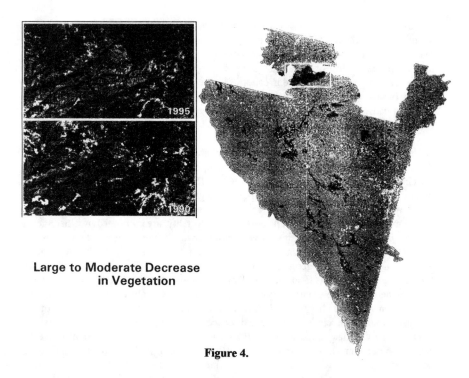

**Large to Moderate Decrease
in Vegetation**

Figure 4.

Figure 4 is an example of an area that burned since 1990 and vegetation loss is clearly evident.

The change map was also overlaid onto a vegetation map of the area to quantify the amount of change by vegetation type. The vegetation map was based on 1990 imagery, and is therefore an effective tool to interpret change between 1990 and 1995. Table 4 below shows these results for scene 2.

					Change by Cover Type (In Acres)									
	Hardwood	%	Conifer	%	Shrub	%	Grass	%	Urban	%	Water	%	Other	%
Large Decrease	2350	1%	6184	4%	92	0%	54	0%	14	0%	7	0%	114	1%
Moderate Decrease	8989	2%	8041	5%	1600	1%	420	0%	83	2%	29	0%	234	3%
Small Decrease	4606	1%	712	0%	1577	1%	1890	1%	87	2%	57	0%	195	2%
Shadow	32310	8%	69671	47%	7912	6%	892	0%	252	6%	789	5%	708	9%
No Change	321772	78%	61805	42%	97385	77%	202613	89%	3108	71%	10817	68%	4644	59%
Small Increase	11221	3%	408	0%	3523	3%	2539	1%	268	6%	655	4%	386	5%
Moderate Increase	24809	6%	484	0%	10206	8%	6420	3%	295	7%	66	0%	752	10%
Large Increase	1978	0%	18	0%	995	1%	1125	0%	188	4%	20	0%	304	4%
Non-Veg Change	4959	1%	164	0%	3911	3%	10553	5%	83	2%	3372	21%	557	7%
TOTAL	412995		147485		127202		226506		4377		15812		7894	942272

Table 4 Change acreage by vegetation type

171

Table 4 shows the change classes, as defined by the change class continuum, by cover type in both acres of change and percent change of the total acres for that class. The table indicates that approximately 15,000 acres of the total amount of hardwoods in scene 2 have decreased. Most of these changes are due to burns and some harvests. 36,000 acres of the hardwoods have increased in canopy closure. This table is especially useful for initially observing changes in landcover. Further work is necessary, however, to determine the exact cause of the change and to identify changes on the map by quantifying corresponding changes on the ground.

At the time of this writing the rest of the project area is being analyzed. It is anticipated that by the end of June 1996 a complete, seamless change map will be finished also marking the end of Phase I of this project. Phase II will begin in early July 1996 and involve a multistage approach to sampling on the ground. The objectives of Phase II are to determine pest caused mortality for the past 5 years and quantify volume loss due to pest damage. In order to determine pest caused mortality, areas changed by other agents will also be identified such as, fire, harvest, blow down, site conversion and preparation, regrowth, housing and road construction, etc. Vegetation maps are available to aid in interpretation as are other geographic information system (GIS) coverages. The National Forest Inventory Analysis Grid (FIA Grid) will be used to estimate productive forestland on all ownerships. The change map will be overlaid in GIS to identify locations where a reduction in vegetation has occured. Causes of change will be determined for all plot locations where productive forestland and reduced vegetation cover occur on the map. SPOT imagery, aerial photography and field work will be used to determine the cause of change. A quick inventory method for 5 or 10 point cluster plots with additional fields for coding salvaged trees will be implemented by trained field crews. Post stratification of plots by change classes, forest types, and mortality types will be applied to refine the information and compile volume estimates. Phase II is expected to be completed by September 1996.

VI. Conclusion

Remote sensing data from 1990 and 1995 have been used to assess changes over 16.7 million acres in the southern Sierra Nevada's. A methodology was developed under a previous cooperative agreement for looking at changes in conifer mortality. Results indicated high correlations between reported changes and conifer mortality. This effort was expanded to multiple landcover types and ownerships. A multi-temporal Kauth-Thomas transformation was employed to detect changes in brightness, greenness and wetness. Images from the two dates were registered and normalized prior to processing. An image segmentation procedure was used which creates regions (polygons) based on spectral and spatial similarities to aid in interpretation. Analysis was undertaken to identify change classes along a continuum of large decreases in vegetation to large increases in vegetation. Two non-vegetation change classes were also reported. A change map has been created for most of the project area which identifies the degree of change in an area marking the end of Phase I. Quantification of the change and information pertaining to what caused the change will be identified in Phase II of the project. Field work will begin in July 1996 to collect information on what has caused the changes.

Results from this first phase are promising and will help to identify trends in changes over time, provide map updates, and assess changes in vegetation extent and composition. This type of large area change detection monitoring is expected to continue on a 5 year cycle through-out the state in cooperation with other Federal and state agences.

Bibliography

Collins, J.B. and C.E. Woodcock, 1994, Change Detection Using the Gramm-Schmidt Transformation Applied to Mapping Forest Mortality, Remote Sensing Environment, 50:267-279.

Collins, J.B. and C.E. Woodcock, 1996, An Assessment of Several Linear Change Detection Techniques for Mapping Forest Mortality Using Multitemporal Landsat TM Data, in press.

Frew, J.E., 1990, The Image Processing Workbench, Doctorate Dissertation, University California, Santa Barbara.

Kauth, R.J. and G.S. Thomas, 1976, The Tassled Cap - A Graphic Description of the Spectral-temporal Development of Agricultural Crops as seen by Landsat, Proceedings of the Symposium on Machine Processing of Remotely Sensed Data, Purdue University, West Lafayette, Indiana, 4b41-4b51.

Macomber, S. and C.E. Woodcock, 1994, Mapping and Monitoring Conifer Mortality Using Remote Sensing in the Lake Tahoe Basin, Remote Sensing Environment, 50:255-266.

Ryherd, S.L. and C.E. Woodcock, 1990, The Use of Texture in Image Segmentation for the Definition of Forest Stand Boundaries, Presented at the Twenty-Third International Symosium on Remote Sensing of Environment, Bangkok, Thailand, April 18-25.

Schott, J.B., C. Salvaggio, and W.J. Volchok, 1988, Radiometric Scene Normalization using Psudoinvariant Features, Remote Sensing of Environement, 26:1-16.

Woodcock, C.E., J.B. Collins, S. Gopal, V.Jakabhazy, X. Li, S.Macomber, S.Ryherd, V. J. Harward, J. Levitan, Y.Wu, and R. Warbington, 1994, Mapping Forest Vegetation Using Landsat TM Imagery and a Canopy Reflectance Model, Remote Sensing of Environment, 50:240-254.

USDA Forest Service, Pacific Southwest Region, 1994, Ecological Units of California, Subsections.

CALWATER 1.2, a geographic information system coverage developed at the California Department of Forestry and Fire Protection, contact Clay Brandow, Watershed Specialist.

MONITORING ASPEN DECLINE USING REMOTE SENSING AND GIS

Gravelly Mountain Landscape, Southwestern Montana

Tim Wirth*, Paul Maus*, Jay Powell*, and Henry Lachowski
USDA Forest Service
Remote Sensing Applications Center
2222 West 2300 South
Salt Lake City, UT 84119

*Pacific Meridian Resources
In Residence at: USDA Forest Service
Remote Sensing Application Center
2222 West 2300 South
Salt Lake City, UT 84119

Kevin Suzuki, Jim McNamara, Pat Riordan, and Ron Brohman
Beaverhead-Deerlodge National Forest
420 Barret Street
Dillon, MT 59725

ABSTRACT

This paper describes a cooperative project between the Remote Sensing Applications Center (RSAC) and the Beaverhead-Deerlodge National Forest to map aspen and aspen decline using remote sensing and geographic information systems (GIS). Current aspen distribution was mapped over a 460,000 acre study area in southwestern Montana using Landsat Thematic Mapper imagery from 1992. Historical aspen populations were mapped for a 6,000 acre area using 1947 aerial photography. Aspen populations for the 6,000 acre area were then compared between 1947 and 1992. Results of this comparison showed that aspen populations decreased by 45% over the 45 year period and that most of the change was due to succession by conifer species.

INTRODUCTION

Quaking aspen (Populus tremuloides), the most widely distributed native North American tree species, plays a very important role in the ecosystems of the Intermountain West. Aspen stands contribute to forest health by increasing plant and animal diversity where they occur. They contribute to forage production for wildlife and domestic livestock. Many species of birds and mammals, including cavity-dependent species, find suitable components for nesting, foraging and roosting in aspen. The amount of aspen in a given area can substantially effect water yields because less aspen means reduced stream flow and lower water yield (Ramsey, 1989). Aspen stands also contribute greatly to the value of scenic and recreational areas.

In the Intermountain West, forests of quaking aspen are considered by most biologists to be subclimax communities (e.g., Mueggler, 1976; Bartos et al, 1983). Mature aspen forests are commonly replaced by evergreen conifers unless the successional process is interrupted by fire or other disturbance. The policy of fire suppression over the past several decades has likely increased the conversion of aspen to conifer (Ramsey, 1989). Other factors such as overgrazing, road-building, and home building have contributed to aspen decline. Historical records indicate an overall decline of aspen populations in the West, and over the last 100 to 150 years, there has been an estimated loss of up to 75% of the aspen throughout the Rockies. The importance of aspen in the ecosystem and its perceived decline over the years has prompted a need for research studies to map current aspen locations and determine where and why aspen populations are being lost.

This study was a cooperative pilot project between the Remote Sensing Applications Center (RSAC) - USDA Forest Service, Salt Lake City, UT , and the Beaverhead-Deerlodge National Forest in Region 1 (Dillon, MT) to map current aspen populations and aspen decline over the last 50 years using remote sensing and GIS technology. The study was funded by the USDA Forest Service Remote Sensing Steering Committee (RSSC) and was conducted during the fall and winter of 1996.

OBJECTIVES

The objectives of this project were to map the current and historical distribution of aspen within a portion of the Gravelly Mountain Range landscape unit, southwestern Montana. The current distribution of aspen was mapped using digital satellite imagery acquired in 1992. Historical aspen was mapped with aerial photographs dating from 1947. The difference in aspen distribution between the current and historical maps was determined by overlaying the two maps in a geographic information system (GIS). The information stored in the GIS was used to determine the extent of change, as well as what changed and what it changed to. An additional goal of the project was the transfer of GIS and image processing knowledge and methods from RSAC to Forest ecologists working on the project. The techniques and results derived from this pilot study will also be used to assist in a larger landscape analysis currently being conducted on the Deerlodge-Beaverhead National Forest. Future goals for this study include updating the aspen component within the entire landscape study area (approximately 2,000,000 acres) and providing a link to large area regional vegetation analysis.

STUDY AREA

The study area is located within the Gravelly Mountain Range of the Beaverhead-Deerlodge National Forest, in southwestern Montana and covers approximately 460,000 acres (Figure 1). Much of this nine quadrangle area is within the West Fork of the Madison River drainage system. The upper end of the Centennial Valley, Ruby River drainage, also falls within the study area. Elevations here range from 5,810 to 10,311 feet. Dominant vegetation types include sagebrush, fescue, aspen, Douglas-fir, lodgepole pine, and sub-alpine fir. Sagebrush and grassland types account for over 50% of the vegetation in this area. Approximately 6,000 acres within the study area were mapped using the historical 1947 photographs. This site is located within the nine quad pilot study area due west of Henry's Lake.

Figure 1. Location of the nine quadrangle study area.

METHODS

Remote sensing data in the form of aerial photographs and satellite imagery were the primary data sources used in the study. Additional information in a GIS was used to improve the vegetation classification. Data used in the study are described below:

- Landsat Thematic Mapper satellite image - 30 meter pixel size
  - Date Acquired: August 7, 1992
- Aerial Photographs
  - 1:16,000 scale Natural Color Photos - Date Acquired: September, 1987
  - 1:30,000 scale Black and White Photos - Date Acquired: August, 1947
- Digital Elevation Models (DEMs) - 30 meter pixel size
- Cartographic Feature Files (CFFs)
- Infra Red Digital Camera Images - Date Acquired: October 9, 1995

The first phase of this project was the creation of a current vegetation map (cover type layer) for the nine quadrangle pilot area. The vegetation classification was derived from a seven band Landsat TM image. Bands 1-5 and 7, and a ratio of band 4 and 3 were used in the classification process (Figure 2). The 4/3 ratio band was developed to assist in enhancing the spectral response of aspen (Ramsey, 1989). A combined supervised and unsupervised approach for image classification was used to develop the cover type layer (see Maus et al, 1995).

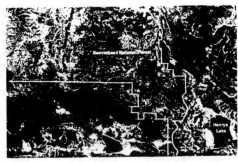

Figure 2. Landsat Thematic Mapper (TM) image acquired August 7, 1992 of the nine quadrangle pilot area. TM Bands displayed are 4/3 ratio (near IR/visible red), band 4 (near IR), and band 3 (visible red).

176

A sixty class unsupervised classification was created using the ISODATA algorithm (ERDAS, Atlanta, GA). Resource specialists familiar with the vegetation identified samples from the unsupervised classes on 1:16,000 scale color aerial photographs. In addition, color infrared digital camera imagery acquired on October 9, 1995 by RSAC was used to determine current aspen sites. As a result of the unsupervised classification, two 1:24,000 scale 7.5 minute quadrangle maps were created and checked in the field to verify class identification. The initial classification scheme consisted of fifteen cover type classes of which eight contained aspen or mixed aspen (Table 1). More effort was put into accurately mapping these categories since the project was aimed at locating existing aspen.

Table 1. Classification scheme used for the 1992 vegetation classification derived from satellite imagery.

Class	Non-Vegetated
1	water
2	rock
3	other non-vegetated
	Vegetated (Pure Aspen)
4	75-100% aspen
5	50-74% aspen
6	25-49% aspen
	Vegetated (Aspen/Conifer Mix)
7	half and half - aspen/conifer
8	1/4 aspen and 3/4 conifer
9	3/4 aspen and 1/4 conifer
	Vegetated (Other Vegetation)
10	conifer
11	aspen/sage mix
12	sparse aspen
13	sage
14	willow
15	other vegetation

In addition to the spectral classes generated by the unsupervised classification, 239 known locations of aspen were delineated on the 1:16,000 scale color aerial photographs by Forest specialists familiar with the study area. These were used as used as supervised training sites and were located and digitized onto the satellite image. A cluster analysis using SPSS statistical software (SPSS Inc., Chicago, IL) was conducted on the digitized training sites. The cluster analysis assisted in locating spectral confusion within the training data set. For example, where aspen training sites were confusing with willow or other vegetation types. Improvements were made by removing and redigitizing problem training sites. After combining the revised digitized training sites with the initial unsupervised (60 class) data set, a second cluster analysis was performed. This process highlighted duplicate spectral classes and training sites as well as identified land cover classes that were not adequately represented by the supervised training sites. Examples of these land cover classes were transferred to aerial photographs for cover type identification. These new training sites were digitized and added to the supervised data set for the final image classification. (For a more detailed description of supervised and unsupervised classification techniques, refer to Maus et al, 1995).

Other refinements used before the completion of the cover type layer included GIS modeling. Classification errors between willow and aspen were common since the spectral response of the two species is similar. Elevation and slope, derived from DEMs, were used to model potential locations of aspen and willow. Other misclassifications (not corrected through modeling) were located on screen and edited to the correct cover type. The combination of the supervised and unsupervised classifications, including modeling and edits, resulted in the final cover type layer (Figure 3).

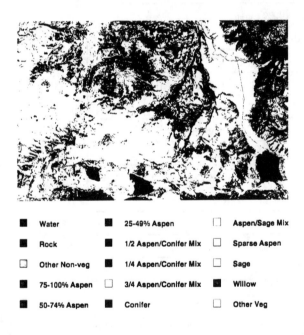

■ Water	■ 25-49% Aspen	▢ Aspen/Sage Mix
■ Rock	■ 1/2 Aspen/Conifer Mix	▢ Sparse Aspen
▢ Other Non-veg	■ 1/4 Aspen/Conifer Mix	▢ Sage
■ 75-100% Aspen	▢ 3/4 Aspen/Conifer Mix	■ Willow
■ 50-74% Aspen	■ Conifer	▢ Other Veg

Figure 3. Classification of cover types for the nine quadrangle pilot area.

Accuracy Assessment

A quantitative accuracy assessment was conducted on the cover type classification. It was determined that the initial 15 class scheme contained more detail than could accurately be mapped within the project's short time frame. Since the classification scheme was hierarchical, it was collapsed from fifteen to eight classes. This collapsed scheme provided enough detail and accuracy to meet the current project objectives. The eight classes assessed for accuracy are listed in Table 2.

For the accuracy assessment, 246 sites were randomly located across the final eight class map. Unique locator numbers were assigned to each site and then transferred to the 1:16,000 scale photos. Photo interpreters from the Beaverhead-Deerlodge National Forest interpreted the accuracy assessment sites in stereo and assigned one of the eight class values to each without prior knowledge of its classified cover type. The photo interpreted data (reference data) were then compared to the classified image data for each of the accuracy assessment sites and the results were tallied an error matrix (Story and Congalton, 1986). The results are summarized in the error matrix in Table 3.

Table 2. Final eight class land cover classes used in the accuracy assessment.

Class	Non-Vegetated
1	water
2	rock/bare
	Vegetated
3	pure aspen
4	aspen/conifer mix
5	conifer
6	sage
7	willow
8	other vegetation

Table 3. Error Matrix

Ref. Data
Row
TM Classification

	1	2	3	4	5	6	7	8	Total
1	15								15
2		23						2	25
3			26	2	1			1	30
4			1	2					3
5			1	8	40	1			50
6		3	4			28		10	45
7			2				9		11
8		8	7			12	11	27	67
Col. Total	15	34	41	14	41	41	20	40	246

Total Accuracy = 170/246 = 69%

Class	User's Accuracy	Producer's Accuracy
1. Water	15/15 = 100%	15/15 = 100%
2. Rock/Bare	23/34 = 68%	23/25 = 92%
3. Aspen	**26/41 = 63%**	**26/30 = 87%**
4. Mix	2/14 = 14%	2/3 = 67%
5. Conifer	40/41 = 98%	40/50 = 80%
6. Sage	28/41 = 68%	28/45 = 62%
7. Willow	9/20 = 45%	9/11 = 82%
8. Other Veg	27/40 = 68%	27/67 = 40%

Table 3 shows the Overall Accuracy for the eight class TM classification to be 69%. The Aspen class (class 3) had a User's Accuracy of 63% and a Producer's Accuracy of 87%. User's Accuracy is based on errors of commission and is the probability that a pixel on the map actually represents that category on the ground. Producer's Accuracy is based on errors of omission and is the probability of a reference site being correctly classified. Further discussion on this the accuracy assessment will be presented later in this paper. (For a more detailed explanation of the Error Matrix and User's and Producer's Accuracy, see Maus et al, 1995).

Interpretation and Transfer to GIS of Historical Aerial Photographs

Black and white 1:30,000 scale aerial photographs acquired in 1947 were visually interpreted in stereo for the 6,000 acre study area. A subset of the final classification scheme (Table 2) was used in the interpretation since only the distribution of aspen was of interest. The land cover categories interpreted from the photographs were aspen, aspen/mix, and conifer classes (classes 3, 4, and 5). Transfer of the photo interpretations into a GIS was difficult due to the photographs having been enlarged resulting in the fiducial marks being cropped. Typically a stereo plotter or transfer scope could have been used to transfer information to a base map. Instead, the photo delineations were reduced to 1:24,000 scale on a photocopier and transferred to two orthophoto quads over a light table. The photo delineations were then digitized and attributed into the GIS (Figure 4). There was no accuracy assessment completed on the historical photo interpreted map.

Once in a GIS, the 1947 cover type map derived from these digitized polygons was overlaid onto the 1992 TM classification. This allowed for the determination of changes in aspen within the 6,000 acre study area between 1947 and 1992.

1947 Photograph of the Henrys Lake Mountains

Figure 4.   The scanned and rectified 1947 photograph with delineated cover types overlay. Aspen and aspen/conifer mix types are shown in yellow and other vegetation types are shown in cyan.

RESULTS AND DISCUSSION

Satellite Image Classification

Analysis of the final cover type layer indicated that mature, highly dense stands of aspen can be mapped by the classification approach used in this study. Although the User's Accuracy in Table 3 for Aspen is only 63%, most errors occurred in classes representing lower aspen densities. Of the 15 errors of commission for the Aspen class, 13 of these were committed to Sage, Willow, or the Other Vegetation class. Most were low density aspen. In general, the cover type layer derived from the TM image over-estimated the sparse aspen areas. Dense aspen however was identified on both the aerial photos and the TM imagery. The difficulty in mapping sparse aspen classes can be partially attributed to the low spatial resolution of Landsat Thematic Mapper imagery (30 meters). The low spatial resolution limits the ability to map some of the patchy and sparse stands of aspen.

Another issue affecting accuracy is the focus of editing efforts. Most of the classification editing effort was spent in the 6,000 acre pilot study area. Since it was critical to accurately map the aspen and time available for the project was short, the focus of the classification effort was in this area. The result was a more accurate map of aspen stands for the 6,000 acre area than for the entire 460,000 acre area. The accuracy assessment however was completed on the entire study area.

Resource specialists from the Beaverhead-Deerlodge National Forest will use the image classification from the nine quadrangle study area to assist in their landscape analysis over the Gravelly Mountain Range. The forest also plans to complete a similar classification for the entire Landscape analysis area. If results prove to be successful, the Forest will use digital satellite imagery in future landscape area analysis and mapping.

Historical Aerial Photo Interpretation

Comparisons of aspen between 1947 and 1992 showed a decline in aspen populations within the 6,000 acre pilot study area (Figure 5). A comparison between the 1947 photo interpretation and the 1992 image classification showed a 45% decline in pure aspen and aspen/conifer mix acreage. During the 45 year period, the acreage in pure aspen and aspen/conifer mix was replaced by approximately 60% conifer, 25% sage, and 15% other vegetation types. Most of the aspen replacement occurred where aspen populations were located adjacent to conifer stands. Many of these 1947 aspen stands where inclusions within conifer stands and stringers along the conifer stands edges. The satellite image classification may have underestimated these aspen areas due to the dominance of conifer in the spectral values.

Some of the areas seen as change between the two maps (1947 and 1992) may also be due to the different methods used for creating the maps. Although one was created from satellite imagery and the other from aerial photographs, the classification schemes used in the interpretation were similar. This ensured to the extent possible that the interpretations were comparable.

Cover Type Layer - 1947-1992 pilot

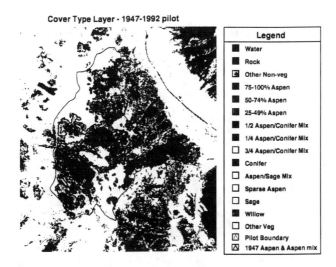

Legend	
■	Water
■	Rock
◪	Other Non-veg
■	75-100% Aspen
■	50-74% Aspen
▦	25-49% Aspen
■	1/2 Aspen/Conifer Mix
■	1/4 Aspen/Conifer Mix
☐	3/4 Aspen/Conifer Mix
■	Conifer
☐	Aspen/Sage Mix
☐	Sparse Aspen
☐	Sage
▦	Willow
☐	Other Veg
◸	Pilot Boundary
◩	1947 Aspen & Aspen mix

Figure 5. 1992 TM classification of the cover type layer for the 6,000 acre pilot area. The 1947 aspen and aspen/conifer mix delineations are shown as a overlay.

CONCLUSION

The results of this pilot study demonstrate that aspen stands can be mapped using Landsat TM satellite imagery. Use of digital satellite imagery and historical photos also proved to be a useful means of determining the location and amounts of aspen decline. Based upon this study, it was determined that the majority of aspen decrease was due to encroachment of conifer species. Data products from this study can now be used within a GIS for further analysis of landscape areas and to address specific research questions on aspen decline.

REFERENCES

Bartos, Dale L., F.R. Ward, and S. Innis. 1983. Aspen succession in the Intermountain west: a deterministic model. General Technical Report INT-153. Ogden, UT: USDA, Forest Service, Intermountain Forest and Range Experiment Station. 60 p.

Maus, P., M. Golden, J. Johnson, V. Landrum, J. Powell, V. Varner, and T. Wirth. Guidelines for the Use of Digital Imagery for Vegetation Mapping - EM-7140-25. USDA, Forest Service, Engineering Staff. Paul A. Maus, Editor. 1995. 125 p.

Mueggler, W.F. 1976. Type variability and succession in aspen ecosystems. In: Utilization and marketing as tools for aspen management in the Rocky Mountains. General Technical Report RM-29. Fort Collins, CO. USDA, Forest Service, Rocky Mountain Forest and Range Experiment Station. p. 16-19.

Ramsey, Robert D. 1989. Modeling leaf area as an input to transpiration models in aspen-conifer forests in Big Cottonwood Canyon, Utah using remotely sensed data and geographic information systems. PhD Thesis. University of Utah - Dept. of Geography.

Story, M. and R.G. Congalton. 1986. Accuracy assessment: A user's perspective. Photogrammetric Engineering and Remote Sensing. 52(3):397-399.

Using Satellite Imagery to Detect and Monitor Forest Change

by

Kass Green
Pacific Meridian Resources

Brian Costentino
State of Washington Dept. of Fish & Wildlife

Abstract

Since the 1940's, land managers have mapped land cover and land use change with aerial surveying methods. In the last 10 years, the proliferation of GIS and satellite technologies has vastly expanded the tools available to land managers for monitoring change . Working under a NASA Earth Observation Commercialization Program (EOCAP) contract, Pacific Meridian Resources investigated the use of multi-temporal satellite and airborne imagery to detect and measure land use and land cover change.

The purpose of the EOCAP project was twofold. First, different techniques for monitoring change were investigated, assessed and compared. Second, the results of the assessments were used to develop easy to use change detection software.

This paper presents the methods and results of an investigative study to determine the usefulness of multitemporal satellite imagery in detecting and measuring forest damage caused by pest outbreaks. The results of this comparative analysis showed that using image differencing of ratios of the near to the mid-infra red bands provided the highest accuracy in measuring forest crown cover change.

I. Introduction

Over the last decade, the western spruce budworm, fir-engraver, Douglas-fir beetle, and other pests have ravaged the forests of Eastern Oregon and Washington and the Sierras of California (USDA Forest Service, 1991). Massive tree mortality and damage have resulted, causing rapid wildlife habitat change and expanding the potential for catastrophic forest fires fueled by increasing levels of dead woody material throughout the forest landscape.

The Blue Mountains of northeastern Oregon have been particularly hard hit by pest morality, A major outbreak peaked in 1986 and declined until 1989 when another infestation occurred. Foresters believe conditions for this latest event, which finally began to subside in 1992, were brought on by a seven year drought and tree overstocking (caused by fire exclusion over the last century). The drought gave the insects a foothold in the forest by lowering natural resistance of the trees. Once the bugs attacked, the close proximity of trees in the overstocked areas made it easy for the pests to spread rapidly from one tree to the next.

The rapidly changing forest landscape presents substantial forest mapping, inventory, and planning challenges for public and private forest management entities. Land mangers need to know the extent and location of forest pest damage so that salvage operations and fire prevention and suppression activities can be planned. Capturing this information in a GIS enables additional analysis of the impact of forest change on harvest scheduling, wildlife populations, range condition, and visual quality goals.

Traditionally, foresters have relied on aerial sketch mapping, aerial photography and field inventories to conduct assessments of forest pest damage. Each method has been effective, but with certain drawbacks. Aerial sketch mapping, ,which involves a trained airborne observer marking tree damage on a paper map and ranking its severity , is difficult to

184

georeference and incorporate into a GIS. Photo interpretation of aerial photography is extremely time consuming and expensive for large area surveys. Field surveys are also expensive and usually cover only a small percentage of a management area.

As part of NASA's EOCAP Pacific Meridian Resources (Emeryville, California) investigated the use of multi-temporal satellite imagery as an alternative means of locating and measuring land use and land cover change. Satellite imagery is a potentially powerful tool because of the large area covered in each scene, the detail of information in each scene, the consistency of spectral response, and the frequent revisit time of the satellites. The overall goal in forest monitoring is to compare and measure spatial representations of forest land at two points in time. Multi-temporal analysis of satellite imagery is effective for forest monitoring because there is high correlation between variation in the satellite imagery and forest cover change.

Change monitoring techniques fall into four categories:
1. Techniques that make the detection of change possible.
2. Techniques for the measurement of extent and magnitude of change.
3. Techniques for updating an existing map or GIS layer to incorporate change.
4. Techniques for estimating the impacts of change on environmental, economic, and political conditions.

This study concentrates on using satellite imagery to detect and measure change.

The Study Area

The project area is located in the central Blue Mountains of northeastern Oregon where some of the most intensive pest mortality has occurred. The area encompasses approximately approximately 24, 7 1/2 minute quadrangles and is located primarily within the Wallowa-Whitman and Umatilla National Forests. The area contains a wide range of land uses including industrial timberlands, municipal watersheds, wilderness, and experimental forests.

The area's topography varies greatly, ranging from broad river valleys to steep ridges at 9000 feet elevation. The most common tree species are grand fir (*Abies grandis*), Douglas-fir (*Pseudotsuga menziesii*, ponderosa pine (*Pinus ponderosa*), lodgepole pine (*Pinus contorta*), western larch (*Larix occidentalis*), subalpine fir (*Abies lasiocarpa*), and Engelmann spruce (*Picea engelmannii*).

II. Methods

Detecting Change

One of the most straight forward uses of satellite imagery in monitoring change is the visual comparison of two images of the same area from different dates. This can be done through side by side comparison or by the combination of multi-temporal imagery to create a change image.

Measuring Change

A change detection image is a qualitative tool. It allows for the manual interpretation of change. Measuring change requires turning the change detection image into a change detection map using image classification procedures. The introduction of digital imagery, including airborne scanners, digital orthophotography, and satellite imagery, has expanded the techniques available for change detection (Jensen, 1981). Image differencing (Maus et al., 1992; Green, et al., 1994), and principal components analysis and/or tasseled cap transformation (Richards, 1984; Byrne et al., 1990; Crist and Cicone, 1984) are the two most often used methods of image-to-image comparison. The major objective of this study was to determine which Landsat TM bands, ratios, or indices most accurately detect and measure tree cover change in a forest.

This study compares the accuracy of using various bands or band combinations for measuring change using image differencing. Image differencing is an automated image processing technique that subtracts the pixel values from one image from the pixels of a second image acquired on a different date. Land cover changes are represented by differences in the reflectance value of each pixel.

The key to image processing is to learn when the image is a good predictor of landcover change, and when it is not. A critical step in any type of change detection analysis is to remove or neutralize any non change image differences that may be misinterpreted as changes in land cover. For example, misregistration of multitemporal images will result in a map of misregistration rather than a map of change. Seasonal differences in sun angle can create shadow in one image that aren't;t in the second and may be mistaken as land cover differences. Variations in atmospheric conditions also can produce radiometric differences between two images acquired on separate dates. Thus, change detection requires that all variation caused by non land cover differences be controlled.

Ideally, multi-temporal images of a study area are acquired during the same season of different years to reduce radiometric differences, However, due to cloud cover, Landsat TM images from the summer of 1989 and the fall of 1993 were used in the Blue Mountain Project. Although the 1989 and 1993 scenes were terrain corrected and in a map coordinate system, they were not in registry. Thus, a second order transformation with nearest neighbor resampling was performed on the 1993 image data sets used in multidate comparisons.

The Classification System

The first step in any mapping project is the specification of the classification system to be mapped. This study focused on the identification and measurement of broad classes of tree crown cover increase and decline. Crown cover refers to the percentage of land area covered by trees. A decline in tree crown cover in a specific area over time suggests loss of trees due to fire, harvesting , pest infestation or other damage.

This project studied the effectiveness of satellite imagery in detecting six categories of tree crown cover change (See Table 1, Crown Cover Change Category Classification System)

Table 1. Crown Cover Change Category Classification System

Class 1	Change in tree crown cover is -71 to 100 percent
Class 2	Change in tree crown cover is -51 to -70 percent
Class 3	Change in tree crown cover is -26 to -50 percent
Class 4	Change in vegetation cover -25 to +25 percent. Considered no change.
Class 5	Change in vegetation cover +26 to +50 percent
Class 6	Change in vegetation cover +51 to +100 percent
Class 7	Water

Forest Service photo interpretation guidelines specify that when a forest canopy changes by more than 25%, a stand should be delineated (USDA Forest Service, 1993). The threshold change value is also near the limit that airborne sketch mapping damage categories can be reliably interpreted from the air (Communication with the Office of Pest Management, Region 6, USDA Forest Service). The vegetation increase categories include increases in all vegetation including trees shrubs, grasses, and herbs.

Field Reconnaissance

After a classification system is specified, the next step is to learn how the landscape varies with respect to the classification system. The purpose of field reconnaissance is to identify the causes of variation in vegetative cover and to link variation in vegetative cover to variation in the satellite imagery. To direct field recognizance, a coregistered multitemporal data set was classified using ERDAS's (Atlanta, Georgia) ISODATA algorithm.

Quadrangle-size hard copy maps (7 1/2 minute) were then plotted. Areas representing forest change and no change were located and field samples delineated on 1:12,000-scale photos. In late October 1993 approximately 50 field sites were visited and estimates of crown cover change were recorded.

Crown cover was estimated for 1993 and 1989 conditions. A portion of the sites had evidence of pest damage prior to 1989 (from the previous pest outbreak) thus making the crown cover change interpretation more difficult. Sites with questionable 1989 crown cover estimates were reviewed and adjusted in the office with color infrared transparencies, other photos, and field notes.

Image Differencing

Image differencing analysis was conducted on a total of nine different muti-temporal bands and band combinations:

        TM 7
        TM5
        TM7/TM4
        TM5/TM4
        TM7/TM4 Calibrated
        TM 5/TM4 Calibrated
        TM7/TM 3
        TM5/TM3
        TM4-TM3/TM4+TM3 (NDVI)

Bands 5 and 7 were emphasized because they are known to be extremely sensitive to plant moisture content, which is an indicator of overall tree health. Image differencing was performed by subtracting the 1993 image from the 1989 and adding a constant to scale the final difference images. The histogram of each difference image was partitioned into change categories. The no change class was estimated first, followed by the forest crown cover decrease classes, followed by vegetation increase. Field data, aerial photography, and other data were used to determine forest change classes from the raw difference images.

Accuracy Assessment

Quantitative accuracy assessment was used to compare the resulting maps. The purpose of quantitative accuracy assessment is the identification and measurement of map errors. It involves the comparison of a map with reference information.

Several researchers have noted the impact of the variation in human interpretation on map results and accuracy assessment (Congalton and Biging, 1992). Gopal and Woodcock (1992) state, "The problem that makes accuracy assessment difficult is that there is ambiguity regarding the appropriate map label for some locations. The situation of one category being exactly right and all other categories being equally and exactly wrong often does not exist." This is especially true in retrospective change detection studies where the analyst has to hypothesize the past condition of land cover. Because of the massive tree defoliation in eastern Oregon, field estimates for 1989 conditions were, at best, difficult, even with a wealth of aerial photography available to the project.

To deal with the impact of variation in human interpretation, several researchers have applied fuzzy logic theory to accuracy assessment (Gopal and Woodcock,1992); Hill, 1993) . Fuzzy logic recognizes that, on the margins of classes that divide a continuum, an item may belong to both classes. Because photo interpretation of tree crown closure typically varies ± 10% (Spurr, 1944), it was assumed that samples within 10% of the correct adjacent reference class are reported as acceptable map labels.

Accuracy assessment site data were collected on plots established on 1:12,000 aerial photos. Along each flight line every twentieth photo was selected and two potential accuracy assessment polygons were delineated in distinct forest or other land cover types. Only sites accessible in the field were used. At field-verifiable sites, a third polygon was delineated on the sample photo and inventoried.

At each accuracy assessment site a field form was filled out detailing stand size structure, crown cover, species, general shrub cover, and non forest cover for 1993 conditions. A form was also filled out for estimated 1989 conditions in a similar manner. The accuracy attribute for each site is the estimated 1989 tree crown cover minus 1993 crown cover. One hundred accuracy assessment reference sites were collected. Approximately 50% of the sites are class 4 (no change) sites.

Results

Accuracy Assessment

Table 2 presents the overall classification accuracy results for the nine maps.

Table 2
Change Detection Map Accuracy Summary

Rank	TM Bands	Accuracy
1	TM7/TM4	83.2
2	TM5/TM4	82.2
3	TM7/TM4 Calibrated	80.2
4	TM7/TM3	79.2
4	TM5/TM3	79.2
5	TM7	77.2
6	TM5/TM4 Calibrated	72.3
7	NDVI	70.3
7	TM5	70.3

All of the bands and band combinations provided acceptable results, with overall accuracies ranging from 70.3 to 83.2 percent. Image differencing using band ratios provided superior classification accuracy over single band analysis.

Unfortunately, ratio images, tend to contain more pixel "noise" than single band 5 or band 7 images. In addition, change detection using only band 5 or 7 may be less expensive and simpler for the image analyst due to less image noise and fewer bands for the analyst to understand and manipulate. However, given the superior classification accuracy achieved with band ratios and the inherent benefits of ratios (e.g., reduction of topographic effects and enhancement of spectral trends) band ratios should be considered in future forest crown cover change analyses.

The superiority of TM7/TM4 over TM5/TM4 may be related to the differences between TM5 and TM7 absorption by foliage water. The TM7/TM4 ratio may possess a greater sensitivity than TM5/TM4 for detecting low levels of coniferous crown cover decrease. Likewise TM7/TM4 ratio provides very good results for monitoring moderate change (class 2) as well as good ability to discriminate no-change (class 4).

Discussion

As mentioned earlier, change detection requires controlling or neutralizing all non change variation. In this study, two sources of non- forest crown cover change were problematic.

- *Grasslands.* Much of Oregon's Eastside forests are a mosaic of grassland and forest, with vegetation types strongly controlled by elevation, slope, and aspect conditions. A large percentage of the study area contains stands that are relatively open and have a grass understory. For satellite remote sensing to be optimally used in crown cover change mapping analyses on forest cover, grassland areas need to be maintained as much as possible within the no-change category. Otherwise, grass spectral response may influence crown cover change estimates. Grass spectral variation can be reduced considerably if satellite scenes are selected for late summer dates when grasses are senescent.

  In this study, wet meadows and seasonally wet grass areas were often classified as substantial vegetation increase (class 6) or as crown cover decrease. These areas vary considerably in the near-infrared during the growing season and from year to year. These areas could be masked out through unsupervised classification procedures to improve map interpretation.

- *Topographic relief.* In areas of high topographic relief, differences in the solar elevation and azimuth angles for the 1989 and 1993 scenes caused unreliable results. For example, mountain summit areas that were sunlit in July 1989 and in deep shadow in September 1993 were incorrectly mapped as change.

Conclusions

The accuracy assessment results form this work indicate that satellite remote sensing data can provide reliable forest change information over a large area with a surprising level of specificity. There is a discernible relationship between Landsat sensor response and forest change. This study demonstrated that creation of reliable forest change detection data depicting categories of crown cover decrease is possible, and that a forest crown cover decrease threshold of 25% is attainable with Landsat data. In addition, forest understory conditions of the characteristically open north-eastern Oregon forests did not significantly complicate detection of crown cover decrease.

Furthermore, the mapping technique demonstrated in this work can provide a cost-effective means to map forest change on a repetitive basis. Because the results are GIS coverages, the change information can also be used to update and existing forest database. In addition, forest change maps and GIS layers can be used to assist silvicultural, fire, and wildlife management activities. For example, the change maps can help fire managers to map fire hazard, and to plan prevention procedures such as prescribed burning and the creation of fuel breaks.

To aid others in utilizing satellite imagery for change detection, Pacific Meridian has developed change detection software in Arc/Info GRID that allows non expert users to perform change detection and to update GIS coverages. Named LUCCAS (Land Use and Cover Change Analysis System), the software steps the user through deciding what type of imagery to use and how to analyze the imagery to monitor and measure land use and land cover change.

Once the change attribute is in the database, it can related to other GIS layers for analysis of the impacts of forest change on other variables such as fire hazard and wildlife habitat. Equally important, the change attribute may provide a rapid means to "flag" stand polygons that may need a boundary or type update. This work could be further extended to include exploration of methods to integrate the evolving "forest practices layer" and change detection information.

From this project it is clearly evident that forest change information generated by satellite remote sensing techniques has great potential for producing cost-effective data for program level forestry activities as well as provide valuable inputs to long-term forest health monitoring programs.

References

Congalton, R. and G. Biging. 1992. How to validate stand maps. *Proceedings of the Stand Inventory Technologies: An International Multiple Resource Conference.* American Society for Photogrammetry and Remote Sensing: Bethesda, Maryland. 9 p.

Congalton, R. and K. Green, 1993. A practical look at the sources of confusion in error matrix generation. *Photogrammetric Eng. and Rem. Sens.*, Vol. 59, No. 5, pp. 641-644.

Gopal, S. and C. Woodcock, 1994. Theory and methods for accuracy assessment of thematic maps using fuzzy sets. *Photogrammetric Eng. and Rem. Sens.*, Vol. 60, No. 2, pp. 181-188.

Maus , P., V. Landrum, and J. Johnson 1992. Utilizing satellite data and GIS to map land cover change. GIS '94 Symposium. Vancouver, BC

Spurr, S. H.. 1947. *Aerial photographs in forestry.* The Ronald Press Company, New York, NY.

USDA Forest Service, 1991. *Blue Mountains forest health report.* Pacific Northwest Region, pp. ll-1 - ll-143.

USDA Forest Service, 1993. *Existing vegetation (EVG) database user's guide.* Wallowa-Whitmann National Forest.

Vogelmann, J. E., 1990. Comparison between two vegetation indices for measuring different types of forest damage in the north-eastern United States. *Int. J. Of Remote Sens.*, Vol. 11, No. 12, pp. 2281-2297.

Vogelmann, J. E. and Barrett N. Rock, 1988. Assessing forest damage in high-elevation coniferous forest in Vermont and New Hampshire using Thematic Mapper data. *Rem. Sens. of Env.*, Vol. 24, pp. 227-246.

Acknowledgments

The authors would like to thank Kevin Corbly for his editorial advise and input.

EVALUATING THE UTILITY OF PARTITIONING REMOTELY SENSED DATA

Bill Cooke
Forest Inventory and Analysis
Southern Research Station
USDA Forest Service
P.O. Box 906
Starkville, MS 39760

ABSTRACT

The utility of partitioning spectral reflectance variation for Advanced Very High Resolution Radiometer (AVHRR) data was tested for the semi-arid and arid regions of west Texas. Comparisons were made between unpartitioned and partitioned data using Bailey's ecoregions as a partitioning device. Woody and non-woody vegetation types mapped by the Texas Parks and Wildlife Department in 1984 was used as reference data and the Jeffries-Matusita signature divergence test was used to quantify separability. In all pairwise comparisons of adjacent ecoregions, separability for woody versus non-woody vegetation improved when the landscape was partitioned. Further tests using mesquite vegetation associations indicated Bailey's ecoregions are more effective than Omernik's ecoregions for partitioning AVHRR spectral reflectance variation in this area.

BACKGROUND

The 1974 Forest and Rangeland Renewable Resources Planning Act (RPA) requires the United States Department of Agriculture Forest Service (USDA-FS) to provide Congress with statistics on current forest land and rangeland conditions. The Southern Research Station, Forest Inventory and Analysis Program (SRS-FIA) has the RPA mandate to conduct forest inventories for all southern states from Virginia to Texas. Except for sparsely forested regions in west Texas and west Oklahoma, forest land in the South has been field inventoried over several cycles in recent history. An assessment of these western woodland resources is needed to meet the SRS-FIA's inventory mandate.

Landform and climate interactions influence phenology and modify spectral reflectance of vegetation across the landscape. Overlaps in spectral characteristics of cover classes can lead to classification errors when satellite data is used to assess resources over large geographic areas. Suitable geographic partitioning schemes that can isolate those ecological processes that influence vegetative distributions should be useful in improving Advanced Very High Resolution Radiometer(AVHRR) classification accuracy. Quantitative determination of optimum spatial partitioning systems for multispectral satellite data will have positive implications both for forest mapping in the U.S. and in global vegetation cover assessments.

The semi-arid and arid regions of west Texas and Oklahoma present a unique inventory situation where remotely sensed data can enhance the effectiveness of field inventories. In most areas surveyed by SRS-FIA, field plots are located on a 3x3 mile grid. The time, effort, and expense involved in using this design in the semi-arid and arid regions of Texas and Oklahoma may be prohibitive. It is hoped that these remote sensing efforts will culminate in the design of a field inventory stratification tool for SRS-FIA. This stratification tool could provide ground inventory personnel with information defining the probability of the occurrence of woody vegetation. Inventory decision makers could then use this information to design a sampling approach in the vast geographic areas of west Texas and Oklahoma with ground plot locations and intensity determined on a "proportion based on probability" approach. If this methodology is accepted for west Texas and Oklahoma, it may also be adopted for use in other areas of the mid-west U.S.

Zhu and Evans (1994) used physiographic partitioning of AVHRR data, as suggested by Loveland et al. (1991), to improve on forest cover identification. In this study classified Landsat Thematic Mapper(TM) data were used to identify sample areas of forest land within each of 15 physiographic regions covering the conterminous U.S. Seasonal AVHRR data that included the original spectral channels and channel transformations were used as predictor variables in a regression procedure. Forest density was predicted from calibration windows of TM classifications co-registered to AVHRR data. These density models provided the input to automated classification procedures used to develop a forest type map. Feedback from Forest Service regional offices, other FIA units, and other cooperators indicated that forest land was overestimated in areas either dominated by non-forest land uses, or with a high degree of forest fragmentation (Zhu 1994). The authors did not fully investigate the contributions that geographic partitioning make to AVHRR classification accuracy. It was recognized at SRS-FIA that research was needed to determine if partitioning is effective in improving AVHRR classifications and if useful, to identify the partitioning scheme(s) that are well suited to the task. These research issues must be addressed before the SRS-FIA can recommend use of AVHRR classifications for pre-stratification of regions for field plot allocation.

Partitioning and Signature Extensibility

Many researchers have alluded to the fact that training data needed for classification of multispectral image data must be representative of the land cover class of interest. Zhu and Evans (1994) partitioned AVHRR image data on the basis of physiographic factors to predict percent forest cover. Loveland et al. (1991) suggest that spectral variations caused by physiographic factors are far greater in data sets covering a large area than a small area. Short (1982) addresses the concept of "signature extension" by suggesting that only a few features such as deep clear water bodies, clouds, and snow may have broad applicability to any scene in a region.

The recognition of the need for partitioning multispectral image data presents the researcher with many choices. Physiography, ecoregions, soils, elevation, climatic data and other land classification systems are often subjectively applied by the researcher. Objective application of any of these systems requires a quantitative test of the viability of a given system for adequately partitioning within class reflectance variation. In essence, finding the land classification system that best divides a spectrally heterogeneous geographic area into homogeneous sub-areas.

OBJECTIVES

The first objective of this study was to obtain quantitative evidence that spatial partitioning of multispectral image data is an effective procedure for removing within class spectral variation prior to automated classification procedures. The second objective was, upon demonstration of the effectiveness of partitioning, to obtain quantitative evidence for choosing either Bailey's (1994)* ecoregions or Omernik's (1987) ecoregions as the optimum partitioning scheme for west Texas and west Oklahoma.

METHODS AND MATERIALS

Three sample areas containing adjacent ecoregions were chosen to provide a representative sample of woody and non-woody vegetation in the sparsely forested areas of west Texas. These areas were chosen on the basis of their inclusion of adequate percentages of woody and non-woody vegetative associations. The Jefferies-Matusita (JM) distance test for signature divergence was used to quantify differences between woody and non-woody vegetation for adjacent ecoregions.

After determining that partitioning was effective, mesquite vegetative associations were used to test the assumption that there is no difference in Bailey's and Omernik's partitioning systems. To test the suitability of one partitioning system versus another, the (JM) distance test was once again employed. The total average separability of mesquite for all ecoregions within Bailey's system was compared to the total average separability of mesquite for all ecoregions within Omernik's system. Additionally, the total average separability for only those ecoregions adjacent to each other was calculated for both Bailey's and Omernik's systems.

Climate and elevation are two variables which may lead to changes in species distribution along a directional gradient. Pairs of adjacent ecoregions in an east/west and north/south direction were chosen to investigate the possibility that directional influences on reflectance due to climate and elevation may occur over the study area.

---

* Refers to Bailey et al. 1994.

RESULTS

<u>Testing Partitioning Effectiveness</u> (woody vs. non-woody)

Partitioning by ecoregion increased the separability of woody and non-woody vegetation for all pairwise comparisons of adjacent ecoregions. In every case, when ecoregions were combined, separability of woody and non-woody vegetation decreased dramatically.

Results of NDVI tests for Bailey's ecoregions indicate that the NDVI improves separability for woody and non-woody vegetation when compared with the best one-channel subset. However, choosing the best two channel subset may be preferable to using NDVI alone.

<u>Comparison of Baileys's and Omernik's ecoregions</u> (mesquite)

Bailey's system clearly increases overall average separability for mesquite associations when compared to Omernik's system. Bailey's system also increased separability for mesquite associations when only adjacent ecorgions were compared. Comparison of average separability for ecoregions which are adjacent in an east-west and north-south orientation, indicate that Bailey's system is more spectrally sensitive to mesquite associations in an east-west direction. Bailey's inclusion of prevailing climate as expressed by mean annual precipitation and mean annual temperature (McNab and Avers, 1994) may explain the dramatic improvement in separability of ecoregions which are adjacent in an east/west orientation. Although Bailey's ecoregions result in greater spectral separability than Omernik's ecoregions, the overall separability is less for the later growing season data than for the early growing season data. This situation may be due in part to the scarcity of deciduous leaf and grass reflectance in the early season data. Juniper and Cedar species in the semi-arid and arid regions of Texas are evergreen and may be more easily separated before green up of grasses and re-foliation of deciduous woody plants occurs.

CONCLUSIONS AND RECOMMENDATIONS

This study indicates the effectiveness of partitioning by ecoregion in reducing within class spectral variance over large geographic areas. Given that partitioning AVHRR data by ecoregions is an effective method for partitioning within class variance, comparisons of Bailey's and Omernik's systems led to the choice of Bailey's system for use in west Texas.

This study should be repeated in other areas. There exists a need for a test of partitioning effectiveness in areas where good reference data is unavailable or difficult to obtain.

LITERATURE CITED

Bailey, Robert G.; Avers, Peter E.; King, Thomas; McNab, W. Henry, eds. 1994. Ecoregions and subregions of the United States (map). Washington, DC: U.S. Geological Survey. Scale 1:7,500,00; colored.

Loveland, T.R., J.W. Merchant, D.O. Ohlen, and J.F. Brown, 1991. Development of a land-cover characteristics database for the conterminous U.S. Photogrammetric Engineering and Remote Sensing. 57(11): 1453-1463.

McNab, W. Henry and Peter E. Avers 1994. Ecological Subregions of the United Staes: Section Descriptions. USDA Forest Service publication WO-WSA-5.

Omernik, J.M., 1987. Ecoregions of the Conterminous United States: Annals of the Association of American Geographers, 77(1):118-125.

Short, N.M. 1982. The Landsat Tutorial Workbook. Basics of Satellite Remote Sensing. NASA Reference Publication 1078.

Zhu, Z. 1994. Forest Density mapping in the lower 48 states: a regression procedure. USDA Forest Service research paper SO-280. January 1994. 11 pp.

Zhu, Z., and D.L.Evans, 1994. U.S. Forest Types and Predicted Percent Forest Cover from AVHRR data. Photogrammetric Engineering and Remote Sensing. 60:(5):525-531.

Efforts to Improve Vegetation Maps
From Landsat TM Imagery Through Subpixel Analysis

Joseph P. Spruce
Lockheed Martin Stennis Operations
Bldg. 1210
Stennis Space Center, Mississippi 39529

William D. Graham
Lockheed Martin Stennis Operations
Bldg. 1110
Stennis Space Center, Mississippi 39529

Dawn R. Gibas-Tracy
Summit Envirosolutions, Inc.
10201 Wayzata Blvd., Suite 100
Minneapolis, Minnesota 55305

ABSTRACT

Producing vegetation maps from Landsat Thematic Mapper (TM) imagery is an inexpensive alternative to producing maps from digital high-resolution airborne imagery, which can cost tens of thousands of dollars. Unfortunately, vegetation maps from 30-meter Landsat data contain classification errors due to what has been referred to as the "mixed pixel" effect. Spectrally mixed pixels commonly occur in small vegetation patches, in spectrally and/or spatially heterogeneous habitats, and along boundaries of surface-cover types. Mixed-pixel classification errors often limit the use of TM data for forestry applications. New image-classification software has recently been developed to improve land-cover mapping of spectrally mixed pixels. Such software may dramatically improve forest-cover-type maps from TM data. In a land-cover study being conducted under the NASA Commercial Remote Sensing Program's Earth Observations Commercial Applications Program (EOCAP), subpixel analysis software was applied in an effort to improve an unsupervised TM classification of forest cover for an area of North Carolina. Preliminary results from this study provide hope that subpixel analysis can help increase overall map accuracy and classification scheme specificity, but more work is needed to confirm this conclusion.

INTRODUCTION

Landsat Thematic Mapper (TM) imagery can be an economical means for conducting general vegetation surveys at the regional scale. For example, a current Landsat scene of about 8.5 million acres can be purchased for around $4,500, whereas 2.5-meter digital airborne multispectral scanner imagery currently costs between $5,000 and $10,000 for acquisition of a square area consisting of 50,000 acres, excluding the cost of ferrying the plane and preprocessing the data. On the other hand, Landsat TM imagery is somewhat limited in its variety of mapped classes and the accuracy of detailed vegetation classifications. Vegetation maps from Landsat TM often have classification errors due to spectrally mixed pixels. Such errors are especially common along habitat boundaries and areas with land-cover patches smaller than the spatial resolution of the imagery.

Subpixel analysis (SPA) is a relatively new technique that enables remotely sensed imagery to be classified according to the percentage of occurrence of a targeted material (Applied Analysis, 1995). In doing so, SPA provides a means to supplement information derived from conventional classification techniques, such as those used to classify Landsat TM imagery. In particular, SPA has potential for reducing classification errors attributable to spectrally mixed pixels of multiple land covers.

To examine whether SPA could improve Landsat TM vegetation classifications by increasing its accuracy and the level classification specificity, NASA's Commercial Remote Sensing Program and Summit Envirosolutions, Incorporated initiated an EOCAP study with the following objectives: 1) explore the use of subpixel analysis software for improving TM-based classification, and 2) optimize signature derivation techniques for SPA classification. This paper reports preliminary results of this study.

METHODS

The study site is located in a 50,000-acre area in north-central North Carolina where Summit Envirosolutions, Inc. was already conducting a variety of environmental surveys. Summit hoped to improve its Landsat TM-based vegetation classifications for these surveys. The research team procured the following remotely sensed data sets acquired during the winter of 1995: 1) digital Landsat TM scene, 2) digital 2.5-meter Calibrated Airborne Multispectral Scanner (CAMS) data, and 3) 1:8,000 color infrared (CIR) aerial photography. The Landsat TM scene was subset to exclude the thermal band and to minimize image data outside the study area. The investigators then performed an unsupervised classification of the Landsat TM subset image, resulting in seven distinct classes:

1) water
2) pine - 3m plus in height
3) seedling pine
4) hardwood - dense (≥ 60% canopy closure)
5) hardwood - sparse (< 60% canopy closure)
6) hardwood/softwood
7) clearing

Qualitative comparisons of this classification to the other unclassified remotely sensed imagery (including CIR aerial photography) indicated that in the Landsat TM imagery, mixed pixel misclassifications occurred in the dense hardwood, sapling and larger pine, seedling pine, and shallow water areas. Subpixel analysis techniques were applied to TM-imaged areas containing these four cover types. The research team performed the SPA using the Applied Analysis Spectral Analytical Process (AASAP) software (Applied Analysis, 1995) developed in a separate EOCAP partnership by Applied Analysis of Billerica, Massachusetts. Previous work by Applied Analysis and others indicates that the AASAP can be used to improve Landsat TM classification of forest species such as slash pine, loblolly pine, tupelo, and cypress (Huguenin, 1994; Huguenin et al., in review).

197

The AASAP software, which runs under ERDAS IMAGINE, corrects raw imagery for duplicate line artifacts, haze, and sun-angle effects. Figure 1 depicts the process for running the AASAP. The software works with IMAGINE to derive and evaluate spectral signatures for SPA. AASAP completes the SPA process by classifying the remotely sensed imagery according to the percentage of material occurring within a pixel.

SPA signatures were generated using IMAGINE software and unsupervised classification, cluster busting, region growing, random sampling, and screen digitizing techniques. When generating signatures, the researchers tried to select training samples from flat areas. Inclusion of training samples with noticeable sun-angle effects was attempted unsuccessfully and subsequently avoided.

Once the training samples for a material were collected, a remote sensing analyst loaded them into AASAP's signature derivation program and then assigned an estimated average material fraction. The analyst then estimated the material fraction confidence level, which Applied Analysis, Inc. defines as the percentage of pixels that actually contain the targeted material. After each signature was derived, it was then subjected to AASAP's classification program, which allows the analyst to set the so-called signature tolerance from 0.1 to 6.0. Increasing the signature (classification) tolerance results in an increased amount of classified pixels said to contain the targeted material. Determining the best tolerance required experimentation: the analyst had to run a series of classifications, each having a different tolerance setting. For spectrally pure signatures, Applied Analysis, Inc. recommends signature tolerances of 1.0 to 2.0.

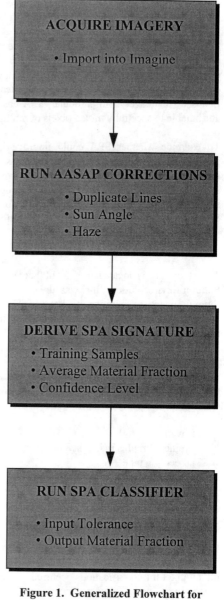

Figure 1. Generalized Flowchart for Using AASAP
(Modified from Applied Analysis, 1995)

Table 1 reports the input SPA signature parameters for each targeted cover type. Note that the hardwood has a material fraction of 0.55, while the other materials were estimated as 0.90, the default for spectrally pure signatures. The confidence level for all four materials was set to the default of 0.80. The classification tolerances for each targeted material were set to either 1.4 or 1.5 as determined by experimentation.

Table 1. Input Parameters for Targeted Materials

Target Material	Number of Samples	Percent Material Fraction	Material Fraction Confidence Level	Classification Tolerance
Dense Hardwood	58	0.55	0.80	1.4
Pine ≥ 3m tall	66	0.90	0.80	1.5
Pine < 3m tall	18	0.90	0.80	1.5
Shallow Water	59	0.90	0.80	1.5

RESULTS AND DISCUSSION

For each targeted material, SPA noticeably underclassified (*i.e.*, omitted) the total area compared to the Landsat TM vegetation map from the conventional unsupervised classifier. Classification errors were generally more apparent in areas with noticeable sun-angle effects and in classifications of spectrally heterogeneous signatures.

The SPA and unsupervised classifications of dense hardwoods both showed commission errors (areas misidentified as dense hardwood), but only the SPA had commission occurring in cleared, bare-soil areas. The SPA result is largely attributable to the input parameters for this material; it is feasible that the signature material fraction and confidence level needed adjustment. Such parameters are difficult to estimate for spectrally heterogeneous pixels, such as those for leaf-off hardwoods. The SPA classification of hardwoods may have improved if the input material fraction and confidence level were lowered or if signature training samples were edited. The hardwood commission in the pine seed-tree cuts may be due to hardwood shrubs.

The SPA classification of mature pine showed negligible commission errors compared to the conventional unsupervised classification. A small amount of commission occurred in the immature seedling pine areas. Omission errors tended to occur in areas having topographic relief affected by the low sun angle. On the positive side, the SPA classification showed some of the pine seed-tree areas containing low proportions of mature pine. Assuming SPA works for mature pine, this is an expected result.

Pine seedlings were generally well mapped, but the SPA classification may have commission errors in some of the seed-tree pine areas. Some of the seedling pine in seed-tree cuts were correctly identified in the SPA classification. Ground truthing would help in interpreting these and other SPA classifications. In this case, stereoscopic interpretation of the 1:8,000 CIR aerial photographs was not a surrogate for field-based verification but was used for qualitative assessment.

Shallow water was poorly mapped on the SPA classification, perhaps due to incorrect user estimates of the signature's average material fraction and corresponding confidence level. The SPA output for this class had both major omission and commission errors. The commission errors evident on both the SPA and conventional classifications were particularly evident in areas affected by topographic shadow. Similar to the leaf-off hardwood, shallow water is spectrally heterogeneous due to water depth, bottom type, water quality, etc. Consequently, a successful SPA classification of shallow water will require more experimentation with respect to selecting appropriate signature training samples and appropriate settings for average material fraction, material fraction confidence level, and classification tolerance.

Optimal SPA classifications require good-quality data, well-defined signatures, and considerable experimentation to optimize the results. According to the AASAP documentation (Applied Analysis, 1995), the training samples used for haze and sun-angle corrections can negatively influence classification results, especially when clouds are selected as bright objects and when impure (*e.g.*, algae-covered) water is selected as a dark object. Signature definition is affected by training sample selection, the input average material fraction, and the material fraction confidence level. The classification tolerance also affects the final results. The amount of experimentation involved with optimizing AASAP SPA results was unexpectedly high, but note that the researchers were not using the most current version of the software. Experimentation levels may decrease when using the updated recent version of the AASAP and as the analysts become more proficient in using the program.

Other factors affecting the AASAP results include the use of georeferenced remotely sensed data, the time of data acquisition, and the use of ancillary data. The effect of using georeferenced Landsat TM data in the SPA classifications is unclear. However, the research team believes that use of georeferenced Landsat TM data did not significantly degrade classification results in this particular study for these reasons: 1) the imagery was resampled with the nearest-neighbor algorithm, which minimizes spectral averaging effects commonly produced by cubic convolution and bilinear interpolation resampling methods, and 2) all signatures were edited with the AASAP to exclude duplicate line artifacts introduced through resampling. Applied Analysis recommends that AASAP users employ non-georeferenced data resampled using the nearest-neighbor algorithm. The results also suggest that the season of data acquisition is an important factor to consider when doing SPA classifications. Early to mid-winter may not be the best time for data acquisition because many of the mapping errors evident on the SPA and unsupervised classifications were in areas affected by sun glare and topographic shadows. Late winter/early spring may be better for distinguishing general categories of pine, hardwood, and mixtures of the two. Leaf-on imagery may also improve the identification of dense hardwoods. The 1:8,000 CIR aerial photography was very useful for assisting signature derivation for targeted materials but unfortunately was available only for a subset of the overall study area. For this study, the SPA signature definition would have benefited more from using medium-scale (*e.g.*, 1:24,000) aerial photography covering the entire 50,000-acre study area.

The AASAP software currently costs around $5500 for a single license and must be run under ERDAS IMAGINE. The AASAP is relatively easy to run, but running the classification program on a 6-band, 1150 by 1050 pixel Landsat image required about 4.5

hours per run. The documentation covers the basics of how to run the software, but detailed documentation on the actual SPA classification algorithm was not found in the older version of the software used in this project.

CONCLUSIONS

The preliminary AASAP SPA classification results for seedling pine, sapling and larger pine, dense hardwoods, and shallow water all contain noteworthy omission errors. In addition, some commission errors occur in SPA classifications of dense hardwood and seedling pine. The SPA classification of shallow water showed both major omission and commission errors. Except for shallow water, the SPA classification results did improve by adjusting signature and classification parameters. In general, SPA produced better classifications for spectrally pure signatures in which the average material fraction equaled or exceeded 0.90. Refined SPA results for spectrally heterogeneous signatures may occur if a more quantitative method can be developed for estimating average material fractions and associated confidence levels. Otherwise, SPA may be cost effective only for materials in which spectrally pure training samples can be collected. SPA classification results may be further improved by using Landsat data that is unaffected by low sun angles. Consequently, the hope remains that SPA can provide supplemental information to previous traditional classifications of study-area vegetation from Landsat imagery. However, more work is needed to optimize AASAP SPA classification results before they can be used to improve vegetation map accuracy and classification scheme specificity. The research team plans to continue its efforts to further optimize SPA classifications of all the targeted cover types using a newer version of the AASAP software and to perform quantitative validation of the best possible SPA classifications.

ACKNOWLEDGMENTS

This work was supported by the NASA Office of Space Access and Technology through the Commercial Remote Sensing Program Office at the John C. Stennis Space Center in Mississippi. Applied Analysis, Inc. contributed AASAP software and technical support throughout the project.

REFERENCES

Applied Analysis, 1995. *Applied Analysis Spectral Analytical Process User's Guide.* AASAP Version 1.1, Billerica, MA.

Huguenin, R. L., 1994. "Subpixel Analysis Process Improves Accuracy of Multispectral Classifications." *Earth Observation Magazine*, July, pp. 37-40.

Huguenin, R. L., M. A. Karaska, D. Van Blaricom, B. Savitsky, and J. R. Jensen. "Subpixel Classification of Bald Cypress and Tupelo Gum Trees in Thematic Mapper Imagery." Undergoing review for publication in *Photogrammetric Engineering & Remote Sensing*.

High Resolution Commercial Satellite Imagery and Data Exploitation for Forestry Applications

Michael E. Bullock
Paul O. DeWolf
Steve W. Wagner

Booz·Allen & Hamilton Inc.
1050 S. Academy Blvd., Suite 148
Colorado Springs, CO 80910

(719) 570-3112 voice   (719) 597-8131 fax

1 ABSTRACT

Beginning in 1997, a revolution in remote sensing will be ushered in with the launch of commercial high resolution multispectral and panchromatic satellite sensors resulting from an investment of over $7 billion from the commercial sector. For the first time in history, commercial satellite imagery having a resolution of 1 meter will be widely available for civilian, environmental, and economic applications. This will create an entirely new paradigm in image exploitation technology for forestry applications. Future users must prepare to have the capability to make maximum use of the information provided by this new generation of sensors. This paper presents an overview of the capabilities of these satellite systems and some important technologies currently under development for the exploitation of this new data. These technologies include: automated map generation, radiometrically accurate image sharpening, and automated change detection.

2 BACKGROUND

Jack Ward Thomas, in a 1994 speech "Future of the Forest Service" said that it was the destiny of the Forest Service to be the world conservation leader. To achieve that goal, he emphasized the need for new sources of learning and the use of the best technology.

"Gifford Pinchot and our predecessors meant the Forest Service to be a guiding beacon for excellence in land management, research, and assistance to others. Our heritage has prepared us to fulfill our destiny as a world conservation leader. In attending to the planet's ecosystems,... we've got to look at new sources of learning and new understanding. We will utilize the best science and management techniques available." [Thomas '94]

Beginning in 1997, new sources of learning and technology will be available to the Forest Service with the launch of three U.S. commercial high-resolution imaging satellites (see Table 1). The imagery provided by these systems will cause a remote sensing revolution on a global scale. It will have profound impacts in the way the Forest Service performs its mission. It will create a new paradigm based on real time access to low cost high-resolution imagery.

For the first time, resource managers will have access to the imagery they need and not just the imagery available to them. High-resolution imagery will be available on demand. There will be a huge increase in the amount of imagery purchased and exploited. High-resolution imagery from multiple sources means being responsive to collection requests at competitive prices. Multiple vendors means competition, which drives competitive prices. And lower prices means larger quantities of imagery may be procured.

Table 1: Companies Launching High Resolution Satellites.

Company	Panchromatic Resolution	Multispectral Resolution
EarthWatch Inc.	3 m 1 m	15 m 4 m
OrbImage Inc.	1 m & 2 m	8 m
Resource 21	–	10 m – 20 m
Space Imaging Inc.	1 m	4 m

Together, these satellites will provide complete coverage of the United States land resources. They will provide redundant sources of image products for any area with re-visit times on the order of a few days. In addition, by tasking the satellite for collection, a new image can be provided in as little as two hours.

Satellite imagery has been an important source of information and understanding for the management of natural resources. The Forest Service makes extensive use of multispectral and panchromatic imagery to assist in the management of the 1.6 billion acres of the National Forests and associated range lands. Multispectral imagery, such as that provided by TM and SPOT, is used by the Forest Service for such applications as resource inventories, map vegetation, map wetlands and wildlife habitats, assess damage, and detect change.

Up to now, commercial satellite imagery has only been available from providers such as SPOT and Landsat Thematic Mapper (TM) with resolutions of 10 meter panchromatic (SPOT only) and 20 and 30 meter multispectral, respectively. Rather than being limited to the low-resolution coverage by these current systems, Forest Service resource managers can access timely, high-resolution images covering any area, virtually on demand. These systems will bring image quality and accessibility to the Forest Service spawning innovative applications well beyond the realm currently envisioned.

One possible application would be the development of a Forest Service forward command center for forest fire management analogous to the way the military conducts battlefield management. In a battlefield environment everything is dependent on timely and accurate information. For instance, the Forest Service could develop a command center that would use data fusion techniques from the National Security Agency to fuse imagery with weather data, DTED (digital terrain elevation data), and crew location information using GPS from such companies as OrbComm. This information would be fused to show the fire manager a terrain truth image with each firefighter location and identity located on the image. The satellite would direct downlink the imagery to the forward command center in real-time. It is feasible that one could conduct fire battle management from a central location and manage and conduct fire storm prediction and crew management and safety.

One of the most immediate uses for accurate, high-resolution images and image processing is in GIS applications. High-resolution imagery will allow GIS applications to evolve to the next level of accuracy. Highly detailed 1 and 4 meter images, for instance, will show resource managers, cartographers, etc., even very small features which can be mapped and included in a vector database.

The Forest Service has been greatly expanding its use of GIS with the procurement of the Integrated Information Management System, known as Project 615. Remote sensing imagery is a critical element of GIS technology. However, the operational use of multispectral imagery does not come without its own difficulties. Some of the major difficulties are:

- The relatively low spatial resolution of the currently used Landsat TM sensor makes some tasks very difficult.
- Accurate interpretation of multispectral data currently requires extensive training.
- The process of processing and interpreting multispectral data requires a significant investment in time and trained analyst labor.

These difficulties have limited the operational use of multispectral imagery. This has resulted in situations where not all of the potential of multispectral imagery is being realized due to the inability or the impracticality of fully exploiting the data to obtain the relevant information. Current GIS technology offers the potential to improve the situation by providing some limited multispectral processing capability. However GIS technology in its present form will not solve these problems. In fact, the availability of geographic information systems can aggravate the multispectral exploitation problem by creating greater demand for new and more specialized image exploitation products. These problems are known within the Forest Service as stated in the Journal of Forestry:

"An often neglected part of (the remote sensing and GIS) technology equation is the person who makes it happen. A shortage of personnel familiar with the technology may hamper extensive and timely integration of GIS and remote sensing throughout the Forest Service. To make the GIS implementation plan succeed, a considerable amount of technology transfer and training must occur throughout the organization." [Lachowski '92]

"The Forest Service does not now have all the skills in our work force to do all the work ahead. We have to have more training and better training. I have given the word out that we can't cut back on training. I was then told, "We can't afford to do the training." I said, "No, we can't afford not to do the training." [Thomas '94]

The Forest Service will soon procure another part of Project 615 with the Remote Sensing and Image Processing Program. The remote sensing revolution will greatly complicate the above problems as well as create a whole new set of problems due to the low cost, increased volume and wide acceptance of high-resolution multispectral. Some of the major problems are:

- The lack of tools to exploit the data and the lack of trained personnel familiar with the tools will hamper the Forest Service's ability to capitalize or even participate in the revolution.
- The task of processing and interpreting the increased amounts of high resolution multispectral data will require additional trained personnel and/or the availability of advanced/automated exploitation tools.
- The process of defining new applications by conducting spectral and spatial trade-offs of current systems such as Landsat and aerial vs. commercial high resolution multispectral imagery is well beyond anything currently envisioned. The different types, sources and resolutions of imagery, will dramatically increase the number of available applications.
- Multiple commercial systems will create multiple imagery archives. How does one identify the needed image, from which archive, and how does one get and exploit the image?

The Forest Service has the opportunity now, while the requirements are still being defined for the 615 Remote Sensing and Image Processing Program, to include requirements for new advanced technologies to help resolve the above difficulties and problems. As the Forest Service moves toward this procurement the implications of this revolution must be made part of the 615 plan. The technology that the commercial world is developing should be of great interest and accounted for in Forest Service future planning. There is opportunity now for the Forest Service to benefit from technology developed to exploit the data provided by these new satellite systems.

The following sections provides a representative set of capabilities offered by the satellite systems being built by Space Imaging Inc., EarthWatch Inc., and OrbImage Inc. as well as some important technologies currently under development by Booz·Allen & Hamilton for the exploitation of this new data.

3 OVERVIEW OF SATELLITE SYSTEM CAPABILITIES

The Forest Service already knows the benefits of an accurate and timely satellite image or aerial photograph. However, until now, the Forest Service was limited to low-resolution imagery available from traditional commercial sources. These new systems will provide new and unique digital imagery products of resolution, quality and clarity previously provided only by U.S. national systems. The advantages of 1 meter panchromatic and 4 meter multispectral provide new opportunities for the Forest Service. Maps will be far more accurate and include rich detail not possible in the past. Resource managers will be able to monitor the different parts of an ecosystem with much greater success.

Some specific capabilities of these systems are provided below as a representative set. Note that not all of the below capabilities are provided by a single satellite system.

1. Imagery will be collected with the following resolutions:
- 1 meter Panchromatic (.45 μm to .9 μm) with a 4 Km wide swath
- 2 meter Panchromatic (.45 μm to .9 μm) with an 8 Km wide swath
- 4 meter Multispectral with an 8 Km wide swath using the following bands:
  - Blue (.45 μm to .54 μm)

- Green (.52 μm to .6 μm)
- Red (.63 μm to .69 μm)
- Near IR (.76 μm to .9 μm)

2. Three-dimensional imaging (Stereo): Up to 30 degrees fore and aft off-nadar pointing provides stereo exploitation with accurate topographic elevation extraction and compilation of digital elevation models.

3. Direct downlink: Potential capability to downlink imagery directly to a Forest Service receiving station for immediate access and exploitation.

4. Re-visit time: An individual satellite can image any area of the Earth within 3 days. Multiple satellites makes it practical to image a requested area in as little as one hour.

5. All-digital data stream: Data is in digital format from the initial collection event through the processing and image exploitation stages to preserve the information and data content.

6. Geo-location: Geo-locate any pixel within 12 meters absolute position. Some systems offer even better geo-location by use of ground control points.

7. Digital Globe™ archive: Access to a digital globe that includes both historic and new imagery. Imagery is priced by the pixel providing a cost saving over buying an entire image footprint.

4 NEW IMAGERY EXPLOITATION TECHNOLOGIES

The Remote Sensing group at Booz·Allen & Hamilton have long understood the importance of aerial and satellite imagery because each plays an important role in Forest Service applications. Our group has been actively developing advanced technologies for the data that is to be the product of this remote sensing revolution. Below are overviews of new technologies being developed by Booz·Allen & Hamilton to exploit high resolution multispectral satellite imagery.

4.1 Semi-automated Map Generation and Map Update

The detection and identification of lines of communication (LOC) is important for a diverse set of applications. The term "lines of communication" applies to both man-made and natural features, such as highways, roads, trails, railroads, waterways, and powerline right-of-ways. The capability to automatically detect LOCs is important for a wide variety of applications such as the automated generation of digital maps and automatic map updating, and the detection of newly built logging roads.

In 1994, the LOCATE (Lines Of Communication Apparent from Thematic mapper Evidence) system was developed to provide the capability to generate vector maps using Landsat TM imagery. Recently, the next generation of the LOCATE system, refered to as LOCATE TNG, was developed specifically to generate maps using high resolution multispectral imagery. Migrating from six band 30 m imagery to four band 4 m imagery required a trade-off of spectral information for spatial information. Therefore, LOCATE TNG was designed to perform much more sophisticated spatial information extraction, while reducing its reliance on spectral information due to having only four spectral bands, all of which are in the VNIR [Bullock '96].

The LOCATE TNG system accomplishes this by combining the available spectral and spatial information for LOC detection. Figure 1 shows the system architecture. First, preprocessing of the multispectral data is performed. This

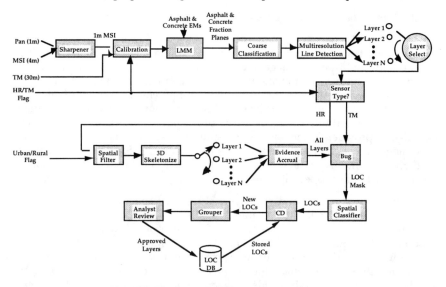

Figure 1: LOCATE TNG System Architecture.

involves multispectral sharpening (see Section 4.2), calibration to units of reflectance, and spectral transformation using the linear mixing model or the Tasseled-Cap transform. This preprocessing maximizes the amount of information available for exploitation.

The next step in the LOCATE TNG system is the detection of all potential LOCs. This is performed by using a hierarchical line detection algorithm, followed by evidence accrual across all the bands being processed. The hierarchical line detector allows for the extraction of various types of LOCs having different spatial properties such as width. The centers of the extracted LOCs are found by use of a 3-dimensional skeletonizer routine that was developed specifically for this system. Next, the detected line and curve fragments are linked together using a *smart bug follower*. This processing step is performed as if a bug were dropped into the plane of the image and were to follow a pre-programmed set of rules in order to connect the LOC fragments.

Spectral information from the multispectral bands is then used for classification to eliminate false LOC detections. This also serves to classify the actual LOCs into application-specific classes, such as concrete road and asphalt road. The final steps within LOCATE TNG further make use of spectral and spatial information to group the LOCs into LOC *nets* [Patterson '94]. Automatic change detection is easily performed by comparing the detected LOCs to those stored in the relational database. The final detected LOCs can then be compared against the LOCs stored in the database. At this point, the user may review these detections and selectively update the database where appropriate. Thus, the user is in a position to *review* the LOC detection process, rather than laboriously *performing* the detections manually.

4.2 Radiometrically Accurate Image Sharpening

The ASIST (Accessory Spectral Investigation & Search Tools) system, developed by Booz·Allen & Hamilton, provides an image data fusion capability by *sharpening* multispectral imagery using higher resolution panchromatic data to significantly improve the spatial resolution of the multispectral image. This system was originally developed for the purpose of sharpening Landsat TM imagery using a SPOT 10 m panchromatic image collected at approximately the same time as the TM image [Bullock '92]. Since its initial development, ASIST has been extended to provide the capability to operate on imagery to be provided by the upcoming generation of high resolution satellite sensors. ASIST can now be used to sharpen 4 m multispectral imagery with 1 m panchromatic imagery to produce a sharpened 1 m multispectral image.

Both panchromatic and multispectral imagery contain different types of information that is important for various exploitation applications. Panchromatic data, due to its higher spatial resolution, offers superior ability to detect smaller features such as man-made objects and lines of communication. Multispectral data provides the capability to obtain spectral information of the features. The ASIST process results in a sharpened multispectral image having 1 m resolution, which offers the best of both worlds: high spatial resolution and spectral information that is suitable for classification and exploitation.

While multispectral image sharpening using panchromatic imagery is not new, our approach is unique in that it is based on a mathematically rigorous pseudoinverse process that models the image formation process of the panchromatic and multispectral sensors. The result is a sharpened image that is radiometrically accurate in a minimum mean-square error sense. By contrast to the "pretty pictures" generated by other ad–hoc approaches (*e.g.*, ISH transforms and FFT approaches), the pseudoinverse approach is designed to maintain the radiometric accuracy of the original multispectral data. As a result, the ASIST sharpening can be used for preprocessing the multispectral data prior to further exploitation by other image processing systems. Recent studies have found ASIST to produce sharpened multispectral imagery having superior quantitative accuracy and higher perceived perceptual quality compared to other standard sharpening techniques [Patterson' 96].

4.3 Automated Change Detection.

As with LOC detection, automatic change detection using multispectral imagery is important for a diverse set of applications. In general, the task of detecting changes in imagery is not difficult. The difficulty sets in when one attempts to detect *only* those changes that are of specific interest, while suppressing those that are not. For example, one user may be interested in automatically finding the amount of actual timber harvested in an area, while ignoring changes due to disease, insect infestation, and biomass quantity. On the other hand, another user may well be interested in changes due to disease and infestation, while ignoring changes due to timber harvesting activity or biomass. The challenge is to develop a system that robustly meets the needs of all users.

The INSPECT (INtegrated SPEctral Change detection Technology) system approaches the problem of change detection by applying several new technologies. Though originally developed to perform change detection in Landsat imagery, the flexibility of this system allows it to also be used with high resolution satellite or aerial imagery. The use of high resolution multispectral imagery provides the capability to detect highly localized changes that can not currently be detected using Landsat imagery.

Our approach to change detection combines both of the major change detection paradigms, pixel level change detection and symbolic level change detection, into one system by taking advantage of numerous tools previously developed for change detection and adding additional spectral exploitation capabilities to customize the system for user specific applications.

First, data preprocessing is done by performing calibration, multispectral band sharpening, and spectral demixing to produce spectral fraction planes. Use of image fraction planes generated by the linear mixing model (LMM) provides the capability to perform change detection that is tailored to specific applications. For example, if the application requires finding natural vegetation changes, then the appropriate green vegetation, dry grass, and wood endmembers would be selected and processed for changes. If, on the other hand, the application requires that man–made changes be found, then a different set of endmembers that include spectra of man-made materials would be employed. This results in the detection of highly selective types of changes, while suppressing changes of other types that are not of interest.

For example, prior work using the INSPECT system applied to two Landsat TM images collected in 1988 and 1991 of the Olympic Peninsula in Washington State has resulted in some interesting results [Bullock '94]. The LMM generated three applicable fraction images, corresponding to the "bright" endmember, "shade" endmember, and the "veg" endmember, respectively. The "bright" endmember corresponds closely to spectrally bright objects in the scene, such as dense forested areas – due to the reflectance off the forest canopy. The "shade" endmember corresponds to the amount of shade in a given pixel, which is high in areas of dense forest and also high in clear cut areas if significant amounts of bare soil are visible. The "veg" endmember corresponds to the amount of biomass in the pixel. This set of endmembers provides excellent features for performing change analysis between the two collected images. These features easily and robustly discriminate between areas that have recently been harvested, areas of replanting, areas of short term growth (approximately a few years), and mature growth (several years). Hence, a map depicting the various classes of change for the forest region is automatically generated.

The INSPECT system performs change detection under one of two possible scenarios: pixel change detection for highly localized changes, and object change detection for changes that are larger in scope. These two scenarios allow the user to address a wide variety of applications, from detection of new construction to detection of gypsy moth infestation in large forest regions.

The INTEREST system has two operation modes, automatic processing and customized. In *Automatic Mode*, we expect novice users will use the imbedded expert system designed to meet the needs of those not having experience in multispectral exploitation. The expert system guides the user through a sequence of questions in order to determine the optimum system parameter sets for the given application. In *Customized Mode*, the system may be optimized for a new application problem domain. For example, if a user needs to track the annual reduction in wetland regions in a particular area of the United States, the user may not find precisely the set of change detection and significance metrics he needs for his problem in the set of built-in applications. This operation mode provides the capability for the user to find an appropriate change projection and parameter set to robustly detect this type of change.

Finally, the interactive review process affords the analyst easy access to change information with changes overlaid on the high resolution imagery. Changes between the imagery and the database are highlighted and relevance ranked for the analyst.

The technologies summarized in this section are just a sample of new technologies that are being developed to process high resolution commercial satellite imagery. New technologies are needed to fill the critical gap between imagery data and mission-relevant *information*. It is not enough to simply await the arrival of high resolution imagery – research and development must be done now to provide the tools that will be needed to effectively use this data as soon as it becomes available.

5 SUMMARY

This paper has described a vision of the pending revolution in remote sensing and the resulting paradigm shift. This paradigm shift is being driven by an emerging U.S. commercial space industry. This revolution will have profound impacts on the way the Forest Service meets its mission goals. For the first time, resource managers will have access to low cost imagery. Remote sensing will gain a much larger role in the ecosystem management process. Commercial competition will drive down the costs of imagery, thereby increasing its use. This will result in a dramatic increase in the amount of imagery the Forest Service must process and exploit. One of the major concerns is the fact that the sensor technology is here and flying tomorrow, but the tools and trained people are not. However, advanced new exploitation technologies are being developed to help transform imagery data into *information* that is relevant to Forest Service missions. The window of opportunity is open now for the Forest Service to address this issue in the Project 615 Remote Sensing and Image Processing Program.

6 REFERENCES

[Bullock '92]   M. E. Bullock, T. J. Patterson, and A. Wada, "Multispectral Band Sharpening Using Pseudoinverse Estimation and Fuzzy Reasoning," *SPIE Int. Conf. on Surveillance Technologies II*, Orlando Fl., April 8-12, 1992.

[Bullock '94]   M. E. Bullock, T. J. Patterson, and S. W. Wagner, "Advanced Exploitation of Landsat Multispectral Imagery for Forestry

Applications," *Fifth Biennial Forest Service Remote Sensing Applications Conference*, Portland, Or., April 11-15, 1994.

[Bullock '96]   M. E. Bullock, T. J. Patterson, and S. R. Fairchild, "Automated Map Generation Using High Resolution Satellite Imagery," *SPIE Int. Conf. on Algorithms for Multispectral and Hyperspectral Imagery II*, Orlando Fl., April 8-12, 1996.

[Lachowski '92] H. Lachowski, P. Maus, and B. Platt, "Integrating Remote Sensing with GIS," Journal of Forestry, Vol. 90, No. 12, Dec. 1992, pp. 16-21.

[Nason '91]   G.P. Nason and R. Sibson, "Using Projection Pursuit in Multispectral Image Analysis," *Computing Science and Statistics: Proceedings of the 23rd Symposium on the Interface*, pp. 579-582, 1991.

[Patterson '94] T. J. Patterson, S. R. Fairchild, and M. E. Bullock, "Perceptual Grouping to Extract Extended Linear Features in TM Imagery," *SPIE Int. Conf. on Algorithms for Multispectral and Hyperspectral Imagery*, Orlando Fl., April 4-8, 1994.

[Patterson '96] T. J. Patterson, R. Haxton, S. Ulinski, and M. E. Bullock, "Quantitative Comparison of Multispectral Image Sharpening Algorithms," *SPIE Int. Conf. on Algorithms for Multispectral and Hyperspectral Imagery II*, Orlando Fl., April 8-12, 1996.

[Thomas '94]   J. W. Thomas, "Concerning the Future Forest Service," Speech given to the Society of American Foresters, Washington, D.C., Sept. 14, 1994.

DIGITAL AERIAL PHOTOGRAPHY USED IN RIVER AND RIPARIAN CORRIDOR ANALYSIS - Blackfoot River Study

Dale Johnson
Julie A. Raine

Positive Systems
250 Second Street East
Whitefish, MT 59937

ABSTRACT

The Blackfoot River in Northwestern Montana provides habitat for a variety of trout species, including the threatened Bull Trout. A variety of human-related river impacts are pushing the bull trout to become another name on the Endangered Species List. The Montana Department of Fish, Wildlife and Parks has undertaken the task of studying bull trout habitat to evaluate river conditions, habitat selectivity, habitat availability and potential restoration possibilities.

The study includes a variety of techniques to monitor the bull trout and their habitat. Researchers have radio tagged fish and are using location information from those radio signals in conjunction with Airborne Data Acquisition and Registration (ADAR) multispectral digital aerial photography to identify trout habitat and migration patterns.

The multispectral ADAR digital aerial photographs are integrated into GIS and Image Processing software where classification routines are used to perform habitat mapping, tree inventory and detection of sediment turbidity. Water depth analysis using the near infrared band of the multispectral imagery defines the boundary between aquatic and terrestrial environments. In addition, ADAR imagery is being used to evaluate the impact of the land surrounding the riparian zone.

Researchers plan to use the results from this study of over 200 river miles to provide information for fisheries management decisions on protecting spawning areas and changing regulations to preserve the threatened bull trout.

This presentation will highlight the various stages of the project: past, present and future.

Evaluation of a Color Infrared Digital
Camera System for Forest Health Protection Applications

Andrew Knapp
USDA Forest Service
Forest Health Protection, Boise, ID 83702

Mike Hoppus
USDA Forest Service
Remote Sensing Applications Center, Salt Lake City, UT 84119

ABSTRACT

In a cooperative project conducted by the Boise Field Office, Forest Health Protection, and the Remote Sensing Application Center (RSAC), a color infrared (CIR) digital camera system was evaluated for use in forest pest management applications. Images were collected over areas affected by various forest insect and disease pests, and areas of recent wildfire activity in Idaho and Utah. Color infrared digital images were compared to visual observations, and to small and medium format aerial photography. Results indicate that the camera system can be successfully used to supplement existing operational remote sensing techniques currently used for monitoring and quantifying forest pest activity.

INTRODUCTION

Forest ecosystems in the Intermountain Region are undergoing serious forest health stress. Fire exclusion, grazing, and past forest management activities have combined with environmental and ecosystem succession changes to create stand conditions which are highly susceptible to attack by forest insect pests and infection by forest diseases (Clark and Sampson, 1995). Numerous forest pests are occurring at record levels throughout host type in the Intermountain Region significantly affecting commercial and recreational values on Federal, State, and private lands (Annual Forest Health Protection Reports).

A major component of forest health management is the detection, monitoring, and quantification of forest insect and disease activity. This information is necessary for conducting suppression projects, vegetative management activities, public information needs, and modeling activities. Forest damage is manifested by changes in the morphology and physiology of living trees which alter the normal spectral reflectance pattern. These differences in spectral reflectance, or signature, form the basis for identification , monitoring, and quantification of forest pest activity. Generally, forest insects and diseases displaying a high signature to background contrast are easy to detect , monitor, and quantify whereas forest insects and diseases with low signature to background contrast are difficult to detect, monitor, and quantify visually or via remote sensing techniques.

Currently, aerial surveys are the primary method used for detecting , monitoring, and quantifying forest insect and disease activity. Three types of surveys are employed: visual sketchmapping surveys, used for initial detection and overall monitoring; aerial photography surveys, used when precise location and quantification is desired; and airborne video surveys, used when rapid image availability is necessary.

213

Color infrared film has been used for numerous forest pest management applications because plants reveal considerably more information in the near infrared portion of the spectrum than in the true color portion (Bobbe, Zigadlo, 1995). Unfortunetly, color infrared film is expensive, aquisition parameters are narrow, and film processing is difficult. Color infrared digital camera systems are a new tool that have the potential to supplement existing operational aerial surveys and improve the temporal, spatial, and quantification limitations of other remote sensing methods. This paper describes initial tests to determine the utility of a color infrared digital camera system to record, identify, and quantify forest pest activity.

EQUIPMENT

The digital infrared camera system used to acquire images of damaged forest stands was the infrared model of Kodak's "Professional DCS 420 Digital Camera". This camera was developed at the behest and with the cooperation of RSAC. Kodak's standard true color digital camera, the DCS 420, was modified to produce color infrared images. The DCS digital camera system consists of a modified Nikon N90 camera which focuses light on a single charge coupled device (CCD). The CCD chip is 13.7 mm wide by 9.1 mm high. A propriatory array of three different filters produces color images. Each cell of the CCD has its own filter which transmits either green, red, or infrared light. The amount of each color assigned to every pixel in the digital scene is derived from mathematical interpolation of the brightness values of neighboring cells. The resulting electronic image is then stored in digital format on a removable hard drive, Personal Computer Memory Card International Association (PCMCIA) card. Disk space required for a single DCS 420 image is 4.5 MB, plus about 0.1 MB for descriptive data, including the date, time, camera settings, and global positioning system (GPS position. The PCMCIA card can store 206 images on 340 MB of usable disk space. The cards are reusable. The camera is capable of taking 5 images at two second intervals, slowing down to one image every three seconds if imaging is continued. In addition, GPS data, date, time, and exposure information is stored with the image data

After the PCMCIA card is removed from the camera and inserted into a PCMCIA port of a personal computer the digital images can be exported into various image processing and GIS software programs where it can be viewed, enhanced, and printed. During image processing each spectral band is assigned a primary color and displayed on a computer monitor. A color conversion process assigns the near infrared spectral band to red, the red spectral band is assigned to green, and the green spectral band is assigned to blue. The resulting color composite images are comparable to high quality film based CIR photographs. A series of overlapping aerial images can also be combined into one digital scene and overlaid onto other digital scenes and/or maps such as a digital orthoquad.

METHODS

Study areas were selected after the completion of annual aerial sketchmap surveys and were based on the presence of current and past pest activity and the variety of pest signatures. The Boise National Forest located in southern Idaho, and the Dixie National Forest, located in southern Utah were selected as primary study areas. These National Forests have extensive forest pest activity and contain diverse forest pest signatures.

Additionally, the Boise National Forest has recently experienced extensive wildfire activity.

Oblique images of forest pest activity were obtained from Cessna 206 and 172 aircraft during August and September, 1995. Flight altitudes varied from 1,000 to 2,000 feet above ground level. The digital camera was hand held and operated in the shutter priority auto mode with a shutter speed of 1/500 second and an ISO setting of 100. A 28 millimeter lens was used for all image acquisition. Exposure compensation was set at -2. When possible, corresponding small and medium format photography was collected concurrently with the digital imagery using a Canon AE-1 camera with Kodachrome 200 film, and a medium format Pentax camera with Ektachrome 200 film. Lighting conditions during test flights ranged from bright sunlight to heavy overcast. Table 1. lists the forest pest or damage agent, the host, and general location of forest pest activity evaluated with the CIR digital camera system.

TABLE 1.

Forest Insect /Disease/ Damage Agent	Host	Location
Western pine beetle (*Dendroctonus brevicomis*)	Ponderosa pine (*Pinus ponderosa*)	Boise National Forest
Douglas-fir beetle (*Dendroctonus pseudotsugae*)	Douglas-fir (*Pseudotsugae menziesii*)	Boise, Dixie National Forest
Fir engraver beetle (Scolytus ventralis)	White fir (*Abies concolor*)	Uinta National Forest
Spruce beetle (*Dendroctonus rufipennis*)	Spruce (*Picea engelmannii*)	Dixie National Forest
Douglas-fir tussock moth (*Orgyia pseudotsugata*)	Douglas-fir, (*Pseudotosugae menziesii*) Subalpine fir (*Abies lasiocarpa*)	Boise National Forest
Dwarf Mistletoe (*Arceuthobium* spp.)	Ponderosa pine (*Pinus ponderosa*)	Boise National Forest
Fire Activity	Ponderosa pine (*Pinus ponderosa*)	Boise National Forest

IMAGE PROCESSING/INTERPRETATION

Image processing was completed by RSAC and hard copies were generated using a Tektronix dye sublimation printer. Individual images were adjusted with Adobe Photoshop image processing software to provide the best possible signature to background contrast to facilitate interpretation. Images were then interpreted by pest management specialists. Interpretation consisted of visually examining the imagery and then comparing it to aerial observations and corresponding small and medium format photography to determine type of forest pest damage, host affected, and extent of pest activity present.

CIR digital images can be reproduced on a number of different types of color printers. RSAC uses a high quality ink jet plotter and a high quality dye sublimation printer to produce hard copies. The ink jet plotter produces copies that have slightly less resolution due to ink overspray and spread. Individual dots in the image matrix are more visible and the colors are not quite as accurate as they are on dye sublimation prints. However, the dye sublimation printer can only produce a 8.5 inch by 11 inch sized print while the ink jet plotter can can make a copy that is up to 33 inches in one dimension. Furthermore, a color copy made on the dye sublimation printer costs about $5.00 compared with about $.50 for a similar sized ink jet copy.

RESULTS

During airborne visual observations or interpretation of color aerial photography, the subtle color variations of dead and dying trees can be difficult to identify. Utilizing the CIR digital camera imagery, identification of individual forest pests, host affected, and quantification of pest activity was enhanced. As with color infrared film, dead and dying trees appear as gray or blue green, while live healthy trees appear in varying shades of red. This signature to background contrast of gray and blue-green against a red background facilitates identification and quantification of dead and dying trees.

Enhancement of identification and quantification of pest activity was greatest with forest pests with low signature to background contrast and least with forest pests with high signature to background contrast. Image resolution was slightly less with the CIR digital imagery when compared to small and medium format photography. However the increased contrast between healthy and dead and dying trees in the CIR digital imagery allowed for easier interpretation and compensated for the decreased spatial resolution.

While no previsual stress in affected trees was observed it was apparent that trees damaged by forest pests and fire which were undergoing a morphological change manifested by a change in crown color were more easily identified on the CIR digital imagery than from aerial observation or interpretation of small and medium format photography. Separation of currently infested , recently killed, and older dead trees was greatly facilitated using the CIR digital imagery.

Imagery was obtained under various lighting conditions. The camera performed well under conditions ranging from bright sunlight to heavy overcast. Usable imagery was obtained under lighting conditions in which standard aerial photography would have been difficult or impossible. Following, summarized by individual forest pests and damage agents, are the results obtained using the camera system.

<u>Western Pine Beetle</u>
In epidemic stages this bark beetle kills smaller second growth ponderosa pine trees, frequently in large dense groups. Because of the high signature to background contrast normally encountered with this pest only slight improvement in pest identification and quantification was observed with the CIR digital imagery when compared to visual observations or standard small format photography.

<u>Douglas-fir Beetle</u>
This beetle attacks primarily large old-growth Douglas-fir trees in groups from a few trees to several thousand trees. Tree mortality was more apparent on the CIR digital imagery than was visible during aerial observation or subsequent aerial photo interpretation. Newly attacked trees, in the early stages of crown color change were easily identified and recently killed trees were easily separated from older tree mortality.

<u>Spruce Beetle</u>
Spruce beetle attacks large, old growth spruce frequently killing up to 90 percent of host type. This bark beetle can be one of the most difficult forest pests to capture via remote sensing techniques because of its very low signature to background contrast and a narrow biowindow for detection and image acquisition. Currently, visual aerial sketchmapping and color infrared photography offer the best opportunity for recording current spruce beetle activity. The increased color contrast between healthy spruce trees and dead and dying spruce trees on the CIR digital imagery greatly facilitated identification and quantification of spruce beetle activity.

<u>Fir Engraver Beetle</u>
The scattered nature of the pest signature of this forest pest where live and dead and dying true fir trees are frequently evenly distributed throughout forested areas makes quantification difficult via traditional aerial survey methodolgy. This "salt and pepper" pest signature has proven extremely difficult to capture with airborne video. Utilizing the CIR digital images seperation of healthy trees and dead and dying trees was accomplished quite easily. Quantification of tree mortality was also enhanced with the camera system.

<u>Douglas-fir Tussock Moth</u>
This insect is one of the most destructive western defoliators and is capable of killing or damaging thousands of acres of trees during short lived cyclic outbreaks. During 1988 to 1993 the largest recorded Douglas-fir tussock moth (DFTM) outbreak occurred on the Boise National Forest. While no current defoliation was present, we were able to record extensive areas of past tree mortality. Because of the scattered nature of the tree mortality, where live and dead trees frequently coexist together in a stand, identification and quantification via aerial sketchmapping, airborne video, or standard aerial photography is difficult. Areas of DFTM caused tree mortality were easily identified on the CIR digital imagery. Even in areas where less than 25 percent of the stand was killed by DFTM tree mortality was easily differentiated from healthy trees.

<u>Dwarf Mistletoe</u>
This parasitic plant infects many species of western conifers decreasing growth rates and predisposing them to insect attack. Comparisons of small and medium format

photography to CIR digital imagery indicated that only a slight increase in pest signature identification was attained using the digital camera. It should be noted that the dwarf mistletoe caused tree mortality we evaluated was an atypical case caused by several years of severe drought which resulted in greater tree mortality, and thus a more visible pest signature than would be expected under normal conditions.

<u>Fire Damage</u>
Images were obtained over areas of recent wildfire activity. Healthy and dead trees were easily identified and separated. Trees damaged by ground fire, but still maintaining a healthy crown appearance could be identified on the CIR digital imagery.

CONCLUSION

Color infrared digital camera systems are a promising tool to supplement existing aerial pest detection and monitoring surveys. Accuracy in pest identification and quantification of forest pest activity can be increased with the use of a CIR digital camera system. The simplicity of the system is an important factor. The CIR digital camera tested was nearly as easy to use as a standard 35 mm camera. Image analysis software is "off the shelf" and can be used by pest management specialists with basic computer skills. Because the data is in a digital format and can be georeferenced by interfacing with a global positioning system it can be quickly integrated into a geographic information system environment.

Resolution of the digital images is nearly equivalent to color transparency film (55.5 lines per inch) but due to the small size of the CCD chip in relation to commonly used film it cannot provide a high resolution image over comparable areas. A single frame of 35mm film has seven times more area; therefore, a 35mm photograph has 2.6 times the resolution of a DCS 420 digital image which covers the same area. Conversely, a DCS 420 digital image, with the same resolution as a 35mm film frame, will cover seven times less area. When compared to operational airborne video systems, the camera system has approximately six times the pixel resolution of S-VHS video. One draw back of the camera system when compared to airborne video is large volume data that is generated. This drawback will likely be less important in the future as new high capacity storage devices are utilized.

Spatial resolution is only one of several image characteristics that allow for accurate interpretation of forest pest activity. The high quality and consistancy of the color infrared response provided by the camera system compensated for the lower resolution when comparing the digital images with 35mm photographs of the same area. When compared to frequently used resource photography the cost of acquisition is far less. The quick turn-around time between image acquisition and a final hard copy of the image is an important consideration when dealing with forest pests because of the short biological windows available for image acquisition and the dynamic nature of forest pest activity. Additionally, images can be obtained under overcast light conditions where standard aerial photography could not be obtained.

Note: The indication of comercial firms or products in this publication is for the convenience of the reader. No endorsement is implied by the USDA Forest Service.

REFERENCES

Clark, L. R., Sampson, R.N., 1995. *Forest Ecosystem Health in the Inland West, A Science and Policy   Reader*, Forest Policy Center, American Forests, 37 p.

Gardner, B. G, and others, 1988-1995, *Forest Pest Conditions in the Intermountain Region*, Forest Pest Management Reports, Ogden UT. USDA Forest Service, Intermountain Region, 26 p.

Bobbe, T. J., Zigadlo, J. P., 1995, *Color Infrared Digital Camera System Used for Natural Resource Aerial Surveys*, Earth Observation Magazine, June 1995, 62. p.

BREAD MAKING AND DESIGNING RESOURCE INVENTORIES: THE GIS CONNECTION[1]
(THE WHOLE LOAF)

H. Gyde Lund and Wanda Wallace
USDA Forest Service
P.O. Box 96090
Washington, DC 20090-6090

William H. Wigton
Agricultural Assessments International Corp.
2606 Ritchie-Marlboro Road
Upper Marlboro, MD 20772

ABSTRACT

The output from a geographic information system (GIS) can be no better than the input. Data sources include resource inventories, maps, and remote sensing. If data sources are to be entered into a GIS, then they must be designed with that use in mind. Considerations for the data collection efforts include appropriateness of the sampling designs, sampling and non-sampling errors, objectivity and quality control of measurements, and methods used for geo-registering field plots and map lines. Using an analogy of the steps needed to make bread in a bread making machine, the authors present an overview of geographic information and recommendations for data collection.

THE BREAD MAKING MACHINE

My wife often works in the evenings so I do most of the cooking. She frequently gives me new kitchen implements to use. I am not sure what message she is trying to convey, but I do appreciate and use them.

This past Christmas, my wife presented me with a bread-making machine. To use the machine, one selects the kind of bread to make, assembles the ingredients, and puts them into the machine in layers. Next we press the proper buttons and two to three hours later, a tolerable "home-made" bread loaf is produced. That is all there is to it! The machine does the mixing, kneading, rising, punch-down, and baking - all automatically.

We can compare making bread in a machine to the process of implementing and utilizing a geographic information system (GIS). We enter data in layers, push buttons, and produce a nice-looking product.

There is an old computer saying - GIGO - Garbage In - Garbage Out. Unlike the bread machine, however, one can dump nearly any data into a GIS and still produce a nice-looking product. And, unfortunately, people are often impressed more by appearances rather than quality. A more appropriate saying may therefore be "Garbage In - Gospel Out!" especially as it relates to GIS, With today's computer technology we can unintentionally make poor data look like gold with the users never being the wiser -- and maps and map-like product appearances do influence people (Monmonier 1991 and Wood 1992).

Many natural resource problems are information and information processing problems. Like bread making, successful information processing is primarily a problem of the proper information system within which data are collected, analyzed, and acted upon by decision makers (Bonnen 1975). This has never been so true as it is today. With the growing fascination in geographic information systems and their implementation, there is the tendency to overlook the design and collection of new data.

Additionally, there is a tendency for analysts to propagate theories, concepts, and models of unknown value because they fail to design and collect data for an adequate empirical test. Many GIS analysts don't look at secondary data before they enter them in a GIS. These analysts receive these data electronically and enter them into the GIS as thematic layers.

---

[1]This paper was originally prepared for AFRICAGIS '95, Abidjan, Cote d'Ivoire. 6-10 March 1995.

Specialization has progressively separated the data collection function from analysis and interpretation.

Inventory designers need to be aware of the analyst's needs and the GIS requirements. Conversely, analysts need to be more conscious of the need for maintaining a common conceptual base for both data and analysis. In addition, accuracy of statistical resource data has two types of survey errors, sampling errors and non-sampling errors. These errors are different from map accuracy that we handle by a GIS.

THE RECIPE

Data are symbolic of some phenomena which we want them to represent. The quality of that representation is only as good as the adequacy of the conceptual base, its operation, or its measurement. Every data system involves an attempt to represent reality by describing empirical phenomena in some system of classes, usually in quantified form. Data are the result of measurement or counting. However, when one sets out to quantify anything, we must first answer "what do we want to count or measure?" If the configuration of data produced is to be internally consistent and have some correspondence with reality, the ideas quantified must bear a meaningful relationship to each other and to the reality we are describing. In other words, there must be some concept of the reality of the population that is to be measured (Bonnen 1975).

The key to success is the "I" in GIS - the information or ingredients going into it. As with a bread machine, there are some rules regarding information that we must follow to ensure a good product from a GIS.

After we decide what kind of bread we wish to make, we then identify the ingredients we need to use. We approach information specification or design by identifying the questions that need answers and by identifying current data and information system problems. Thus, we work back toward the data specification needed to answer our resource management questions. We call this process an Information Needs Assessment or Analysis (INA).

Through the INA, we identify information needed for an integrated GIS by sizing up the data requirements and setting data preparation priorities. Keep in mind that data are not information. Data can be considered tables and output from various sources. These data become information when decision makers can use these data. It is the decision maker who changes the data to information.

The steps for doing an INA are as follows:

1. Determine what decisions the administrator needs to make (what kind of bread he or she wants). This process includes conducting GIS awareness sessions.

2. Identify the analyses we need do and what information needs we must have to meet those needs (the ingredients and amounts). This includes identifying the data needed for use in project work and developing GIS product descriptions. Like making bread, the linkage between desired results, the bread making machine and needed ingredients, must be maintained. Keep in mind that improvements in statistical methodology and data-processing techniques cannot offset failures at data needs identification. For no matter how well we can manipulate numbers, we may still be measuring the wrong attributes. Furthermore, when decision makers want to have relationships between variables for their decisions, they are usually need statistical relationships, correlations and regressions, rather than the logical relationships between information layers in a GIS. These correlations can be calculated when data are collected from identical units. For example, we can relate agricultural production of land and natural resource depletion when we make measurements and counts on the same units. In another example, height and weight are commonly related for use in anthropometric surveys. Obviously, the height and weight measurements must come from the same child. To obtain information on many of the relationships useful to decision makers, we must design data requirements in advance of data collection activities.

221

3. See what is available (check your shelves for the ingredients) and determine the suitability of existing information (check your ingredients for quality, quantity, and expiration date).

4. If the data are not available, then you collect the needed information (go to the market). Identify the data sources needed and develop a plan for data acquisition and preparation. Maintain objectivity and quality control in the data collection process (acquire, measure, and handle ingredients carefully and according to the recipe). Create and test the data bases and establish maintenance procedures.

5. Enter data into the computer (put the ingredients into the bread machine) and perform the analysis (start the machine). Most analyses required by decision makers are in written form. For example, land tenure data, as we collect them in the field, may not help decision makers with their critical decisions. Usually data collected must be analyzed by experts at the Land Tenure Center (LTC). Commonly, the analysis shows probable results if certain courses of action are taken. Again, the GIS may help in the analysis but will not provide all the analysis.

6. Evaluate the results (taste the bread) and change as needed.

Society has grown more complex and specialized. Demands are not just for more data and greater accuracy in the articulation of detail, but also for data in a learning or development mode (Dunn 1974). In such a case, we may not be able to specify completely the decision making goals. One purpose of an information system is to help the decision maker in specifying the goals in a progressively more complete form. In a developmental mode, goals end problems may continue to change as learning takes place and thus may never be completely specified. In this situation, one is not well served by data that are static.

In the learning or developmental mode, the information system that perceives and acts on data is itself changing in structure and behavior in response to new input. Thus, the information system must be capable of perceiving changes not only in the environment but in itself, even under conditions in which such changes themselves become goals (Dunn 1974).

THE INGREDIENTS

The first step in developing any information system is to determine the information needs. This conference theme is "Integrated Geographic Information Systems Useful for Sustainable Management of Natural Resources in Africa." 'Integrated' is an important word. In the past, information systems were carried out by sectorial agencies each with their own system and needs. Modern managers require information on many sectors. For example, a forestry project may impact the environment, women in development, health aspects of rural populations, agricultural systems and thus modern managers need multi-sectorial data.

In addition, the term 'sustainable' here implies some very specific information needs. By sustainable, we mean that we can maintain or increase the current production of the natural resources. To achieve sustainability, we may need not only information about the resources themselves but also the factors that can impact the resources. Factors, for example, which contribute to deforestation are grazing, fuel gathering, and shifting agriculture. Factors contributing to fuel wood gathering and shifting agriculture, in turn, are increasing populations and urbanization. To halt deforestation, one must mitigate population growth and increase the productivity of lands already devoted to agriculture. To expand the forest resource, one needs to know the capability of the non-forested lands. Thus to sustain the forest resource, the decision maker not only needs information about the forested lands but about non-forested lands, as well, including those devoted to food production and urbanization.

Table 1 provides a listing of anticipated environmental information needs for the Year 2000 Global Assessment by accumulated ratings from four international forest experts meetings (Malingreau, da Cunha, Justice 1992, FAO/ECE 1993, UNEP/FAO 1993, WRI 1993). The highest rating shows the most demand for the information. Table 1 can serve as a shopping list for information resource managers need to address emerging global and national environmental, ecological, and economic issues.

Table 1 - Anticipated future information needs

INFORMATION NEEDS	RATING
Land Cover	15
Biodiversity	15
Biomass	12
Land Use	12
Forest Health	12
Socio-Economic	11
Vegetation Type	9
Carbon Storage	9
Ownership	7
Soils	7
Ecofloristic Zone	6
Soil Productivity	6
Fragmentation	5
Watershed/Hydro.	3
Topography	2
Fauna	1
Accessibility	1

THE CUPBOARD

Now that we have identified what information we need, we next check the data we have on hand. In doing so, we must remember that we owe it to our employers and publics to provide them with the best information available. Unfortunately as we are populating our GIS data bases, we often grab any data we have residing on our shelves to save money and with the hope it will meet our needs. The reason for this can best be explained by another saying "If all you have is a hammer, then all problems become nails." This may or may not be appropriate depending on the decisions we want to reach, how we initially collected the data, and what control we used in that process.

If the data were collected before you got your GIS, chances are it may not be what you need. In all probability, the data were collected with a specific purpose in mind and not designed to be combined with other data sources. The acquisition and mapping may have been nothing more than a person's perception of the status of a resource crudely draw on a map using a grease pencil. Since the primary user was the person collecting the data, this sufficed his or her needs.

Converting data to digital form and building the related attribute data base is a new cost that can be as much as 75 percent of the cost of implementing a GIS. This makes data conversion a very expensive ingredient! It is very important to identify what information is relevant for populating your data bases.

You could be spending huge sums on useless or outdated information. Keep in mind that two or more data source will be combined in a GIS. By doing so, general information tends to override specific, bad drives out good, and errors in data tend to multiply.

Data bases and maps concerning natural resources are especially complex. Boundaries, although gazetted as site specific, are often fuzzy. Mapped polygons usually contain many inclusions. Data recorded and stored are often averages for the site. When such information is overlaid with more specific or higher resolution data, error in interpretations can be made.

Furthermore, you will find a considerable data are available in national statistical data bases such as those found in the United States Department of Agriculture (USDA). These include Forest Service data on the Nation's forest lands, the National Agricultural Statistics Service (NASS) with data on all the agriculture in the 50 states, the Natural Resource Conservation Service (NRCS) with the National Resource Inventory (NRI) data and the Foreign Agricultural Service (FAS) with data on export and import requirements of nearly any country in the world. The Food and Agriculture Organization (FAO) of the United Nation has additional large data bases such as those for global forest assessments. These data bases are extremely useful without having a GIS component. If we want sound information, we must learn to work with statistical data bases as well.

INGREDIENT QUALITY

Once we have identified what we have on hand, we need to check the existing ingredients for quality, suitability, and expiration date. This step is very important. One cannot make the right decisions with the wrong data. Through an analysis of systems problems we can begin to model the various assumptions as to organizational structure, environment, objectives, and other dimensions. All these efforts help us move toward the urgent objective of identification and conscious management of our data systems as systems and as part of a still more comprehensive set of information systems (Bonnen 1975).

Suitability
Assuming our needed data are available and are geo-referenced, there are several factors to look for in evaluating the existing information. These include the information source, objectivity of data collection methods, standards followed, and when the information was gathered (Lund and Thomas 1995). Data quality is more important than the data amount.

In looking at the source, make sure it is credible - done by a group that has the expertise and resources to carry out data collection and summary activities. Have there been any problems noted about the data or the data source? Next look at how the data were collected. Is the method statistically valid? Is there an accuracy assessment, what is the "resolution" of the data or mapping? Why were the data collected, who is using it and for what purpose? When were the data obtained or the maps produced? Have there been any updates? Source materials need to be up-to-date and complete. Resolve any inconsistencies and errors before digitizing.

For statistical data, ask about the sampling frame and observe the completeness. Identify the sampling units and compare these units to the reporting units. Ask about data collection and supervision. Ask about the sampling and non-sampling errors. Keep in mind that statistical data are not as easily entered into a GIS as map data.

Statistical data become more accurate as the area covered and the sample size becomes larger. Conversely, as the area gets smaller, the sample becomes smaller and the accuracy becomes less and less accurate. For example, an estimate of planted wheat acres at the national level may be close to 1 percent sampling error while an estimate at the state level is 5 percent sampling error. An estimate at the county level from the same survey may be more than 50 percent sampling error. Subjective guesses are more accurate!

GIS applications can be expensive. If decisions about the natural resources are required, it seems logical to ensure that the resource data are at least as good as the GIS system. In some applications the natural resource data are the most important data. Let's ensure that the data are up to the use.

Shelf-life
There are four ways data may be obsolete - those dealing with changes in the resources themselves, those dealing with changes in resource management policies, those dealing with changes in how data can be collected, and those caused by changes in the organization itself (Bonnen 1975).

1. When the represented phenomenon changes rapidly, we must modify the conceptual base of the information system to keep up with the change in the reality represented and the problems studied. If the change rate is high enough, the need for redesign becomes nearly continuous. This is the basic problem in the information design for natural resource management. Failure to keep up with the changes in problems and in reality leads to significant obsolescence. The system begins to lose its capacity as an accurate guide for problem identification and solution or management.

When the world continues changing rapidly, the need to redesign the system eventually becomes continuous. Therefore the capacity for redesign must be a normal information system function. If the designer does not become part of the system in this situation, the system's capacity to produce useful information will deteriorate (Bonnen 1975).

2. Changes in questions and policies - When the questions change, you may find that the existing data are no longer appropriate. Issues facing us today that we did not consider 10-20 years ago include global warming, carbon sequestration, and biodiversity conservation. Our existing data may not address such needs and indeed may not even be appropriate.

If one is to address these new issues, then it follows the information needs to be in the data base. Additionally, if we are to use a GIS in the analysis, then data about these sectors have to be geo-referenced. Converting from a manual system to an electronic system shifts the cost from time consuming manual data analysis and map preparation tasks to performing more complex analysis.

3. Changes in data collection capabilities - Some resource data are more accurate today than before. Satellite-based remote sensing, for example, can provide quick estimates and updates of forested land extent. Thus, timber and fuelwood production estimates could be far better today than they were in the past if we used the new technology.

4. Changes in institutions - Institutional changes also may dictate changes in information systems and information flows. Similarly, changes in basic statistical measurement techniques, unmatched by an implementing organizational adjustment, also can create another form of institutional obsolescence and inefficiency. As a result of institutional obsolescence or reorganization, current administrative structures often do not bring the necessary information together at the time and places in the structure where it is most needed by decision makers (Bonnen 1975).

THE MARKET

Table 2 groups the information identified in table 1 by source and difficulty of obtaining. Information at top of table are the easiest to get. Information toward bottom is most difficult.

In addition to sustainability, this conference focuses on 'integrated' geographic information systems. By integration, we mean the parts have to come together into a whole. This means there has to be some continuity and commonality between the parts. A GIS will inevitably combine and use different fields of knowledge. Therefore, the concepts underlying the information system will be derived from different disciplines.

If such a system is to produce useful data and, in the process, manage its own continuing redesign, a general theory of social information processing or a theory of theories or a 'meta-theory' is needed. In other words, a means must exist for synthesizing concepts from different bodies of knowledge into meaningful relationship to each other (Dunn 1974).

Table 2 - Environmental data by source and ease of obtaining.

<---------------------Remote Sensing--------------------->

ANCILLARY INFORMATION	RESOURCE INVENTORY	MODELS
Topography	Land Cover	Accessibility
Land Use	Fragmentation	Carbon Storage
Ownership	Vegetation Type	Hydrological Cycle
Socio-Economic	Biomass	Soil Productivity
	Biodiversity	Ecofloristic Zone
	Fauna	Forest Health
	Soils	

<--------------------Field Surveys-------------------->

In most GIS applications, data sources have to be geo-referenced to the same base, scale, and resolution. Using standard cartographic techniques for manuscripting and digitizing will help to ensure a high level of accuracy and consistency. In resource inventory, this may mean we collect data through multiple resource effort that is unbiased and un-interpreted, and which data are made available for all sectors to use. Data are essentially collected once and used by many. Guidance on how to design integrated inventories may be found in Lund (1986).

Despite conventional wisdom, we cannot view information problems as merely a matter of inadequate measurement techniques. The inadequacy lies in the design and conceptual base of the information-processing structures (Bonnen 1975). With GIS, there is more likelihood that someone else will be using the information that we gather. Therefore the data residing in a GIS should be objective and unbiased. Store data in a format that is readable by others. We call format specifications data standards. The more people who will be using the data, the more "correct" it needs to be. Additionally, one needs to document the meta-data (data about data). Meta-data includes: the information source, who was responsible, why and for what reason, how the data were collected, and what was the quality assurance and control.

If you need to collect new information, use this opportunity to seek partners in the venture. Partners would be those who need similar information at the local, nations, regional or global level including both government organizations and non-government organizations. These partners may be potential sources of funds, expertise, and perhaps additional data.

There are three areas for consideration for resource inventories - 1) the objectivity by which the samples were chosen, 2) the objectivity and quality control in the actual field measurements, and 3) the methods used to geo-register the location where the observations were obtained with the coordinate system used in the GIS.

Objectivity of inventory sample selection
Use statistically-valid or scientifically-based sample designs. Avoid curbside cruises, windshield surveys, and other subjective techniques. A variety of sampling designs applicable to inventorying natural resources may be found in Lund and Thomas (1989) and other sampling texts.

Measurements
"The important thing is not what you do, but how you measure it" (Horwood n.d.). Make measurements as precise as possible. In the inventory process, focus on collecting basic data (that which is directly measured or inventoried) not interpreted or derived data. Then, if the application definitions and models change, the basic data will still be

226

useful. Record and store actual measurements instead of interpretations or classes. For example if measuring forest canopy coverage record the actual measurement (e.g. 15 percent) instead of a class such as "open."

Geo-referencing
   Register plot locations, stand boundaries, etc. to a coordinate system that will be used in the GIS. Global positioning systems are a preferred method of getting coordinates. A fall back includes annotating plot locations or stand boundaries on aerial photographs and then either registering the photographs electronically or transferring the information using stereo-plotters to a control base. The last resort is to "eyeball" transfer data from aerial photographs to base maps and then digitize that information.

   Taking plot data and geo-referencing it implies that the plot data are an accurate area representation around the plot. That may not be the case. It takes many plots to represent the resources in a given area. Be careful here!

THE KNEADING AND THE BAKING

   Once you have the ingredients, measured them correctly, and dropped them into the bread-making machine, you turn it on and let the kneading and baking beginning. With natural resource information, however, the task is not quite so simple. Industrialization and development increase the demand for information. Development leads to functional specialization and organization. This increases the need for coordination and, thus, the social returns to and the demand for information. It also causes a change in the kind of information demanded (Bonnen 1975) and how it is handled.

   An electronic environment requires a new organization in its management and use. Common legal and administrative characteristics must be agreed upon up front and each inventory must be modified to conform. You need to have a program of quality assurance and control and follow national or international standards. Spot check field crews and photo interpreters to see that they are following standards and procedures correctly. Have a penalty or reward system for assuring quality.

   Calculate the reliability of the inventory or resource mapping activity. Most sampling texts such as Lund and Thomas (1989) give procedures for calculating sampling errors in resource inventories. Similarly, many recent mapping and GIS texts such as Congalton (1994) have procedures for determining the spatial information accuracy. The detailed steps in GIS data preparation are very time consuming, but do help to point out many problems with data consistency and redundancy.

   Include the information on the reliability of the inventory or maps with the documentation for the GIS. Document the source, date, and quality of data entered in the GIS. Provide this information to the users and decision makers through your meta-data documentation.

   Once you have a well-organized corporate set of resource data in a GIS, the maintenance and operation responsibilities must be put in place. You need to determine who is authorized to make changes to the corporate data files and how these changes can be transmitted to users.

THE TASTING

   The final step in bread making is the tasting. Only through such an activity can one determine if the outcome is suitable and what changes have to be to improve in future baking efforts.

   The capacity for renewing any system must involve feedback or learning loops within the information system itself. As a minimum, any major data system should have a group of professionals working continuously on the conceptual base, definitions, measurement, and data quality. This might be characterized as a statistical system design and quality control shop. There would have to be a similar organization at the information system level. Such organizations would monitor, stimulate, and perhaps contribute to conceptual development in the disciplines upon which the data and information systems are dependent. These groups must maintain close relationships with the data users. They also would provide a place in the system which could be the common ground on which information and data users, statistical methodologists, and disciplinary methodologists met (Bonnen 1975).

THE FINAL PRODUCT

In this paper, we have briefly looked at some potential problems with using existing information in a GIS for sustainable development of natural resources. We have looked at how to evaluate existing information and the steps to go through in collecting new data for use in a GIS.

Information is an expensive and valuable commodity. The cost of poor decisions and subsequent lack of appropriate information is extremely high. The foundation of effective information management is careful data and information design. Appropriately designed information allows one to reduce uncertainty and to manage its undesired consequences. A meta-theory for information system design may be an impossible goal, but the logic of its need is valid. The design of data and information systems is not a job we can assign to any but the best minds (Bonnen 1975).

Failure to invest in redesign of data and analytical capability is a cost imposed on future generations. Administrators and decision-makers need good data now. Peoples' health and welfare depend on it. As such, appropriate resource information becomes a "Bread of Life."

REFERENCES

Bonnen, James T. 1975. Improving information on agriculture and rural development. Presidential address presented at the American Agricultural Economists Meeting. Michigan State University. 11 p.

Congalton, R.G. ed. 1994. Unlocking the puzzle. Proceedings International Symposium on the spatial accuracy of natural resource data bases. 1994 May 15-20; Williamsburg, VA. Bethesda, MD: American Society for Photogrammetry and Remote Sensing. 271 p.

Dunn, Edgar S., Jr. 1974. Social information processing and statistical systems-change and reform. New York: John Wiley & Sons.

Food and Agriculture Organization/Economic Community of Europe (FAO/ECE). 1993. Report. FAO/ECE Meeting of Experts on Global Forest Resources Assessment 2000. 3-7 May 1993, Kotka, Finland. 22 p.

Horwood, Edgar M. n.d. 10 Laws of data processing and information systems. Seattle, WA: University of Washington. 1 p.

Lund, H. Gyde. 1986. A primer on integrating resource inventories. Gen. Tech. Report WO-49. Washington, DC: U.S. Department of Agriculture, Forest Service. 64 p.

Lund, H. Gyde; Thomas, Charles E. 1989. A primer on stand and forest inventory designs. Gen. Tech. Report WO-54. Washington, DC: U.S. Department of Agriculture, Forest Service. 96 p.

Lund, H. Gyde; Thomas, Charles E. tech. coords. 1995. A primer on evaluation and using existing information for corporate data bases. Gen. Tech. Report WO-62. Washington, DC: U.S. Department of Agriculture, Forest Service. 168 p.

Malingreau, J.P.; da Cunha, R.; Justice, C. 1992. World Forest Watch (WFW) Proceedings, 27-29 May 1992, Sao Jose Dos Campos, Brazil. EUR 14561 EN, Joint Research Center, Commission of the European Communities. Ispra, Italy. 84 p.

Monmonier, Mark. 1991. How to lie with maps. Chicago, IL: The University of Chicago Press. 176 p.

United Nations Environment Programme/Food and Agriculture Organization (UNEP/FAO). 1993. Report of the UNEP/FAO Expert Consultation on Environmental Parameters in Future Global Forest Assessments. Nairobi, Kenya, 1-3 December 1992. Earthwatch Global Environmental Monitoring System GEMS Report 17. Nairobi, Kenya. 262 p.

Wood, Denis. 1992. The power of maps. NY: The Guilford Press. 248 p.

World Resources Institute (WFI). 1993. Results of the Workshop on Future Environmental Information Needs for Decisionmakers. 16-17 December 1992. World Resources Institute, Washington, DC. 13 p.

TECHNOLOGY TRANSFER, AN IMPORTANT ROLE:
GPS IN URUARA, BRAZIL

Carlos D. Rodríguez-Pedraza
International Institute of Tropical Forestry
USDA Forest Service
PO BOX 25000
Rio Piedras, PR 00928-5000

ABSTRACT

Uruará is a small city in the western part of the Amazon Basin in the State of Pará, Brazil. The municipal government of Uruará in collaboration with Brazilian government (state and federal) and non-government organizations put together a global plan for the development of Uruará. This plan seeks to address the social and economic problems of the city within the theme: "Desenvolver sem Devastar" (development without devastation).

As one of the principal institutional participants in this plan, the Superintendencia do Desenvolvimento da Amazonia (SUDAM) in Belém, had the responsibility of developing thematic maps of soil, geology, vegetation and geomorphology in conjunction with Landsat Images of the area of Uruará. Since 1993 the International Institute of Tropical Forestry has been assisting SUDAM (among several state and federal agencies and universities in northern Brazil) in the use and applications of the Global Positioning System (GPS) as part of its mission of technology transfer. Brazilian colleagues learned (on the job) to collect, process, and output GPS data. They have realized the importance of GPS technology for natural resource applications and ecosystem management.

INTRODUCTION

"The establishment of the Tropical Forest Experiment Station, located on the grounds of the Insular Agricultural Experiment Station at Rio Piedras, Puerto Rico, is proceeding slowly and surely. A laboratory and office building to house the station is under construction on a hill overlooking San Juan and the island to the west. It should be completed and ready to occupy in September of this year. It is hoped that full use of the Station and its facilities will be made by all countries and persons interested in tropical forestry. ... For those countries not able to send anybody to the Station, but who are anxious to carry on experimental work and who request technical collaboration and assistance, the Station expects to have a technical staff to furnish such a service." This is part of the companion letter of The Caribbean Forester of April of 1940 written by Arthur Bevan, first director of the Station. With an initial staff of 5 people, and a budget of some $25,000 (Wadsworth 1995), the Station started its mission.

Today the International Institute of Tropical Forestry (IITF, initially Tropical Forest Experiment Station, and former Institute of Tropical Forestry, created in 1939 and made international by Congressional mandate in 1992) continues its commitment to research and technology transfer in the tropics. To address the aspects of physical, social, and economic issues of managing tropical forests, the IITF has a 57-year tradition of interdisciplinary research. The staff, consisting of foresters, wildlife and terrestrial ecologists, geomorphologists, biologists, botanists, zoologists, and atmospheric scientists, work both independently and jointly with outside collaborators. They are involved in long-term studies in Puerto Rico, Brazil, China, Costa Rica, Venezuela, and elsewhere in the Caribbean and Latin America (IITF 1992-93 Annual Letter). The principal IITF-International Cooperation goal is to exchange knowledge critical to the sustainability of tropical ecosystems and their contribution to humankind through activities that include demonstrations, training, science exchange and technology transfer.

Since 1993, as part of its mission, the IITF has been assisting, the Superintendencia do Desenvolvimento da Amazonia (SUDAM, and other federal agencies and universities in northern Brazil) in the use and applications of GPS for research and management of natural resources. SUDAM is a federal Brazilian agency created in the late 60s and based in the city of Belém, in the northern state of Pará. It is responsible for the planning, promotion, and control of the federal actions in the Amazon region.

In 1994, SUDAM and the municipal government of Uruará in collaboration with other state and federal Brazilian government agencies and non-government organizations put together a global plan for development of the city of Uruará. After several meetings, all findings and needs were grouped in five general projects, and the institutions responsible were identified and requested to help on the endeavor. This plan seeks to address the social and economic problems of the city within the theme: "Desenvolver sem Devastar" (development without devastation). From the plan, SUDAM was requested to prepare maps for the city to address the lack of basic information on natural resources, actual land use, and urban and agriculture structure.

LOCATION

Uruará lies in the western part of the Amazon Basin in the state of Pará, Brazil (Fig. 1, Lat. 3° 43' 19.548" S; Long. 53° 44' 15.020" W WGS-84 Datum). With elevations inferior to 200 m, Uruará has a humid tropical climate with an average annual temperature of 27° C. The rainy season has its peak between February and May, with a mean annual precipitation of 2400 mm. The agriculture structure consists of 3740 farms of 100 ha, 186 farms of 500 ha, and 27 farms of 3000 ha for a total of 3953 agricultural farms. Close to the urban area, these farms have been subdivided in to small lots, called "chácaras", with areas of 1 to 10 ha.

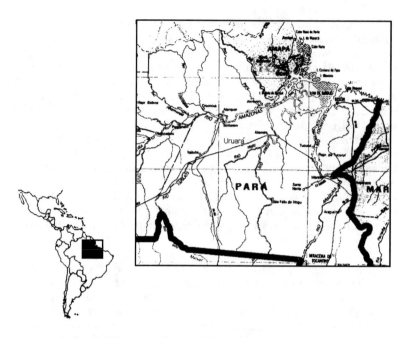

Figure 1. State of Pará, Brazil.

The road system is made up by of the Transamazonica highway (BR-230) running from east to west, and cut from north to south every 5 km by transversal roads. None of them are paved. Hauling timber products has damaged many of these roads, and during the rainy season transportation is a problem.

MAPPING

Our first task was to establish two control points to be used as base references. One group was working with Ashtec Dimension GPS units, and the rest were using Trimble's PRO XL and Geoexplorers. One team was surveying the control points established by the Instituto Nacional de Colonização e Reforma Agraria (INCRA), which is responsible for the colonization and agriculture reform in the Amazon region. INCRA has maps of the agricultural structure of the city. These maps were done with traditional surveying methods, and were not tied to the Brazil South American Datum of 69 (SAD-69). SUDAM, with the data from the control points, will georeference these maps and use them as layers of information in combination with Landsat TM images.

The other groups were collecting information on actual land use, urban area, and road systems. With the guidance of personnel from the Comisção Executiva do Plano da Lavoura Cacausira (CEPLAC) who provide technical

232

support to farmers, we visited farms that produced the principal cash crops of the region. In each farm we collected ground truth data for the crops. As seen in Figure 2, exact location for these crops is now available and its use for the vegetation assessment will be of extreme value.

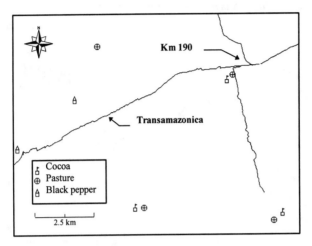

Figure 2. Actual land use.

This information will aid SUDAM to perform a more specific classification of the vegetation. With the growth and development of the city, farms close by were subdivided into small lots, called "chácaras", with areas of 1 to 10 ha which were not surveyed, nor updated in the INCRA maps. The survey of these lots was initiated as part of this project. Also the municipal government lacks accurate maps of the city, so one group surveyed the residential and commercial lots, and all of the streets (Fig. 3).

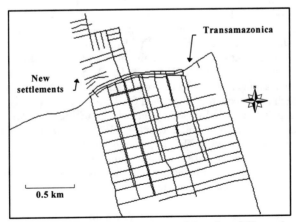

Figure 3. Partial road system of the city of Uruará.

Road alignments are not always visible on the satellite images, and since they are not paved they show the same composite color as cleared areas. We surveyed the Transamazonica highway from east to west within the limits of Uruará, and the transversal roads (every 5 km), about 5 kms from each side perpendicular to the Transamazonica highway (Fig. 4).

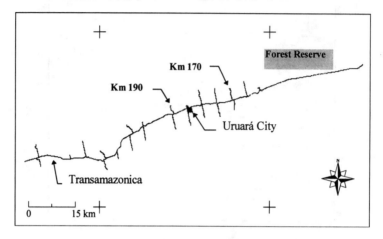

Figure 4. Transamazonica highway (BR-230) within the Uruará boundaries (Geodetic tics shown- 30').

CONCLUSION

Brazilian colleagues learned throughout the project to collect, process, and output GPS data. Combining the GPS information and the existing satellite data, SUDAM will develop a groundcover classification with high specificity for the region. Tropical forests are known for their complexity and variations, and GPS has proved to be an excellent mechanism to gather ground information for vegetation assessment and classification (Rodríguez-Pedraza *et al.* 1994), among others. Field information is essential for forest inventory since it creates the connection between the satellite-sensed data to the actual ground conditions (IUFRO 1994). Annual deforestation rates for Latin America and the Caribbean were estimated at 7.4 million ha (FAO 1990), of which 84% occurred in tropical South America. Also, it is likely that perennial crops with closed canopies have been included in forest classes in previous remote sensing applications. GPS information will play an important role in improving the vegetation classification, and in so doing should allow for improved estimates of deforestation rates in Brazil (Walker 1994). The lack of basic information on the resources available represents additional threats to these ecosystems.

This is where technology transfer plays an important role. The USDA Forest Service, through the IITF, has made available new technology for use and application in these important ecosystems. What has been important for our Brazilian colleagues is that this new information was made available in their own language. Another important aspect is that universities (professors and students) were included in the working groups. This way future professionals will develop with current knowledge on new technologies available. They realize the importance of GPS technology for natural resources applications and ecosystem management. This year we have been invited by the University of the Amazon in Manaus to teach the use and applications of GPS in forestry activities. Field assessments in the Amazon can now be performed more accurately and faster.

ACKNOWLEDGMENT

This work was possible thanks to the coordination and the cooperation of Pedro Mourão de Oliveira Director of the Basic Research Division of SUDAM. And to the rest of the participants, Adelaide Pereira Nacif, Geraldo Pereira Da Silva, and João Girard De Almeida of SUDAM. Maria Graciete Do Amaral Torres and Vicente Raimundo De Azevedo from the Instituto Brasileiro de Defensa dos Recursos Naturais e do Meio Ambiente (IBAMA), Flavio Altieri from the Secretaria de Ciencia e Tecnologia e do Meio Ambiente (SECTAM), Ana Cristina Machado from the Instituto de Desenvolvimento Economico e Social do Estado do Pará (IDESP), Diemerson Correia Barile from the Comisção Executiva do Plano da Lavoura Cacausira (CEPLAC), Ulisses Cunha from the Universidad Federal do Amazonas (UFAM), Waldemar Manoel Pereira from the Instituto Nacional de Colonização e Reforma Agraria (INCRA), Jailsom Rocha Brandão Mayor of Uruará, Pedro Birro Rosa special assistant of the Mayor, and the LaSalle center for their hospitality and lodging facilities.

REFERENCES

Bevan A. 1940. Companion Letter of The Caribbean Forester. April, Vol. 1 (3).

FAO. 1990. Forest Resources assessment 1990: Tropical Countries. FAO Forestry Paper 112. 102 p.

International Institute of Tropical Forestry. 1992-93. Annual Letter. United States Department of Agriculture Forest Service, Rio Piedras, PR. In press.

IUFRO World Series. Päivinen, R., H. G. Lund, S. Poso, and T. Zawila-Niedzwiecki, editors. 1994. IUFRO International Guidelines for Forest Monitoring. IUFRO Working Party S4.02-05. Vienna, IUFRO. Vol. 5, 102 p.

Rodríguez-Pedraza C.D., R. Walker, and P. Mourão de Oliveira. 1994. Use of GPS with LANDSAT TM images to identify main crops in farms along the Transamazonica Highway, Brazil. *In* International Society for Photogrammetry and Remote Sensing, Proceedings ISPRS Commission VII Symposium: Resource and Environmental Monitoring; September 26-30, 1994. Rio de Janeiro, Brazil. 30(7a):35-42

Wadsworth, F.H. 1995. A Forest Research Institution In The West Indies: The First 50 Years. *In* A.E. Lugo and C. Lowe, editors. Tropical Forests: Management and Ecology. Springer-Verlag, New York. 33-56.

Walker, R. 1994. A remote-sensing, survey-based approach to the identification of sustainable land use. Research proposal submitted to the National Science Foundation

LASER SYSTEM REMOTE SENSING APPLICATIONS

Jeffry E. Moll, P.E.
San Dimas Technology and Development Center
444 E. Bonita Ave., San Dimas CA 91773

ABSTRACT

The laser system provides a ground-based survey method for corridors and sites that results in topographical, attribute, design, and monitoring information. Three-Dimensional (3-D) data are quickly measured, downloaded electronically to a data recorder, and formatted for automated import to CADD, road or site design software, or to data bases. The system supports remote sensing and works in concert with GPS, photography, and other data gathering and display technologies. It works well in many situations where other technologies do not, and is adaptable to a multitude of resource applications and disciplines. Further development effort is required to make a universal survey platform and to address limitations of the laser instrument, mainly the local attraction and operational problems related to the fluxgate compass.

INTRODUCTION

At present, the laser survey system provides a user friendly ground based method for corridor and site surveying with many applications in engineering and the natural resource disciplines. The system is comprised of the Laser Technology, Inc., (LTI) Criterion 400 survey instrument, the LASERSOFT survey software platform, and any one of several MS-DOS data recorders or laptop computers. The system uses a point-and-shoot routine and results in spherical coordinates represented by the ordered triple of slope distance, horizontal angle, and vertical angle. An instrument person may work alone in clear areas, while in brush or heavy vegetation a rod person with a reflector is required to ensure measurement only to the desired point. The end product of a laser survey is a topographical strip with ground attributes for each point measured-to, and can be used for site layout, design, data bases, and monitoring of resources.

The system supports remote sensing and works in concert with GPS, photography, and other data gathering and display technologies. It works well in many situations where other technologies do not, for example, in caves, steep sided canyons, dense vegetation, or polar regions, and is adaptable to a multitude of resource applications and disciplines.

INSTRUMENT DESCRIPTION AND SPECIFICATIONS

The Criterion 400 laser survey instrument uses an infrared semi-conductor laser diode for slope distance (SD) measurement. A vertical tilt-sensing encoder provides vertical inclination (VI), while a fluxgate electronic compass measures magnetic azimuth (AZ), completing the data required to establish a point's three-dimensional location in space. Instrument dimensions are 89 mm by 165 mm by 216 mm, and weight is 2.77 kg plus yoke, staff, and battery.

Manufacturer specifications give the instrument a slope distance measurement range of 1.5 to 9,150 meters (5 to 30,000 feet) when sighted on a retrodirective prism, and from 1.5 to 450 meters (5 to 1500 feet) when sighted on a "non-cooperative" target, such as a tree or the ground. A common 7.5 centimeter (3 inch) diameter plastic automotive reflector--with hundreds of tiny prisms--carried by a rodperson, is used in conjunction with a filter on the instrument to ensure measurement only to the desired point. With filter and reflector, the instrument will measure through heavy vegetation, reducing the amount of clearing required compared to existing survey methods. See figure 1 for the survey laser field hardware configuration.

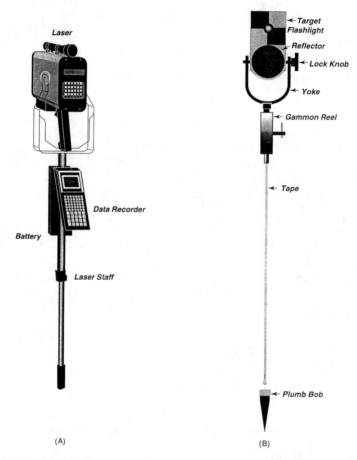

(A)

(B)

(A) Instrument, battery, and data recorder mounted on telescoping rod

(B) Yoke-mounted reflector assembly with gammon reel, tape, and plumb-bob

Figure 1.—Laser survey system hardware.

Accuracy specified by LTI for slope distance measurements is plus-or-minus 0.1 meter (0.3 foot); for vertical inclination, plus-or-minus 0.2 degree (0.35 percent slope); and for azimuth, plus-or-minus 0.3 degree. Results of controlled tests performed by SDTDC show average error to be within these values.

Instrument limitations and accuracy problems are mainly attributable to the fluxgate compass. Compass readings are made only when the instrument is oriented within a window of plus-or-minus 15 degrees with the horizontal; a routine requiring an additional sighting of the instrument and two extra keystrokes provides azimuth readings in these situations. Local attraction can cause error in azimuth measurements resulting in 5 or 10 times the magnitude of error in slope distance measurements.

Eye safety for the instrument meets FDA Class 1, (CFR 21), which means no measurable eye damage results after three hours of constant exposure to the laser beam. The laser beam contains only five percent of the energy of the average TV remote control.

LASERSOFT DESCRIPTION

The LASERSOFT field survey platform is MS-DOS software for corridor surveys made up of a baseline traverse with sideshots. Site surveys may also be performed, although alternative platforms are being designed to optimize site survey field operations and address specific needs. Traverses may be foresight, backsight, or both. Side shots may be assumed to lie on the angle bisector in the traverse-which restricts required data collection to slope distance and vertical angle--or they may lie in any direction, in which case the azimuth must also be collected. Side shots may be tagged for right or left of centerline, may be designated as turning points, and can account for "boot heights," which occur when the target is raised or lowered to provide line-of-sight for the instrument person through brush or around obstacles.

LASERSOFT is flexible in terms of data gathering. The traverse may be established first and side shots taken later, or side shots may be taken at a traverse point prior to traversing ahead. Established 3-D coordinates may be input for any point measured-to; the software will calculate coordinates for every other point in the survey based on collected data. Height-of-Instrument (HI) and Height-of-Reflector (HR) can be input, as can boot heights. Approximately 10,000 side shots may be made at a traverse point; 100 lines each with up to 100 side shots. Each side shot line may have multiple turning points--also known as instrument points--from which additional side shots are taken.

Short notes consisting of 16 characters may be input for each traverse point and for each side shot. Long notes having up to 76 characters may additionally be input for traverse points. An "auto points" mode is available for automatic incrementing of data screens. This "quick key"

239

allows data entry of successive points without touching the data recorder. Other quick keys perform various functions such as scrolling up or down data screens, entering the traverse angle bisect mode, inserting or deleting shots, and toggling between traverse and side shot modes.

SURVEY TYPES AND DATA USAGE

Corridor

Corridor surveys include roads, trails, caves, streams, and other applications conducive to a traverse baseline with cross sections or radial side shots. Traverse length should generally be limited to approximately 200 points to ensure manageable files. Corridor width is usually dictated by application needs and may be affected or restricted by sight distance, although use of turning points allows almost any desired width. Data is used in such design software as LUMBERJACK, RDS, FLRDS, Eagle Point, Softree, and AUTOCAD, and can be incorporated into databases.

Site

LASERSOFT has been used for facilities, archeological, and recreation sites, as well as lakes, timber stem mapping, landslides, and vegetative grid studies. An all-purpose site survey system has been conceived that could operate in an interactive mode with AUTOCAD. Laser data would be converted to 3-D coordinates and downloaded into AUTOCAD; any point previously measured-to could become an instrument point simply by pointing or clicking. A field going computer capable of running AUTOCAD would of course be required.

The system would allow rapid construction of contour maps, an invaluable tool in the field for optimizing data gathering. Also desired of the system is a triangulating routine by which slope distance and vertical angle are used along with a reference azimuth and the law of cosines to result in 3-D coordinates without reliance on the fluxgate compass. This system would be useful for a majority of survey applications needed to support applications in engineering and the natural resource disciplines.

CONCLUSIONS

The laser survey system provides a user friendly ground based method for corridor and site surveying with many applications in engineering and the natural resource disciplines. The system uses a point-and-shoot routine and results 3-D coordinate information. The end product of a laser survey is a topographical strip with ground attributes for each point measured-to, and can be used for site layout, design, data bases, and monitoring of resources. Further development effort is required to make a universal survey platform and to address limitations of the laser instrument, mainly the local attraction and operational problems related to the fluxgate compass.

Submeter GPS Positioning Over Long Baselines

Carl W. Sumpter
and
Leslie J. Gross
USDA Forest Service
Medicine Bow-Routt National Forest
2468 Jackson Street
Laramie, Wyoming 82070

INTRODUCTION

This paper discusses two methods for obtaining submeter positions over baselines exceeding 100 kilometers. Trimble Navigation's Phase Processor (carrier-phase) and MCORR400 (code-phase) differential correction post-processing software is used to obtain sub-meter horizontal positioning accuracies over baselines up to and exceeding 100 kilometers (km) in distance from the reference receiver.

Currently, in version 3.0 + of Trimble's Pathfinder software there are two differential correction engines, MCORR300 and MCORR400. MCORR300 is an intergal part of the PFINDER program and is used to provide submeter horizontal positioning accuracies to a distance of 50 km from the reference station and the mobile receiver.

For distances greater than 50 km the accuracy using MCORR300 decreases on the order of 10 parts per million (ppm) resulting in differential positioning accuracies of 1-5 meters depending upon the distance from the reference station (Biacs and Bronson, 1995).

MCORR400 is a stand alone DOS utility in PFINDER and is used to provide positioning accuracies of less than 1 meter to a distance of 500 km. The positional accuracy decreases on the order of 0.5 ppm, resulting in submeter positions over distances greater than 50km from the reference station and the mobile receiver (Biacs and Bronson, 1995).

The Phase Processor, a Windows based software, is an another method of obtaining submeter positioning to a distance greater than 50 km and in some cases greater than 100 km.

This type of postioning technology is very significant to Forest Service applications, considering these accuracies are obtained using "resource grade" or "mapping grade" GPS receivers.

The GPS receivers necessary to obtain submeter positions include: Trimble Pro-XL 8-channel (Maxwell chip) mobile receivers and Pro-XL 12-channel (Maxwell chip) reference receivers using carrier-phase or code-phase measurements. Other Trimble 4000 series receivers may be used as reference receivers. These include the 4000SE Land Surveyor II and 4000SSE Geodetic receivers, and others which are Maxwell chip carrier-phase technology.

The Geo-Explorer, (a small, lightweight 6-channel) "Gauss" based technology is capable of receiving carrier and code-phase measurements to provide submeter positioning accuracies consistently on distances greater than 30 km up to 50 km.

The use of the optional Phase Processor is required for obtaining submeter positions using carrier-phase measurements with either the Geo-Explorer or the Pro-XL. It should be noted the use of MCORR400 will improve the positions obtained with the Geo-Explorer, but will not produce submeter postions on distances over 50 km from the reference station consistently using code-phase measurements.

PROJECT DESCRIPTION

The use of sub-meter positioning technology was utilized on the Big Horn National Forest for establishing geodetic control points for Digital Orthophoto Quads (DOQ) during the fall of 1994.

The project area for 1994 was the north half of the forest including areas outside of the forest boundary in order to have complete USGS Quadrangle coverage. This resulted in a project area encompassing an area expanse of approximatly 70 miles East-West and 100 miles North-South covering portions of Southern Montana and Northern Wyoming. The area included lands managed by the USDA Forest Service, Bureau of Land Management and Bureau of Indian Affairs.

FIELD PROCEDURES

The first step was to identify National Geodetic Survey horizontal and vertical control stations and their location relative to the project.

The require accuracy of the DOQ's was 1 meter horizonatal and 2 meters vertical with the coordiates in Universal Transverse Mercator (UTM), a vertical datum of NAVD 1929, and a horizontal datum of NAD 1927 with a reference ellipsoid of Clarke 1866. Given these parameters, it was determined horizontal, vertical, ellipsoid and geoid separation would be determined by Wyoming's High Accuracy Reference Network (HARN) with supplementary classical NGS control stations for vertical elevations and the connection to NAD27. The vertical values were consistently less than one meter for the stations. The following geodetic control stations were selected to meet the criteria:

242

<u>Wyo. HARN Stations:</u>

Sheridan
Owen
Lake
Battle (new station)

<u>NGS Vertical Stations:</u>

Red
Ranchester NW

The second step, once the station was reached, the aerial photography was interpreted and correlated to the selected field identifyable position. The photography was then "pin pricked" for spatial position relative to the quad corner, a rebar with aluminum cap was set, the observation was done and the accessory data was noted in a field book.

The design of the project was not to get a position on every quad corner, but to establish a geodetic position within 1 mile of selected quad corners. The selection criteria for these points was based on several factors. These included:

1. Proximity to roads or trails

2. Ground points easily identifiable on existing resource photography such as
    a. road intersections
    b. prominate features
    c. tangent intersection of curves on roads

3. Distance between DOQ horizontal control stations and reference stations

Step three was the data processing using the stand alone DOS utility,MCORR400 and the Phase Processor.

Step four involved the data evaluation and final output. The Phase Processor was chosen since it provides an output showing the integrity and quality of the data. MCORR400 doesn't do this without first importing the data into GPSURVEY. Once that is done then the data has integrity and quality control. However, unless the end user has access to GPSURVEY this step is not possible.

DATA COLLECTION PROCEDURES

The data collection field procedure set forth utilized Trimble 4000SSE geodetic receivers on two HARN stations and two Trimble 4000 survey receivers on two classical NGS vertical stations for reference control during each observation period. The DOQ stations observations were done with 2 Pro-XL 8-channel receivers resulting in redundant baseline vectors between reference and rover stations.

HARN station Owen was utilized throughout the project as a tie between the two halves of the project. HARN station Lake was used on the west half and HARN station Sheridan was used on the east half with one new station set in Montana at the Little Bighorn National Battlefield Monument.

The receivers were configured with the following parameters:

1. 4000 series receivers
    a. 5 second logging rate
    b. 13 degree elevation mask
    c. DOP mask 6
    d. 3D position solutions

2. Pro-XL receivers
    a. 1 second logging rate
    b. 15 degree elevation mask
    c. DOP mask 6
    d. SNR mask 6
    e. 3D position solutions
    d. decimeter, aka phase mode
    f. 10 minute occupation times

Once the reference receivers were setup and secured the survey crews started the sometime long drives to get to the DOQ stations. Even though the survey crews were equiped with a variety of maps, aerial photography and had allotted adequate time to reach the stations. There were problems reaching the stations due to flat tires, getting the truck stuck in a creek, locked gates, new gates, closed roads, wet roads and roads that on a map looked like the station access was achievable. When the station access was not good, it forced the finding of a new access route thus causing a delay by vehicle or several hours of walking to reach the station.

The sheer expanse of the project and the time between DOQ stations was enormous resulting in only four to six, 10 minute occupations per rover a day being logged. This resulted in the reference receivers occupying the control stations for up to 12 hours while logging data. The long occupation times allowed for a precise verification of the station accuracies and establishment of new geodetic postions and ellipsoid references on the classical NGS vertical stations.

The project was accomplished in a two week period, but in reality, in only 6 days (3 days per week) as the survey crew had an 8 hour drive to and from the project site on Mondays and Fridays.

RESULTS

The field observations were processed with both MCORR400 and the Phase Processor to determine DOQ positions and the relative difference between the two methods.

RESULTS OBTAINED FOR BASELINES USING DECIMETER PROCESSING COMPARED AGAINST MCORR400 PROCESSING, FOR BASELINE LENGTHS FROM 7.5km TO 106.8km.

STA	BASE STA. PROCESSED AGAINST	PHASE PROCESSING LATITUDE / LONGITUDE / HEIGHT	EXPECTED ACCURACY	MCORR400 PROCESSING LATITUDE / LONGITUDE / HEIGHT	BASELINE LENGTH	DIFF. IN LATITUDE	DIFF. IN LONGITUDE
1	SHERIDAN	44-37-01.1814 / 107-15-10.4591 / 2575.107m	<0.3m	107-15-10.467 / 2576.590m / 431/595	32,200m	0.000m	0.175m
2	SHERIDAN	44-37-51.9608 / 107-30-00.9906 / 2714.014m	<0.3m	44-37-51.952 / 107-30-01.014 / 124/596	48,200m	0.278m	0.502m

???/??? POSITIONS

STA	BASE STA. PROCESSED AGAINST	PHASE PROCESSING LATITUDE / LONGITUDE / HEIGHT	EXPECTED ACCURACY ???/??? POSITIONS	MCORR400 PROCESSING LATITUDE / LONGITUDE / HEIGHT	BASELINE LENGTH	DIFF. IN LATITUDE	DIFF. IN LONGITUDE
1	BATTLE	44-37-01.181 / 107-15-10.4744 / 2574.624m	<1.0m	44-37-01.1935 44-37-01.186 / 107-15-10.451 / 2576.507m / 590/595	106,800m	0.247m	0.502m
2	BATTLE	44-37-51.9893 / 107-30-01.0224 / 2716.200m	<1.0m	44-37-51.944 / 107-30-00.995 / 2714.768m 2716.646m / 549/596	104,400m	1.389m	0.589m

245

SHERIDAN

3

44-45-16.5180	44-45-16.492	BATTLE 44-45-16.4988 44-45-16.507
107-37-56.2991	107-37-56.337	107-37-56.338 107-37-56.326
2496.241m	2498.285m	2496.865m 2497.642m
<1.0m	602/602	<1.0m 499/602
	54,900m	91,900m
	0.803m	0.247m
	0.829m	0.262m

LAKE

4

44-46-53.1303	44-46-53.142	OWEN 44-46-53.1314 44-46-53.141
107-58-02.4958	107-58-02.487	107-58-02.5007 107-58-02.519
1897.410m	1898.341m	1897.229m 1898.325m
<0.3m	603/603	<0.3m 603/603
	30,000m	38,400m
	0.370m	0.309m
	0.196m	0.393m

OWEN

5

44-44-33.1091	44-44-33.105	LAKE 44-44-33.1068 44-44-33.098
108-15-49.4487	108-15-49.528	108-15-49.4313 108-15-49.433
1369.832m	1375.116m	1369.904m 1371.803m
<1.0m	612/612	<0.3m 612/612
	61,000m	9,100m
	0.123m	0.278m
	1.724m	0.044m

SHERIDAN

6

44-52-30.6322	44-52-30.636	OWEN 44-52-30.6297 44-52-30.650
107-15-23.8048	107-15-23.838	107-15-23.8200 107-15-23.802
1183.635m	1186.018m	1183.655m 1185.156m
<0.3m	601/601	<0.3m 601/601
	26,200m	27,500m
	0.123m	0.617m
	0.720m	0.393m

246

7   LAKE

44-52-31.3346   44-52-30.734   OWEN 44-52-31.3369 44-52-31.332
108-22-05.7558   108-22-05.278   108-22-05.7669 108-22-05.739
1194.375m   1184.615m   1194.700m   1196.016m
<0.3m   598/598   <1.0m   598/598
           8,500m   71,800m
          18.553m   0.154m
          10.429m   0.611m

8   SHERIDAN

44-59-28.7412   44-59-28.730   OWEN 44-59-28.7418 44-59-28.740
107-22-16.4347   107-22-16.451   107-22-16.4232 107-22-16.416
1267.470m   1268.715m   1267.508m   1267.921m
<0.3m   611/611   <0.3m   611/611
          39,900m   34,200m
          0.340m   0.062m
          0.349m   0.153m

9   OWEN

45-00-21.9165   45-00-21.932   SHERIDAN 45-00-21.9512 45-00-21.930
107-36-40.2943   107-36-40.271   107-36-40.3099 107-36-40.289
1310.333m   1310.775m   1308.783m   1310.023m
<0.3m   591/591   <1.0m   307/591
          35,600m   57,500m
          0.494m   0.648m
          0.502m   0.458m

10 OWEN

44-58-35.4938   44-58-35.491   SHERIDAN 44-58-35.4809 44-58-35.486
108-00-28.5692   108-00-28.577   108-00-28.5725 108-00-28.572
1774.336m   1775.070m   1774.208m   1774.611m
<1.0m   597/597   <1.0m   597/597
          51,000m   86,400m
          0.093m   0.154m
          0.175m   0.000m

11 OWEN

45-07-25.8857   45-07-26.635   SHERIDAN 45-07-25.8857 45-07-25.928
107-52-51.7972   107-52-51.872   107-52-51.7865 107-52-51.826
2092.059m   2141.515m   2091.639m   2093.693m
<1.0m   604/604   <1.0m   604/604
          56,300m   82,300m
          23.122m   1.297m
          1.636m   0.873m

12 LAKE

45-07-44.7260   45-07-44.724   OWEN 45-07-44.7286 45-07-44.719
108-11-54.2811   108-11-54.291   108-11-54.2852 108-11-54.273

13 LAKE	1316.122m <0.3m	1316.693m 600/600 38,300m 0.062m 0.218m	1316.078m <1.0m	1317.134m 600/600 73,300m 0.309m 0.262m
14 OWEN	45-15-07.2601 108-06-09.3423 1652.870m <1.0m	45-15-07.261 108-06-09.332 1654.288m 601/601 53,700m 0.031m 0.218m	OWEN 45-15-07.2558 108-06-09.3440 1652.149m <1.0m	45-15-07.256 108-06-09.324 1654.632m 601/601 78,000m 0.000m 0.436m
15 LAKE	45-13-49.1104 108-23-03.5712 2094.381m <1.0m	45-13-49.123 108-23-03.610 2098.483m 589/589 91,700m 0.401m 0.851m	LAKE 45-13-49.1095 108-23-03.5692 2094.360m <0.3m	45-13-49.116 108-23-03.585 2095.751m 589/589 47,900m 0.185m 0.349m
16 LAKE	45-22-23.3441 107-52-11.7360 1005.845m <5.0m	45-22-23.346 107-52-11.721 1006.419m 596/596 73,800m 0.062m 0.327m	OWEN 45-22-23.3429 107-52-11.7385 1005.995m <5.0m	45-22-23.345 107-52-11.718 1007.045m 523/596 80,700m 0.062m 0.436m
	45-21-57.6369 108-08-01.2761 1356.221m <1.0m	45-21-57.603 108-08-01.266 1358.545m 598/598 65,000m 1.050m 0.218m	OWEN 45-21-57.6365 108-08-01.2975 1356.409m <1.0m	45-21-57.583 108-08-01.240 1358.834m 598/598 89,700m 1.636m 1.265m
17 SHERIDAN	45-29-33.5963 107-14-19.7416 1030.234m <1.0m	45-29-33.597 107-14-19.730 1032.303m 609/609	BATTLE 45-29-33.5658 107-14-19.7332 1030.746m <0.3m	45-29-33.624 107-14-19.708 1031.965m 609/609

SHERIDAN 18

45-29-51.0603   45-29-51.059   80,200m   BATTLE 45-29-51.0550 45-29-51.063   17,500m
107-30-55.4530   107-30-55.450   0.031m   107-30-55.4512 107-30-55.443   1.790m
1041.227m   1043.066m   0.262m   1041.421m   1042.614m   0.545m
<1.0m   619/619   <0.3m   619/619
                                       89,500m   10,200m
                                       0.031m   0.247m
                                       0.065m   0.175m

SHERIDAN 19

45-37-22.3365   45-37-22.341   94,200m   BATTLE 45-37-22.3399 45-37-22.343   16,200m
107-14-29.7479   107-14-29.748   0.154m   107-14-29.7535 107-14-29.753   0.093m
1024.551m   1024.984m   0.000m   1024.482m   1025.242m   0.000m
<1.0m   621/621   <0.3m   535/621

SHERIDAN 20

45-37-19.8759   45-37-19.874   100,900m   BATTLE 45-37-19.8844 45-37-19.885   7,500m
107-29-39.3165   107-29-39.327   0.062m   107-29-39.3229 107-29-39.323   0.031m
938.916m   940.316m   0.240m   938.730m   939.673m   0.000m
<1.0m   10/621   <0.3m   621/621

249

SUMMARY

This type of postioning technology is very significant to Forest Service applications, considering these accuracies are obtained using "resource grade" or "mapping grade" GPS receivers. The use of Trimble PRO-XL 8-channel resource grade GPS receivers with MCORR400 or Phase Processing software for obtaining submeter positions over long baselines a very cost effective alternative where high accuracy geodectic surveys are not needed. The use of resource GPS equipment keeps the cost down, the learning curve low and the production high. The use of MCORR400 and the Phase Processor now extends the effectiveness and makes better use of the Forest Services extensive network of Community Basestations for the collection of data in support of GIS now and in the future.

Note: The mention of product names or corporations is for the convenience of the reader and does not constitute an official endorsement or approval by the US Government of any product or service to the of others that may be suitable.

REFERENCES

Biacs, Zoltan and Bronson, R. 1995, <u>Wide Area DGPS results with MCORR400 and PRO-XL receivers</u>, Trimble Navigation Ltd., Sunnyvale, Ca.

USING GPS IN WILDLAND FIRE MANAGEMENT

Dick Mangan
Fire Program Leader
Missoula Technology and Development Center
Building 1, Fort Missoula
Missoula, Montana 59804-7294

ABSTRACT

Hand-held Global Positioning System (GPS) units can help wildland fire managers by providing relatively accurate locations quickly in the field. That information is needed both for wildfires and prescribed fires. Although GPS units are rapidly improving and becoming less expensive, they are still not widely available at the field level for firefighters. Once they are, GPS units can be integrated into the incident command system, allowing more accurate locations when fires are detected, allowing firefighters on the ground and in aircraft to more readily find specific areas on a fire, and allowing planners to more accurately determine the boundaries and area of a fire.

INTRODUCTION

Wildland fire, by definition, occurs in wild areas; oftentimes off roads or in roadless areas with no clear lines of ownership; often wildland fires are in areas that lack clear terrain features to help establish location.

Since the turn of the century, wildland firefighters have relied on the compass, (hand held or configured as an "Osborne Firefinder" in lookouts), to determine their own location through triangulation, and to find their way to remote fire locations. There was also a heavy dependence on experienced "locals" who had been brought up in the area, spending much of their lives hiking, hunting, and gathering firewood on the ground. Even this familiarity with the ground didn't always work, however; many of us, have spent long hours (including some overnights) looking for an elusive lightning-caused fire in the Western mountains.

Not getting to a wildfire quickly was more accepted then than it is today. The transportation system was primitive, resource values were low, and there were few if any homes in the woods. Suppression equipment often consisted of men with hand tools, horses and trucks. Media attention was scarce on all but the largest fires.

All that has changed over the past few years. Higher resource values, threatened and endangered species habitat, escalating fire suppression costs, the urban-wildland intermix, and intense media scrutiny require us to conduct our fire management activities in a timely, efficient, and professional manner. A new tool—the hand-held Global Positioning System (GPS) receiver—is available to help wildland fire managers meet these goals in wildfire and prescribed burning activities in the wildlands.

FIRE MANAGEMENT APPLICATIONS

Important applications of hand-held GPS occur in most stages of our fire management activities:

Preplanning

GPS can be used prior to prescribed fires and wildfires to locate and map critical resources such as cultural resource sites, bridges and trail improvements, range structures, wildlife snags, critical habitat such as elk wallows, designated superior trees, remote homes and cabins, and section corners. In addition, specific hazards such as mine shafts, power lines and substations, well sites, and pipelines can be mapped accurately.

Knowing the specific locations of critical resources and specific hazards can be invaluable for dispatchers and initial attack fire crews on wildfires or when scheduling fuel treatment activities.

Detection and Dispatching

The use of GPS in wildfire detection and dispatching can be especially valuable, since accurate reporting of location and timely initial attack are often closely related to a fire's ultimate size and the suppression costs. Using GPS can greatly increase the accuracy of a fire report when fixing the location of a wildfire discovered by ground personnel or aerial detection, especially in areas lacking good reference points (such as roads, streams, or recognizable facilities), and when fires are reported by individuals who are unfamiliar with the terrain. Smokejumper aircraft flying to a fire location often discover other fires. The use of GPS in a smokejumper aircraft can help pinpoint the location of multiple ignitions in a small area, and help ensure that sufficient ground crews are dispatched for each ignition.

With more accurate locations of reported wildfire ignitions, dispatchers will be better able to determine ownership and protection responsibility, and to dispatch suppression forces by the most effective route. Crews that have to leave a road and travel cross country to a known fire location can use GPS to give them a bearing and distance that is constantly updated en route to the fire, eliminating much of the time that might otherwise be lost finding their way. The decreasing cost of hand-held GPS units makes it feasible for each engine crew and smokejumper team to have its own unit.

Suppression

There are numerous opportunities for GPS to increase the efficiency and effectiveness of our fire suppression activities, especially on large fires managed by incident command teams. Some specific applications of hand-held GPS on these large fires include:

Field Observer—Hand-held GPS can increase accuracy in locating and reporting critical control points, the fire perimeter, potential helispots and water sources, and safety hazards.

<u>Suppression Overhead Personnel</u>—Overhead personnel, such as crew supervisors, strike team leaders, and division supervisors, can use GPS to accurately report their positions throughout the shift, identify specific hazards (such as burning snags), accurately identify spike camp and "coyote" camp locations, and direct medivac crews to line locations if needed. Safety is enhanced if firefighters become entrapped by fire; retardant and helibucket drops can be delivered with pinpoint accuracy if both the aircraft and the firefighters on the ground have GPS. The use of GPS when requesting air tanker and helibucket support on hot spots can result in major cost savings. Not only will the use of GPS increase the accuracy of the drops, (possibly reducing the fire spread and ultimately the final fire cost), but it can also reduce the flight time on aircraft that may cost more than $100 per minute (such as large Type 1 helicopters). At those rates, a hand-held GPS unit could easily pay for itself in one 12-hour shift.

<u>Planning Section</u>—Planning personnel are able to use GPS to:

• Map a fire's perimeter and area (from either an airborne platform or on the ground)
• Identify division breaks and drop points
• Locate isolated improvements needing special protection efforts.

The use of GPS can be especially valuable to the planning section when line personnel accurately report either the location of the fire's edge throughout the burning period, or pinpoint where line construction has actually progressed during a shift, rather than where it was expected to be. This allows subsequent incident action plans to be updated to more closely reflect real-time conditions on the incident. In preparing the medical unit plan (Incident Command System-206), the latitude and longitude of predesignated medical facilities and burn centers can be specified so that medivac transport can easily use the most direct flight route.

MECHANIZED EQUIPMENT

The use of GPS with mechanized equipment (engines, bulldozers, water tenders, lowboys, helitenders, crew carriers) on wildfires offers valuable opportunities to:

• Track the location of mechanized equipment—often a difficult task on large wildfires, especially when units are assigned to an unfamiliar area. A GPS unit placed with each of these key pieces of equipment can increase the coordination needed between maintenance and fuel support vehicles
• Identify specific resource protection objectives, and ensure that equipment is properly positioned for implementation
• Give single resource leaders (such as strike team leader dozers) the ability to pinpoint the locations of resources they find within their area of responsibility
• Expedite recovery and repair of disabled vehicles.

AIRCRAFT

The use of GPS in aircraft operations on wildfires offers many advantages, both for fixed and rotary wing aircraft:

- Air detection flights can accurately record fire locations, reducing the time needed for initial attack forces to reach a fire
- Rappel helicopters and smokejumper aircraft equipped with GPS, and dispatched to a specific fire located with GPS, can reduce response time, flight hours, and costs
- Pilots, especially those unfamiliar with an area, can be directed to accurately located helispots
- Ground forces equipped with GPS can more accurately request air tanker or helibucket drop requests
- Helicopters and air tankers equipped with GPS can more accurately respond to the requests of ground forces
- Any aircraft over a large fire (helicopter, air tanker, lead plane) can report the location of spot fires and hot spots in a timely and accurate manner
- Helibucket fill locations can be plotted and used to reduce flight time, especially for new helicopters on an incident
- Air medivac operations can be enhanced by pinpointing the location of injured firefighters needing air transport
- The location of smokejumper jump spots in remote areas can be accurately recorded, an important safety consideration if the smokejumpers have to be retrieved later.

POST-FIRE

After a wildfire has been controlled, GPS can continue to play an important role in post-fire activities. As incident management teams begin leaving a fire, specific locations identified by GPS can be extremely useful:

- Newly arriving personnel can be given accurate locations to help them check hotspots and spot fires
- Caches of tools, pumps, hoses, and other equipment can be referenced for retrieval at a later date
- Critical rehabilitation needs (for skid trails, creek crossings) can be identified for timely work
- The location of damaged resources (fences, buildings, trails and bridges, and cultural sites) can be identified in a timely and accurate manner
- Severely burned areas or hydrophobic soils can be accurately mapped for rehabilitation teams
- The fire perimeter can be easily mapped and the fire area easily calculated. This can be especially important when determining cost shares on multijurisdictional fires.

PRESCRIBED FIRE

Hand-held and aircraft-mounted GPS units have numerous applications on both planned ignitions and prescribed natural fires in wildernesses. As large-scale eco-system burns (such as the 15000-acre "Buchanan Burn" in New Mexico) are planned more often, GPS units can be used to:

• Help control lighting patterns
• Identify critical protection needs such as cultural resource sites, superior trees, sensitive wildlife habitat, locations of threatened and endangered plant species, range improvements and property corners
• Identify water sources for both ground-based and helibucket operations
• Identify time-sensitive control points that will influence firing patterns
• Locate critical areas for holding and patrol crews
• Maintain accurate locations of assigned resources.

Many of the uses of GPS identified for wildfires and planned ignitions are also appropriate for prescribed natural fires (PNF) in wilderness. In addition, GPS allows the prescribed natural fire manager to coordinate specific on-the-ground locations with geographic information system-(GIS-) based information sources to:

• Ensure that actual fire behavior is consistent with predictions
• Quickly and accurately map the fire perimeters
• Help assure the safety of monitoring teams by closely monitoring their locations relative to the fire's location.

AVAILABILITY OF GPS

While technological advances in GPS are extremely rapid and the price of hand-held receivers is dropping quickly, the availability of GPS units at the field level is not yet widespread. If GPS is to reach its potential for use on large-scale prescribed burns or wildfires, the units must become more available. Fire overhead (down to the crew supervisor), field observers, line scouts, weather observers, ground support personnel, helicopter managers, and engine captains will need GPS units assigned to them if GPS is to be integrated into management. Any crews working in high-risk environments (such as interagency hotshot crews) should also be equipped with hand-held GPS.

One potential method to increase the availability of GPS units on a wildfire incident or prescribed burn is to include GPS kits in the fire cache system, similar to the procedure for personal portable radios. This could help ensure full utilization of a limited resource, provide accountability of a sensitive item, and offer skilled mainte-nance capability after use. Limited instruction to overhead and other potential users at the incident base camp will provide the skills needed for basic operations in the field.

Some Federal agencies now have access to the military GPS units designated as "PLGRS" (called "Pluggers"). Those GPS units have computer programs available to increase their accuracy by compensating for the intentional error in the publicly broadcast GPS satellite signals. The Missoula Technology and Development Center in

Missoula, MT, programs and updates these units for Federal natural resource agencies. Several of these units are out in the field in Florida and New Hampshire for evaluation during the 1996 fire season.

CONCLUSION

As we enter the 21st century, fire remains one of the most significant forces of nature affecting our natural resources, yet the individual firefighter and fire manager still remain the essential human link between the fire and the ground.

The two constants of fire and man interact to shape the land, just as they have for the past 100 years. By reaching into the sky to access the power of the satellite, 21st-century fire managers will be able to use GPS to more efficiently manage fire on the land.

INTEGRATING GLOBAL POSITIONING SYSTEM (GPS) DATA INTO CARTOGRAPHIC FEATURE FILES (CFF)

Carol A. Brady
Forest Service
Geometronics Service Center
Salt Lake City, Utah 84119

ABSTRACT

Global Positioning System (GPS) data has been collected by national forests for their use in accurately locating roads and trails within forest boundaries. This data has not previously been used to update Cartographic Feature Files (CFF) used by the Forest Service's Geometronics Service Center (GSC) to produce the 1:24,000-scale hard-copy maps. These same CFF's are used as base information by forests for building GIS files for various planning and tracking purposes. This paper gives a brief overview of incorporating Pike National Forest GPS data into CFF's from initial processing in Pathfinder software to merging the data with the CFF. It will cover creation of a CFF compatible file and the manipulation necessary to make the data usable for CFF files and hard-copy map production. The presentation will include a discussion of proposed guidelines/standards for GPS data collection and how these standards can affect integration of the data into CFF as well as the effect on the 1:24,000-scale hard-copy maps.

INTRODUCTION

Global Positioning System (GPS) data has not previously been directly incorporated into Cartographic Feature Files (CFF). GPS information has been used by the forests to create correction guides, but only to the extent of depicting the location of roads on hard-copy maps or plotting files for use in hand-collecting data from those plots.

The Pike National Forest submitted a disk containing GPS files with their 1995 correction guides. These files were differentially corrected but unedited and contained point features as well as road and trail locations.

The files were plotted when they were received and forwarded to GSC's Photogrammetry unit with the correction guides. Photogrammetry identified those roads that could be seen on orthophotos and marked all other roads requested on the correction guides as location approximate. The project was then forwarded to GSC's Digital Mapping unit.

INITIAL PROCESSING AND IDENTIFYING PARAMETERS

GPS data contains more points than necessary or desirable for both CFF and 1:24,000 hard copy production, partly due to the way it is collected and partly by the nature of the trails and roads most often collected. Most map symbology takes up more space on the map than the feature does on the ground. Cartographers have to decide how and where to place the symbology to best represent features with as little positional distortion as possible. In the case of roads and trails, this means that too much road

257

curvature (like switchbacks) has to be altered in order to keep the map intelligible.

The GPS files were loaded onto Pathfinder in GSC's Cadastral unit. The data was examined and an attempt to thin the number of points was made. This involved compressing the point interval from 5 seconds to 15 seconds by averaging all the values of each point collected between 15 second intervals into one value. The resulting files showed a distinct loss of curvature definition. At this point, thinning the data by linear feet seemed the better option.

The GPS files were processed though Pathfinder using coordinate translate to change the data to NAD27 and State Plane zone 0502 (Colorado Central zone). Once the data had been changed, it was exported to TONTOCADD where it was converted to an Integraph-compatible CFF.

The CFF files containing the GPS information were plotted prior to any thinning or editing. The plots were made on a Versatec raster plotter with the same symbology as the 1:24,000 scale hard copy maps. The plots were reviewed for how the number of points affected the symbology on the plot (Fig. 1). It was evident that the number of points in the file needed to be reduced. An initial two quads were run through a thinning program at 5, 10, 15, and 20 foot intervals. The two quads were reviewed on screen using the original files as background. The 15 foot interval was found to be the optimum number of points removed while maintaining road position and curvature.

The thinned GPS files were plotted again to identify areas that needed additional editing (Fig 2). Most of these were areas of dense point collection or "clusters", often to the side or at the end of the feature. These "clusters" may have occured when the people collecting the data failed to stop collection of linear data while going off the road or trail to collect a point feature. These were marked on the plots for identification in the graphics edit. Other areas marked for edit were narrow switchbacks that couldn't be adequately represented at a 1:24,000 scale (Fig. 3)

MERGING THE GPS DATA INTO THE CFF

All files, the original GPS file, the CFF, and the thinned GPS files were all processed into graphics (design) files. These files could then be directly overlaid for comparison and editing of the thinned GPS data.

Removal of the off-road point clusters along the length of the road was the first step in editing the GPS data. Since these were associated with point features, they were removed entirely. Road-end clusters posed more of a problem. A hard copy map containing elevation contours was used to assist the editing process for these clusters. When point clusters occured at the junction of another road and were associated with little or no elevation change, they were eliminated. Some, however, were associated with a sharp change in elevation. Where these occured, it was necessary to examine the cluster to determine if the point collection was distinguished by any pattern or direction. If so, that might indicate switchbacks or obstacles that could have caused the people collecting data to double back or attempt different routes. Whenever possible, this information was edited and kept in the files.

GPS data is collected without road network connectivity. Any road or trail incorporated into a CFF must be connected by a single node or point to the existing network. This was done after editing as the GPS data was copied into the CFF.

Where the GPS data overlaid an existing road or trail, the GPS data was used. In some cases, the GPS road or trail closely followed the alignment derived from orthophotos or photography. This happened more often on dirt or gravel roads than trails. Where GPS trails overlaid existing trails, both the amount of curvature detail and the alignment differed greatly from the CFF design file.

Where roads or trails conflict with older physical obstacles such as lakes or rivers, the data was incorporated and notes made on the plot to check the location of the obstacle on the orthophotos. One of the files involved a trail that wandered into and over a river. Examination of the orthophoto showed that the course of the river had changed sufficiently to account for the position of the trail.

DATA COLLECTION STANDARDS

In the Geometronics Service Center's Draft Guidelines for Digital Map Updates, Minimum standards for GPS data to be added to CFF's lists Positional Dilution of Precision (PDOP) should be no greater than 8 and the signal level mask setting should be not less than 6. Roads and trails collected using 2-D will only be used if there is 40 feet or less relief along the feature collected.

There are additional collection parameters that should be considered. The point collection interval should be changed for the type of feature being collected. A 5 second interval produces a good alignment for gravel roads, but can be too dense for dirt roads or trails.

Simple things such as stopping point collection on roads or trails prior to moving away to collect a point feature such as a culvert or picnic table, will eliminate one type of point cluster problem.

Once GPS data has been forwarded to GSC for incorporation into the CFF, the decisions In the Geometronic Service Centers's Draft Guidelines for Digital Map Updates, minimum standards for GPS data to be added to CFF's lists Positional Dilution of Precision (PDOP) should be no greater than 8 and signal level mask setting shoud be not less than 6. Roads and trails collected using 2-D will only be used if there is 40 feet or less relief along the feature collected.

There are additional collection parameters that should be considered. The point collection interval should be changed for the type of feature being collected. A 5 second interval produces a good alignment for dirt or gravel roads, but can be too dense for trails or dirt roads.

Simple things such as stopping point collection on roads or trails prior to moving away regarding thinning and elimination of clusters will become GSC's. Editing by the forest before sending the data would allow the people who collected information to make decisions and eliminate extraneous data based on first-hand knowledge of the roads and trails.

CONCLUSION

It is possible to incorporate GPS data into CFF's and generate 1:24,000 scale maps. The time involved in modifying this type of data is little or no greater than that required to hand-collect the information from orthophotos and then digitize that into the CFF.

REFERENCE

Boeder, A., 1995, <u>Draft Guidelines for Digital Map Updates</u>, available from U.S.D.A. Forest Service, Geometronics Service Center, Salt Lake City, Utah 84119

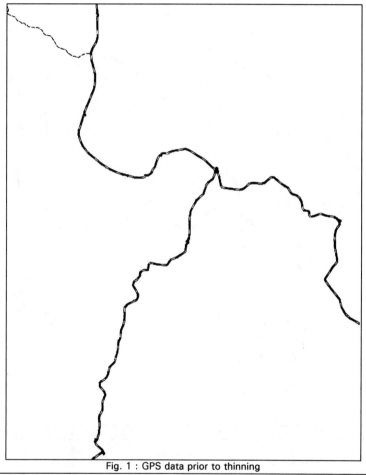

Fig. 1 : GPS data prior to thinning

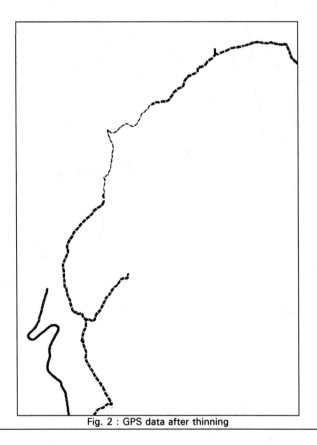

Fig. 2 : GPS data after thinning

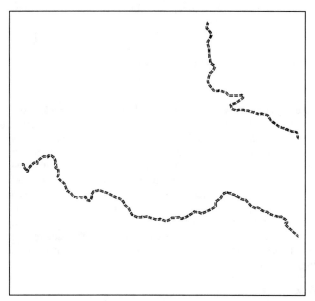

Fig. 3 : Thinned GPS - showing narrow switchbacks

A REAL-TIME GPS SYSTEM FOR MONITORING FORESTRY OPERATIONS

William R. Michalson - Assistant Professor
Joshua Single, and Michael Spadazzi
Department of Electrical and Computer Engineering
Worcester Polytechnic Institute
100 Institute Road
Worcester, MA 01609

Michael Wehr - Mechanical Engineer
Forestry Sciences Laboratory
410 MacInnes Dr.
Houghton, MI 49931

PROJECT OVERVIEW

Abstract

A GPS-based system has been developed at Worcester Polytechnic Institute to research and evaluate developing technologies for continuous, automated acquisition of geographic position information and navigation. Knowledge of geographic position is important in planning, managing, and conducting numerous forestry operations. Applications of this type could include mapping, controlling the rate of herbicide application, or developing accurate space/time plots of equipment such as harvesters, skidders and forwarders. Initial system testing will be conducted on herbicide application equipment. The initial prototype system consists of a portable differential GPS reference station which will work in conjunction with mobile display units which present the vehicle track, land boundary, and spray coverage density to the equipment operator.

This paper will describe the complete system and accuracy and performance achieved in initial testing.

Problem Description

The National Forest Service requires a vehicle tracking system which will allow field supervisors to accurately keep track of areas which are sprayed with an herbicide. These areas are one square kilometer or less in area, and are deforested. Herbicide is applied to these areas, and the effect of the herbicide is examined over time. Hence, it is important that the areas sprayed with the herbicide are accurately known.

Additionally, this vehicle tracking system needs to assist the driver of the application vehicle in applying the herbicide consistently over the entire work area. Therefore, the system must also provide the operator with some sort of indicator that shows where he has or has not yet sprayed.

Since the system will be used in the field, inside a moving vehicle, a keyboard is not an acceptable means of user input. Therefore, a touch screen needs to be used for this system. The touch screen interface must be uncluttered and simple to use, since most of the operator's attention should remain focused on the task of driving the vehicle.

264

In order to achieve these goals, the project will use a differentially corrected GPS system

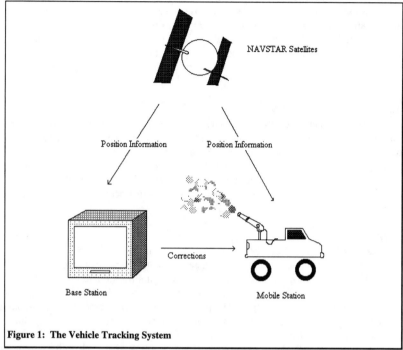

Figure 1: The Vehicle Tracking System

for positioning information. The GPS system, and differential GPS are described below.

THE GLOBAL POSITIONING SYSTEM

The NAVSTAR GPS system is an full-time, all-weather world wide radio-navigation aid. The system is operated by the United State Department of Defense. The system is composed of three segments, the Space Segment, the Control Segment, and the User Segment. We will discuss each one individually.

Space Segment

The Space Segment of NAVSTAR/GPS consists of 24 satellites in earth orbit in six orbital planes inclined 55 degrees to the equator. The orbital period is approximately 12 hours, and the satellites orbit at an altitude of approximately 10898 nautical miles. Each satellite transmits a signal encoded on two frequencies, L1(1575.42   MHz) and L2(1227.60 MHz). These satellites transmit several different messages encoded on the carrier. First is a pseudo-random noise code. The pseudo-random noise code is a long term repeating code which is broken down in to short segments. Each of these unique segments is then assigned to a different satellite. There are two kinds of this code, the coarse acquisition(C/A) code, and the precision(P) or encrypted precision(Y) code. The C/A code is the code that civilian receivers use to determine position. The signal is a code which is at 1.023 MHz. The way that the receiver figures out the time for the signal to travel from the satellite is discussed later. The P code is a higher frequency code at

265

10.23 MHz which allows better accuracy in measuring the transit time of the signal, and therefore better range measurements. The Y code is the encrypted version of the P code which is in use today now that the system is fully operational. To use the Y code you must have an authorized receiver with the appropriate cryptographic key.

Also transmitted on the carrier is a satellite status message, as well as ephemeris data updates which are the mathematical approximation of the orbit of the satellite. Included in this message is room for additional information to be transmitted at a later date if it becomes necessary or desirable.

Control Segments

The Control segment of the NAVSTAR/GPS is headed by the master control station at Falcon Air Force Base(AFB) in Colorado. At this and other monitoring stations at Diego Garcia, Ascension Island, Kwajalein, Hawaii, and Onizuka AFB, accurate measurements of the satellite orbits are made, and translated into ephemeris data, which is then uploaded to the satellites regularly, along with corrections to the onboard system clocks. Also, if the satellites fail for whatever reason, the control stations instruct the satellites to transmit a signal that indicates bad health.

User Segment

The user segment consists of all of the GPS receivers on the planet today. They operate on the same basic principle of pseudorange measurements. A pseudorange measurement consists of the time shift need to correlate the actual signal transmitted by a satellite, and the replica generated by the receiver multiplied by the speed of light which is the propagation speed of the radio signal. The pseudoranges are measured with a correlation detector which controls a delay lock which aligns the two signals, actual and replicated This distance measurement is called a pseudorange because of the errors introduced into the system by relativistic effects, ionosphere delays, and Selective availability introduced to degrade system performance by the United State DOD. When one satellite pseudorange is determined, the distance describes a sphere on which the receiver could lie. When two satellites are included in the solution, the intersection of the spheres is the solution, which is a circle. When a third satellite is included it resolves to two points. The fourth satellite not only resolves the ambiguity but also determines the system time which is needed for accurate measurement of the signal travel time. The receiver will than compute a transformation of the position which is in an earth centered earth fixed frame of reference to a useful system such as latitude and longitude or UTM grids for example.

DIFFERENTIAL GPS

Differential GPS is a system addition which allows better than P code accuracy without the need for a P code receiver.

Differential GPS is based on adding a base station which placed at a known point to the system. As discussed above, the receiver calculates the distance to each satellite. Since the base station knows both the position of the satellite and receiver, it can calculate the distance to the satellite precisely. The difference between the actual and measured signal is the differential correction for the pseudorange to that satellite. These corrections can either be transmitted over a data-link for real time use, or saved for later post-processing of satellite ranging data. In our system these corrections are then transmitted to the mobile station receiver. The mobile receiver applies these corrections to the same satellite's ranging data, and then solves the navigation solution with all satellites for which correction are available. Our system transmits a header to tell the GPS card what the data is, and also a GPS week and seconds into the week and two correction components for each satellite, the correction to the satellite ranging data, and a rate of change for the satellite correction data. The rate of change data is present so that the mobile receiver can update the correction data in between correction input, since the card generates 60 times more navigation solutions than the number of corrections sent.

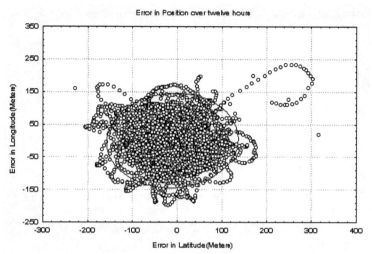

Figure 2: Uncorrected GPS fixes

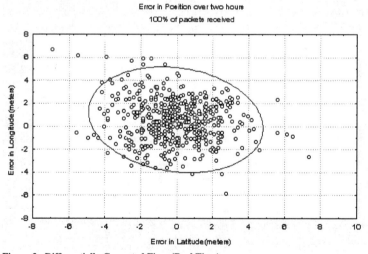

Figure 3: Differentially Corrected Fixes (Real Time)

The accuracy of the system is degraded by several things, such as distance between base station and mobile station, number of satellites in view shared by both stations, and any latency in the system. Both the distance between the base and mobile station, and the

267

system latency affect the corrections by delaying the application of the correction to the system. The degradation is caused by the increase in time in which the receiver must extrapolate actual correction data. Since accuracy is also degraded by then number of satellites in view, when the number of shared satellites in view is low, accuracy suffers.

PERFORMANCE GOALS

Once completed the system developed was expected to meet the following design criteria. Differentially corrected fixes should be accurate to at least 3 meters, while non-corrected fixes should be accurate to within 100 meters. The system should be able to fix and plot its position several times per second, but since the fixed positions will be stored on a floppy disk, the frequency of position fixes will be limited two fixes per second. This will allow the system to operate for eight hours at a time without interruption.

The mobile portions of this system are configured such that they stand up to the rugged outdoor environment in which they will be used.

The software is well-engineered. It is as simple to install and use as possible. It is maintainable and lends itself to further development.

Overall Package Description

The hardware and software package that will be delivered to the Forest Service will be made up of two sets of hardware and four software programs. Together, these items will make up three "stations" which will together will comprise an entire vehicle tracking system. These three stations are the Base Station, the Mobile Station, and the Supervisor Station.

The Mobile Station. The Mobile Station is, as its name implies, a moving GPS receiver. It is placed on the vehicle which is doing the spraying, and creates for the driver a display which tells him where he has sprayed and where he has been. The Mobile Station can be used as a normal GPS receiver, or as a Differential GPS receiver. In can be configured to display an area on the screen which guides the driver and shows him exactly where to spray, or it can be set to display a radius of operations, allowing the driver to drive anywhere within that radius. Whether or not the driver stays on the displayed map, the Mobile Station will always keep track of where the vehicle has been, how accurate each fix was, and the status of the vehicle's sprayer (on or off).

The Base Station. The Base Station (commonly called the reference receiver) is what makes the system a differential system. This station consists of a GPS receiver, a computer, and a radio. When it is placed in a known location, the Base Station will compare its location to the location it receivers from the GPS system, create a data packet which expresses the difference or error between these two positions, and transmit this error packet to the Mobile Station. When the Mobile Station receives this radio message, it can apply the errors encountered to the location it receives from the GPS system, and therefore radically increase the accuracy of its own position determination.

The question that remains: How do we determine exactly where the Base Station is? Two options exist. The first way to tell the Base Station where it lies is to put it at an accurately surveyed point. By placing the Base Station's *antenna* over a VABM (Very Accurate Bench Mark), for example, the antenna's position can be determined by finding the VABM on a topographical map and picking off its coordinates, being careful to be accurate down to one hundred-thousandth of a degree (0.036 seconds), better if possible.

Another option is by allowing the Base Station to figure out its position by itself. This is done by taking a *time averaged position* of the Base Station over a period of several hours. Since most of the error in GPS is cyclical, the Base Station can come up with a reasonably accurate idea of where it is by taking an average of all the positions it obtains over a long period of time. The system user will be allowed to choose how long to let the Base Station run, though the maximum (and default) run time will be twelve hours.

The Supervisor Station. The Supervisor Station, physically, is any modern desktop IBM-compatible PC, which is located hopefully within driving distance of the work site. It is at the Supervisor Station where the system is configured for each spraying job, and where the data collected from each job can be displayed, scrutinized, printed, and archived for future reference.

The first supervisor software program is the job setup program. This program produces a command file which is used by the Mobile Station to initialize itself. The job setup program will ask the user a series of questions about the job, such as the size of the vehicle sprayer, the type of map the user has, and the size and shape of the work area.

As briefly discussed earlier, the user has two ways to determine exactly what the work area is. The simplest way is by specifying a radius of operations to the job setup program. If, for instance, the user decides to work within a 100 meter radius, the Mobile Station will start the vehicle in the center of the display, and allow the user to wander at will anywhere within 100 meters of his starting point. A second way of determining the work area is by specifying a polygon called a perimeter. When the user inputs the latitude and longitude of each corner of the perimeter, the Mobile Station will display this perimeter and plot the vehicle when it is within the general area of that perimeter. If the user cannot supply the coordinates of the perimeter corners, he will be able to first drive the Mobile Station around the perimeter, pressing the "Mark" button as he reaches each corner. The job setup program will then be able to translate these marked positions into a displayable perimeter.

The second supervisor software program is the data reader program. This piece of software will take the floppy disk from the Mobile Station once a job is completed and recreate the display map seen by the driver. The user will be able to make a hard copy of the screen image and store the data for future reference.

CONCLUSIONS

This project provides the user a short range highly accurate system that meets the need of having a precise positioning data in various areas of operation. The need to move from location to location with minimum set-up effort is key to this systems' utility. Even with the P/Y code receivers currently authorized to the National Forest Service, this system will provide anywhere from 4 to 10 times greater accuracy. This prototype system can be reduced significantly in both cost and size. This reduction in cost and size along with a minimal learning curve will provide many end users with a system that provides need accuracy in a more timely and effective manner for many real-time needs.

Fixed-base large scale aerial photography applied to individual tree dimensions, forest plot volumes, riparian buffer strips, and marine mammals.

Grotefendt, R.A.[1], Wilson, B.[2], Peterson, N.P.[3],
Fairbanks, R.L.[4], Rugh, D.J.[5], Withrow, D.E.[5], Veress, S.A.[6] and Martin, D.J.[7].

[1]P.O. Box 1794, North Bend, WA 98045.
[2]U.S. Forest Service, Alaska Region, Regional Office—Timber Management, P.O. Box 21628, Juneau, AK 99802-1628.
[3]Simpson Timber Co., 700 South 1st St., Shelton, WA 98584.
[4]Foster Wheeler Environmental Corp., Suite 1300, 10900 NE 8th St., Bellevue, WA 98004-4405.
[5]National Marine Mammal Laboratory, Alaska Fisheries Science Center, NMFS, NOAA, DOC, 7600 Sand Point Way NE, Bin C15700, Seattle, WA 98115-0070.
[6]College of Engineering, Civil Engineering, University of Washington, Seattle, WA 98105.
[7]2103 North 62nd Street, Seattle, WA 98103.

Abstract

A fixed-base large scale (1:1,000 to 1:2,200) aerial photography system was developed to estimate tree, forest plot, riparian,buffer, and marine mammal dimensions and characteristics. Photo scale, independent of flying height and ground control, was obtained by mounting two 70mm metric cameras a fixed distance apart on an aircraft fuselage or boom. Development of a 40 ft boom, 22 ft longer than any other boom previously used in aerial photography, allowed a larger area of stereoscopic coverage with the necessary amount of overlap needed for accurate measurement of vertical and horizontal distances. Horizontal errors ranged from 0.20 to 1.76 percent and vertical errors ranged from 1.16 to 2.61 percent on a control field. Correlation of photo measurements of individual tree dimensions to tree volume in old growth forest predicted average plot volume within 10.9-13.4% of actual volume. This is approximately a four-fold improvement over high altitude aerial photography methods that used averaged photo dimensions to predict volumes rather than individual tree dimensions. The larger image sizes obtained by the fixed-base large scale method result in more accurate estimates of tree dimensions even under dense, old-growth canopies. The increased measurement accuracy justfies the use of fixed-base large scale aerial photography even though the cost of implementation was approximately 7% higher than high altitude methods.

Introduction

The accuracy improvement that large scale photography should provide was tested on measurements of natural resource characteristics in the following studies: (1) individual tree dimensions and forest plot volumes; (2) riparian buffers; and (3) marine mammals. Normal aerial photography is collected in scales around 1:12,000 to 1:15,840 for land management uses that range from habitat mapping to detailed forest measurements, as well as use of the photography for navigation. Accuracy of measurement tends to be limited because of the small scale and ground control. A photo with a scale of 1:15,840 means one inch on the photo equals one fourth of a mile or 1,320 ft on the ground. Photography taken at lower altitudes provide scales of 1:1,000 to 1:2,000 where one inch on the photo equals 83 to 167 feet on the ground. These lower altitude images magnify what would be seen on the high altitude photography by 8 to 16 times and yield much better accuracies in measurement and easier photo interpretation (Photo Plate Pair 1). The reference section includes examples of research spanning over 42 years that have utilized this type of large scale or close range photogrammetry (Avery, 1958, 1959; Biggs, Pearce, and Wescott, 1989; Biggs, 1990, 1991; Bonner, 1975; Bradatch, 1979; Browlie and Firth, 1989; Erlandson and Veress, 1975; Firth, Hensel, and Carson, 1989; Folke, 1992; Gagnon and Agnard, 1989; Lyons, 1967; Paine and McCadden, 1988; Rhody, 1977, 1981; Riser, 1991; Sayn-Wittgenstein and Aldred, 1972; Setzer and Mead, 1988; Veress and Tiwari, 1976; Withrow and Angliss, 1992, 1994; Young, 1954, 1955)

Instrumentation and Camera Support System

A dual camera system mounted on a boom and carried by a helicopter was used to take the photography in all studies except for marine mammals where a fixed-wing aircraft was used. The fixed-wing aircraft flew at speeds of approximately 90 knots, and the helicopter flew at approximately 3 knots to reduce image motion or blurring and to ensure precise flying over the area of interest. In the more common high altitude aerial photography which produces 9 by 9 inch negatives, the scale is determined by the relationship of flying height to camera focal length or the relationship of a known object's ground dimension to the object's photo image size. With low-altitude fixed-base photography, the scale is determined by the relationship of distance in space between two parallel cameras' axes ("fixed-base"), the cameras' focal lengths, and camera format. This eliminates the need for ground control and increases possible applications.

In the fixed-base system, the distance between the two cameras is constant and the relationship to each other is maintained. Experimentation and testing at the University of Washington Civil Engineering laboratory facility demonstrated that the cameras' optical axes could be maintained closely parallel (within 0.02 degrees). In the fixed-wing aircraft this was accomplished by mounting the cameras in the nose and tail of the fuselage.

Collection of photo imagery for precise measurements demand strict tolerances are followed to prevent introduction of untraceable errors. The cameras, film, platform supporting the cameras, and equipment used to measure the images all must be designed with specifications that allow repeatable measurements. The cameras utilized in this study were metric Rolleiflex SLX cameras with 70mm bulk backs that provided 60 to 70 frames per roll and had calibrated elements known to within 1 micron (0.001mm). Film used the majority of time in Alaska was 70mm Ektachrome ASA 200 and in Washington was 70mm Agfachrome ASA 200. Very little software is developed for comparators which measure the imagery from the fixed-base paired camera system. We were fortunate to find such software newly developed by Ward Carson of Carto Instruments for the AP190 comparator. Although this comparator 's precision of 7 to 15 microns was less than other comparators, it was more than adequate for these studies, where often the inability to see the correct ground level introduces considerably more error than the 7 to 15 micron instrument tolerance.

An advantage of using stereo-photography is that actual contours and individual elevation points, as well as vegetation and stream characteristics, can be digitized directly from the imagery. DXF files can be exported from the comparator software and plotted by a variety of mapping programs.

Instrumentation Accuracy Assessment

Prior to field implementation, control targets were placed near the airport control tower at Olympia, Washington. Low-altitude photographs were collected from 3 flying heights (300, 400, and 500 ft) and included a parking lot and control tower as well as the control targets. A total station or digital theodolite was used to survey the coordinates of the control targets and fence corners. A tape measure was used to directly measure the height of the control tower and check total station distances and a clinometer was used to measure the control tower height and some nearby vegetation. Horizontal errors ranged from 0.20 to 1.76 percent, and vertical errors ranged from 0.26 to 5.46 percent. These results indicated the prototype system was performing as designed (see Table 1).

Forest Plot Volume Study

Southeast Alaska contains vast areas of roadless forest land that make ground sampling a lengthy and difficult endeavor. Collection of sufficient samples to make statistically valid predictions of stand characteristics is costly. Improved methods were needed to obtain forest inventory information more easily and economically. Low-altitude photography enlarges the tree image size and reduces shadows thus improving accuracy of measurements as well as providing better species identification. A systematic photo sample of forest plots can be obtained in one day from aircraft and then measured in the stable environment of an office.

The Forest Service needed to develop a more accurate methodology of volume determination to comply with the proportionality requirement of the Tongass Timber Reform Act (TTRA) (Public Law 101-626, November 28, 1990). The primary objective of this study was to evaluate using low-altitude fixed-base aerial photography as an alternate method to improve the accuracy and precision of proportionality analysis for timber sale projects at a reasonable cost. A secondary objective was to consider whether this alternative method could be used to improve the quality of other resource analyses as well.

The Project Area is located about 70 miles northwest of Ketchikan, Alaska, in the central portion of Prince of Wales Island immediately west of the community of Thorne Bay. It includes the watersheds of Control Creek, Cutthroat Creek, Rio Roberts Creek, Rio Beaver Creek, Rush Creek, and Goose Creek. The Project Area covers an area of 53,370 acres, which is mostly forested, and ranges in elevation from 50 to 2,809 feet above sea level just west of Cutthroat Lake.

Individual Tree Dimensions

Tree height and crown area were measured for all visible trees and plot crown closure,

Table 1. Control field test measurements at Olympia Airport.

Object separation distances or heights	Measuring tool	Measurement (ft)	Photo Estimate (ft)	Percent Error (%)
Fence post to fence post	Tape measure	100.00	98.95	1.05
Fence post to fence post	Tape measure	100.00	99.80	0.20
Control field target	Total station	183.46	181.57	1.03
Control field target	Total station	183.46	180.23	1.76
Control field target	Total station	50.30	49.98	0.64
Control tower height	Clinometer	62.10	61.94	0.26
Control tower height	Tape measure	61.23	61.94	1.16
Tree height	Clinometer	47.60	45.00	5.46

tree crown closure, and branch density percent were estimated on the fixed plots. Ground heights compared closely with photo heights. The average height error from actual was 5.6 ft (SD=4.1 ft, n=109). To check the camera system, horizontal ground distances of over 100 ft were measured and yielded accuracies within 0.2 to 0.4% of actual. These accuracies of within 0.2 to 0.4 % corresponded to pre-field testing of the system at the airport in Olympia, Washington. Diameter at breast height (dbh) is highly correlated to stem volume. A regression equation was developed that used crown area, plot crown closure, and height from the photos to predict dbh (R^2=.746, S.E. = 3.34 in).

Several difficulties encumbered collection of individual tree data from the low altitude photos. First, incorrect locations caused plot centers to be missed on 4 plots of 55 photographed, thus reducing the sample number to 51. Second, the end of the roll of film was sometimes reached prematurely because the film magazine had advanced extra frames unknowingly to the operator. High humidity effects on the sensitive electronics were suspected as the cause. Third, some trees were hidden on the photos. This occurred if the crown of a suppressed tree was growing under a dominant tree. The effect of this on the total plot volume turned out to be minimal because those hidden trees had a very low percent of the plot volume. Fourth, finding the plot center was sometimes difficult if the target was partially or entirely obscured because of the angle from which the photograph was taken or if bears had destroyed or modified the new target replacements. This never prevented eventual correct placement because video footage was also taken at the same time as the still photography, aiding location in these situations. Fifth, tree height was determined from the elevation difference of locating the top of the tree and the same ground level as the tree's base. If the tree was on a slope and the ground level of the tree could not be seen, this process was done uphill and downhill of the tree, and the average height was used. Sixth, if the crown was over-exposed less detail was visible, and the crown top was hard to distinguish. Western redcedar that was chloritic tended to be the most difficult. Still, the range of height errors was acceptable.

Overall, the difficulties in measuring or observing tree characteristics did not have a significant effect on prediction of plot volume. Also using a refined ground sampling method will eliminate some of the above problems.

One thing not known before the study commenced was the clarity with which crown defect could be seen from the low-altitude photography. Broken and gnarly tops that were rotten and twisted could be clearly seen. It is possible that a defect index could be determined for each tree that would be correlated to actual percent bole defect.

Forest Plot Volumes

In the original study design, ground truth plots were laid out to test different photographic sampling techniques for predicting volume. In 1994, 40 ground targets had been placed in visible locations close to the center subplot of clusters of 5 subplots. The average of the 5 subplots was used to determine each plot's volume (see Fairbanks et al. 1995). These were to be re-visited to verify that the targets were still intact and to replace them if needed prior to the low-altitude photography. White flagging that was one foot in width was spread on the ground in the form of a cross with the long axis pointing to the north. White or yellow plastic plates (9 inches in diameter) were also nailed down to wood or the ground.

Of 21 ground variable plots re-visited to check target integrity, 17 required complete or partial replacement due to animal damage or litter accumulation. Even after target replacement, animals frequently came back and again damaged the flagging prior to the photo flight. Most damage appeared to be from bears.

Full ground measurements were completed on plots at 10 points during this study (Table 2); in addition several spruce tree were added. Ground volumes ranged from 620 to 26,900 board feet per acre (GBF32) on the ten 0.1-acre fixed plots. Volumes at these same points ranged from 1,509 to 46,409 board feet per acre (GBF32) based on the average of 5 variable plots. Ground volumes and individual tree measurements at these 10 points are summarized in Table 2

Table 2 shows some relatively large differences between the different cruising methods. Because the variable plots use a prism to determine whether to tally a tree or not, differences occurred in the number of trees chosen for measurement between variable and fixed plots. Also differences exist in Superstand determined volumes between trees of similar size. This introduces additional variability when predicting individual tree volume. Half of the large difference in plot 349 volume was because a large cedar was "in" the variable plot but just barely "out" of the fixed plot.

In this study, 0.1 acre plots were visited on the ground for tree measurement and stem mapping in September, 1995. A 37.2 ft radius circular fixed plot was sampled for all trees above 9 inches in diameter at breast height (dbh).. Height and dbh were measured for each tree, and species was recorded. No defect was determined because a regional average is applied to the gross volume based on fell and buck study data.

High altitude photography provides less detail of individual trees so the plot volume was predicted from the average height of the 5 tallest trees occurring in a 1-acre circular fixed plot and other stand and tree characteristics (crown closure and crown width). In the original work 40 plots were sampled. (see Fairbanks et al. 1995)

Tree height, crown area, plot crown closure, tree crown closure, and percent branch density were measured or estimated from the low altitude photos of 10 plots and additional spruce were added outside of plots to increase the sample size. A stem map which locates all the trees on the fixed plot was drawn for each sample. This enabled the photo interpreter to easily identify each tree when photo measurements were made. The volume of each tree was predicted based on these measurements. Then the sum of each tree's predicted volume on the fixed area plot provided the plot volume.

Plot 395 is seen in the photographs of Photo Plate Pair 1. This plot contained 15,380 gross bd ft volume (32) and was a medium volume plot. The approximate flying height and scale were 498ft and 1:1,897, respectively. Species present are western hemlock, western redcedar, and spruce. At the confluence of the streams are targets and yellow plates. The range of heights and dbh's were 49-110 ft and 11-32 in, respectively.

Equation development

We developed our own equations to relate individual tree photo measurements to individual tree ground volumes occurring on the 0.1-acre fixed plots. Ground measured, individual tree volumes were determined from equations in Forest Service Superstand computer software. To determine which photo measurements explained the greatest amount of the variation in ground-measured volumes, we used stepwise multiple regression. Gross board feet volumes (32 foot scale) was used as the dependent variable. Tree height, crown area, plot crown closure, tree crown closure,

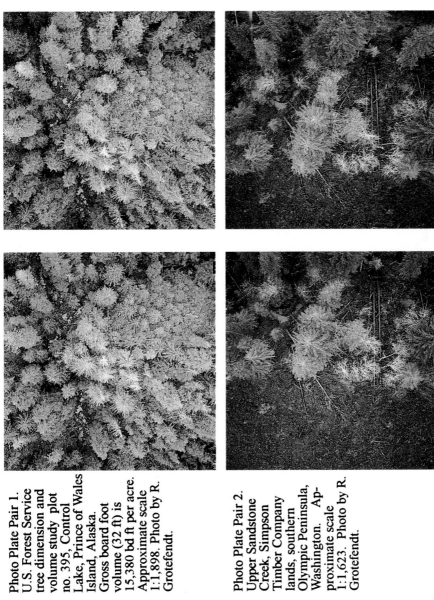

Photo Plate Pair 1. U.S. Forest Service tree dimension and volume study plot no. 395, Control Lake, Prince of Wales Island, Alaska. Gross board foot volume (32 ft) is 15,380 bd ft per acre. Approximate scale 1:1,898. Photo by R. Grotefendt.

Photo Plate Pair 2. Upper Sandstone Creek, Simpson Timber Company lands, southern Olympic Peninsula, Washington. Approximate scale 1:1,623. Photo by R. Grotefendt.

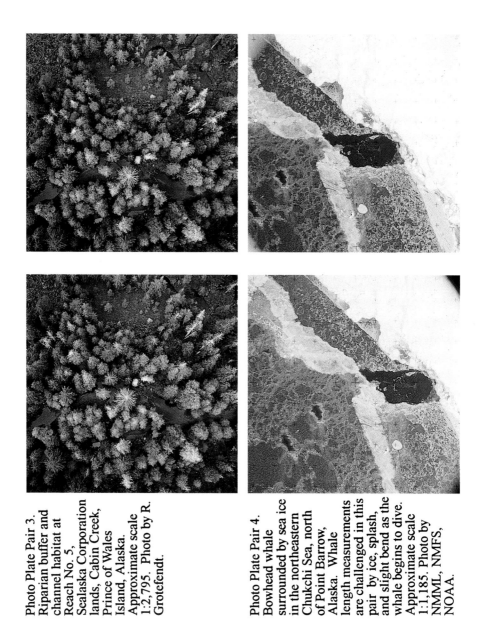

Photo Plate Pair 3. Riparian buffer and channel habitat at Reach No. 5, Sealaska Corporation lands, Cabin Creek, Prince of Wales Island, Alaska. Approximate scale 1:2,795. Photo by R. Grotefendt.

Photo Plate Pair 4. Bowhead whale surrounded by sea ice in the northeastern Chukchi Sea, north of Point Barrow, Alaska. Whale length measurements are challenged in this pair by ice, splash, and slight bend as the whale begins to dive. Approximate scale 1:1,185. Photo by NMML, NMFS, NOAA.

Table 2 Ground and predicted volumes from low and high altitude photography and percent errors.

Plot #	Low Altitude Ground volume	Equation Ht only Photo volume	Equation Ht only Bdft error	Equation Ht only Percent error	Equation Crown Area and Ht Photo volume	Equation Crown Area and Ht Bdft error	Equation Crown Area and Ht Percent error	High Altitude Ground Volume	High Altitude Equation Photo volume	High Altitude Equation Bdft error	High Altitude Equation Percent error
54	26,900	19,956	-6,944	-25.8	19,141	-7,760	-28.8	27,396	26,295	1,101	-4.0
65	21,450	22,258	808	3.8	21,344	-106	-0.5	22,893	21,336	1,557	-6.8
101	620	678	58	9.3	713	93	15.1	4,299	-2,500	6,799	-158.2
274	11,520	12,668	1,148	10.0	11,685	165	1.4	11,861	29,339	17,478	147.4
311	2,940	3,717	777	26.4	4,033	1,093	37.2	1,509	7,267	5,758	381.6
349	20,670	21,005	335	1.6	21,418	748	3.6	46,409	28,569	17,840	-38.4
361	1,220	1,188	-32	-2.6	1,232	12	1.0	1,955	3,066	1,111	56.8
379	4,700	4,752	52	1.1	5,009	309	6.6	7,196	12,818	5,622	78.1
395	15,380	16,866	1,486	9.7	18,014	2,634	17.1	17,929	21,689	3,760	21.0
418	2,800	2,974	174	6.2	4,341	1,541	55.0	7,829	9,894	2,065	26.4
average	10,820		1,181			1446		14,928		6,309	
% error			10.9	9.7		13.4	16.6			42.3	

277

and percent branch density were measured or estimated from the photos and used as independent variables with various transformations. The most important variable identified in the all-species-combined equation was total tree height. Height always is the highest correlated variable, because it is the best to field measure, not the best variable. The problem with relying on just height is that we know tree volume is also strongly correlated with dbh and in aerial photography dbh is correlated with crown area. Therefore a more robust equation would incorporate additional variables besides just tree height (personal communication, Dr. K. Rustagi) A second equation was developed for individual species that incorporated the product and transformation of crown area and tree height.

Table 2 contains a comparison of the accuracy of volumes for 0.1-acre plots predicted from the two equations and volumes predicted for 1.0-acre plots predicted from a stepwise regression equation using high altitude aerial photography (see Fairbanks et al. 1995).

Although the equation with height only produced the best estimate of plot volume, we feel the species equations incorporating crown area should eventually be used (see Table 2). Low-altitude fixed base aerial photos were collected which have no fixed area ground plots. For a small amount of effort these ground plots could be measured and a wider range of stand densities and tree sizes would be added to the data base and it is likely the equations with crown area would improve and define the individual tree volume better and more reliably than the equation just relying on height. Also note one reason these equations produce much better results than the high altitude results is the tree parameters are measured much more accurately.

The all species equation gave a slightly better prediction of plot volumes than the individual species equations with an absolute average error difference from actual of 10.9% . The high altitude equation had an absolute percent error difference from actual of 42.3%. If the low and high altitude estimated average absolute bd ft difference from actual are compared, they are 1,181 versus 6,309, respectively (see Table 2). This indicates that when detailed measurements from the low-altitude photography are used to predict volume, improvement in accuracy is almost four-fold. The ability to identify individual species was greatly enhanced by the use of low-altitude aerial photography.

Riparian Buffer Study

Simpson Timber Company and Sealaska Corporation have monitoring programs that normally quantify stream and watershed characteristics from ground based surveys. Because the riparian zones are complex and difficult to quantify the ability to obtain a large number of samples is time consuming, expensive, and limited by environmental conditions. Low-altitude aerial photography that can measure various habitat characteristics and facilitate this type of monitoring was investigated in pilot studies.

Photographic limitations exist when vegetation obscures ground characteristics, but field work can fill in missing information. The ability to have an infinite number of points from which measurements can be made as well as the ability to tie those points to specific geographic locations weighed strongly in favor of photogrammetry as a possible method to improve characterization of streams and riparian buffers.

In 1995, imagery was collected on approximately 3 miles of Cabin Creek, Prince of Wales Island, Alaska and a total of 2 miles on Sandstone Creek, Canyon River, and Wildcat Creek, southern Olympic Peninsula, Washington (see Photo Plate Pairs 2 and

3). Each photo pair overlapped the previous pair by approximately 50%. The Cabin Creek imagery began at salt water. Flying heights were in the range of 420 to 750 ft. This imagery was expected to provide the following:

- Multiple measurements of numerous riparian attributes.
- Documentation for a historical review of stream changes.
- Detailed photographs that show effects of land management practices.
- The ability to repeat measurements now, as well as in the future.

The following characteristics were investigated: distance of root ball to stream edge; fall distance of tree from channel; down tree perpendicular distance from channel edge; location of blowdown in stream along 400m reach; diameter of downed trees; heights of standing trees; crown closure percent; channel width; fall direction of tree; did fall tree hit channel; is down tree position in or over channel; side of creek down tree is on; down tree count; stump counts; live tree counts; tree species; predicted dbh using height, crown area, and plot crown closure as independent variables; log debris position maps; all visible downed logs map; stem map.

Photo measurements

Measurements or estimates were sought from the low altitude imagery that could be easily distinguished in all riparian conditions. The paired images taken by the two cameras were used to make the measurements or estimates of tree height, crown area, down tree dimensions, and digitization of stream borders and other features. Round plastic plates that were 9 inches in diameter were placed approximately every one hundred feet on both sides of the stream being photographed in Washington.

Normal aerial photography is best applied when 4 or more control points are widely distributed in the stereo pair formed by two overlapping photos. Although not needed for photo measurement, these were deployed to provide an extra means for checking the performance of the system in the future when time allowed.

In Washington, horizontal and vertical distances were measured on the photography between control points surveyed with a Topcon total station. Down and standing trees were photo and ground measured. Ground measurements were made with a clinometer and tape measure for heights and the down trees were measured directly with a tape. Diameters of down trees were measured with calipers. Several photo test measurements were made for comparison (see Table 3). More photo test measurements can be made in the future from the existing imagery and ground data.

In the Washington field tests at Sandstone Creek, the most reliable field vertical elevation was a comparison between two control points surveyed with the total station and the error was 2.61%. Small elevation differences may be more difficult to quantify with comparators of lower accuracy. Tests on standing Doug fir timber had errors of 1.72 to 2.72%. The height of a spruce and alder had errors of 7.3 to 12.86%. The alder top was especially difficult to delineate because of the many fine branches. The ability to judge correctly the ground level by the tree and discern the top, as well as ground instrument errors or tree tilt, make the comparison between ground and photo estimates less concrete. The error in diameter measurement of a downed tree was 5.7% or within 3.6 cm of a 63.2 cm diameter log (1.4 inch of a 24.9 inch log). Down tree total length from butt to tip had errors ranging from 3.28 to 3.78%.

The fixed base system produced very accurate results with errors of only 0.10 to 1.76% when comparisons were made between photo measurements and ground measurements

Table 3. Objects field and photo measured at Sandstone Creek, Washington test area.

Object	Measuring tool	Measurement	Photo Estimate	Percent Error
Control point A to B	Total station	156.76 ft	157.23 ft	0.30
Control point B to D	Total station	233.27 ft	232.84 ft	0.18
Control point B to D	Total station	233.27 ft	233.04 ft	0.10
Control point D to E	Total station	214.98 ft	213.20 ft	0.83
Control point D to E	Total station	214.98 ft	215.56 ft	0.27
Control point E to B	Total station	259.49 ft	258.35 ft	0.44
Elevation A to D	Total station	30.23 ft	31.02 ft	2.61
Diameter downed tree	Calipers	63.20 cm	59.60 cm	5.70
Spruce tree height	Clinometer	102.60 ft	110.10 ft	7.30
Doug fir tree height	Clinometer	155.40 ft	152.70 ft	1.70
Doug fir tree height	Clinometer	172.80 ft	168.10 ft	2.72
Alder tree height	Clinometer	103.40 ft	90.10 ft	12.86
Down Doug fir length	Tape	158.40 ft	153.20 ft	3.28
Down Doug fir length	Tape	158.80 ft	164.80 ft	3.78

of round targets on the ground. Placement of the comparator's floating mark on round targets is much more accurate than on natural objects. This indicates the fixed-base system was producing accurate results

Utilization of maps digitized from photo pairs on Cabin Creek and Sandstone Creek allowed estimation of some stream characteristics. Table 4 contains these measurements or counts that were obtained directly from the AP190 comparator or

Table 4 Stream characteristics obtained directly from the AP190 comparator or from manually measuring and counting map features that occurred within 66 ft of full bank edge.

Stream characteristic within 66 ft of full bank edge	Cabin Creek Reach #5	Sandstone Creek, Washington
Area full bank stream (ha, acres)	0.25, 0.62	0.04, 0.09
Full bank length (m)	165	104
Minimum channel width (m)	10.4	0.6
Maximum channel width (m)	32.3	7.3
Stem count (all species)	153	46
Snag count	6	3
Stump count	0	0
Distances of visible root wads to stream edge (m)	none digitized	1.5, 8.5, 0.0, 9.8, 9.8, 4.3, 4.0, 9.1, 9.1, 9.1, 9.1, 4.6, 4.9
Blowdown count on land only	7	8
Blowdown count in stream	28	16
Total blowdown count	35	24
Stem count in stream per 100 m	17	15
Area covered by stereo pair (ha, acres)	1.96, 4.85	0.66, 1.64

MAP 1

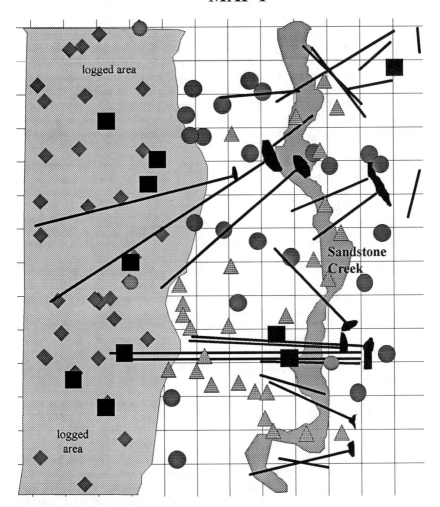

Map 1 was digitized from Photo Plate Pair 2 taken at upper Sandstone Creek, southern Olympic Peninsula, Washington. Circles are conifers, triangles are alder, squares are snags, diamonds are stumps, heavy black lines are blowdown, irregular black polygons at the ends of blowdown are root wads that were visible, control targets are smaller circles, and the creek was digitized at full bank. The background grid is in 20 ft intervals. Tables 3 and 4 contain several items displayed on this map.

from manually measuring or counting features from maps digitized from photos (see Map 1). Presently, no field measurements to calibrate photography have been made at Cabin Creek.

Some problems were encountered in collecting the stream characteristics. The vegetation density and orientation, as well as the stream size affect visibility. A map made from the photography should be carried in the field to fill in missing information. If the stream is not centered properly in the photo pair, part of the riparian buffer may not be in stereo and photo measurements can't be made. Stem locations may be in error by approximately 15 ft horizontal (based on U.S. Forest Service crown width data) due to foliage hiding the stem. The best methodology to determine stem locations is still being evaluated. Map 1 stem centers were placed by using the top of the tree. This method was tested because it would be more repeatable by different operators and on subsequent photography. Delineation of the 66 ft buffer on the map was done by hand with calipers, however a computer method could be developed that would digitize this automatically.

Preliminary testing with the U.S. Forest Service study data produced an equation that predicted diameter at breast height with a R^2 of 0.75 and SE of 3.34 in. This equation used the dependent variables of $\log_{10}$ of crown area, the square of plot crown closure, and total tree height.

The results that compare photo versus field measurements indicate the accuracy obtained from the low-altitude fixed base aerial photography system enable an accurate evaluation of many stream characteristics. Measurements taken from the digitized maps indicate that many stream and riparian characteristics can be quantified and evaluted. This system is expected to supplement ground based stream surveys and provides information not reasonably available from standard ground survey techniques. Overhanging vegetation can obscure some dimensions that must be measured on the ground. A map can now be generated of all visible objects and utilization of this when conducting field measurements will improve efficiency and additionally allow photo obscured dimensions to be field collected. The accuracy of the instrument or comparator used to make the photo measurements directly affects the errors, but results were still comparable to field values.

Preliminary cost estimates indicate that for a 200 by 200 ft area covered by a stereo pair, the total cost per mile to collect and measure photography and produce summary tables, maps, and reports may range from $4,900 to 6,600. (This includes $2,500-3,000 per mile to collect the photography and $2,400 to 3,600 for photo measurement, comparator costs, analysis and map production on various combinations of stream characteristics). Increased application and development may reduce these cost ranges. Smaller stream channels will require that fewer of the stream characteristics be collected and thus reduce costs. Although the application of this photogrammetric principle requires more rigor in technique, the additional new information obtainable, as well as the greater quantity of standard information, makes this system appear to be a useful way to monitor and evaluate riparian buffers and is cost effective when compared to field costs.

Marine Mammal Study

Age structure of marine mammals, as with terrestrial mammals, is useful in determining reproduction rates that are needed for population management. In 1985, no anatomical part of baleen whales were known to provide animal age. In 1983 and 1984 studies were conducted using single and dual camera systems to estimate sizes of bowhead whales (Davis et al. 1983; Cubbage et al. 1984). In 1985, a fixed-base system was implemented to improve the accuracy of size measurements (see Photo Plate Pair 4).

Total length, fluke width, three girth measurements and the length from the tip of the rostrum to the center of the blowhole were taken for each whale that was adequately photographed (lacking excessive motion blur, ice or splash over end points, or extreme bending of the whale when diving..

Software to handle the fixed-base system was non-existent during the 1985 and 1986 studies. Therefore the cost of measuring the stereo pairs became prohibitive and required utilizing the simple relationship of camera focal length to altitude to determine scale. Additionally, the ability to synchronize camera shutters was more difficult then. Currently, camera shutter synchronization has improved considerably.

Even though cumulative type errors could increase by not using the fixed-base pair, accuracies were still considered sufficient to identify three major size classses: calf (<6m), immature (6-13m), and adult (>13m). The length structure of the bowhead whale population migrating past Barrow, Alaska, during the springs of 1985, 1986, and 1989-92 (Withrow and Angliss 1992; Withrow and Anglis 1994) were established using aerial photogrammetric techniques. In addition to the standard 70mm camera system, a large format aerial reconnaissance camera was tested in 1991 and used extensively in 1992. When photogrammetric data from both years were incorporated with data from 1985, 1989, and 1990, the average percent frequencies of bowhead whale calves, immatures, and sexually mature adults are 4.1, 57.7, and 38.2% respectively. A significantly higher proportion of immature whales (64.1% and 65.5%) and a corresponding lower proportion of adult whales (31.8% and 33.7%) were observed in 1991 and 1992 than in previous years. This may be the result of variable recruitment, temporal differences in sampling, and/or ice conditions. (Withrow, D.E. and Angliss, R.P., 1994)

Acknowledgments

This work could not have been accomplished without the skill and advice of pilots, machinists, and professionals. These include: Michael Bucove of Temsco Helicopter Inc.; Joe DeMarco of Fly Wright Helicopters; Dan Clark of Northwest Helicopters; Ken Knowlan, Dick Terry, and Dr. K. Rustagi of the University of Washington;. John Fuller and John Johnson of Timberline Forest Inventory Consultants Ltd.; Xiaoping Yuan, Joe Nemeth, Mr. Bradatch, and John Wakelin, Province of British Columbia, Ministry of Forests; Ward Carson, Oregon State University; Stephen E. Reutebuch and Bob McGaughey of USDA, FS, PNW Research Station; Paul Walker; and Dennis Riley of Riley Mapping. Funding was provided by Simpson Timber Company, United States Forest Service, Sealaska Corporation, National Marine Mammal Laboratory, and Richard A. Grotefendt.

References

Avery, G. 1958, Helicopter stereo-photography of forest plots, Photogrammetric Engineering, Vol. 24, No. 4, pp 617-624.

Avery, G., 1959, Photographing forests from helicopters, Journal of Forestry, Vol. 57, No. 5, pp. 339-342.

Biggs, P.H., 1990, New approaches to extensive forest inventory in Western Australia using large-scale aerial photography, Australian Forestry. Vol. 53, No. 3, pp. 182-193.

Biggs, P.H. 1991, Aerial tree volume functions for eucalyptus in Western Australia, Canadian Journal of Forest Research. Vol. 21, No. 12, pp. 1823-1828.

Biggs, P.H., 1989, Pearce, C.J., and Westcott, T.J., GPS Navigation for large-scale photography, Photogrammetric Engineering and Remote Sensing, Vol. 55, No. 12, pp. 1737-1741.

Bonner, G.M. 1975, Cluster sampling with large-scale aerial photography in forest inventories, Information Report, Forest Management Institute, Canada. No. PMR-X-80.

Bradatch, H.1979, Application of large-scale fixed-base aerial photography with helicopters to forest inventory in B.C., .in Kirby, C.L.: Hall, R.J. (Compilers): Practical applications of remote sensing to timber inventory,.Proceedings of a workshop held September 26-28, in Edmonton, Alberta.

Brownlie, R.K. and Firth, J. G., 1989, A photogrammetric method for the volumetric assessment of wood chip stockpiles, Harvest Planning Research, Forest Measurement and Resources Division, Forest Research Institute, Ministry of Forestry, Rotorua, NZ, July 25.

Cubbage, J.C., Calambokidis, J., and Rugh, D. J. 1984. Boywhead whale length measured through stereo photogrammetry. Report to National Marine Fisheries Service, 7600 Sand Point. Way NE, Seattle, WA. 67 p.

Davis, R. A., Koski, W. R., Miller, G. W. 1983. Preliminary assessment of the length-frequency distribution and gross annual reproductive rate of the Western Arctic bowhead whale as determined by low-level aerial photography, with comments on life history. Report for National Marine Fisheries Service, Bldg 4, 7600 Sand Point Way NE, Seattle, WA .91p.

Erlandson, J.P. and Veress, S.A., 1975, Monitoring deformations of structures, Photogrammetric Engineering and Remote Sensing, November.

Fairbanks, R.L., Boyce, J.A., Grotefendt, R.A., 1995, Evaluation of Photo-point Inventory Methods For the Estimation of Timber Volume and Proportionality in Southeast Alaska, Final Report to U.S.Forest Service.

Firth, J.G., Hensel, R., and Carson, W., 1989, Better planning with aerial photography - What's New in Forest Research, Forest Research Institute, Ministry of Forestry, Rotorua, New Zealand, No. 172

Folke, E. 1992. Large scale photogrammetry with the transverse heliborne twin camera boom, Master's Thesis in Civil Engineering, University of Washington, Seattle.

Gagnon, P.A. and Agnard, J.P., 1989,Twin-camera fixed base photography of tree plots: possibilities and Accuracy of a system, Canadian Journal of Forest Resources, Vol. 19, pp. 860-864.

Grotefendt, R. A. and Fairbanks, R. L., 1996, Estimation of Timber Volume in Southeast Alaska using Low-Altitude Fixed Base Aerial Photography, Final Report to U.S. Forest Service.

Grotefendt, R. A., 1996, A Pilot Study Utilizing Low-Altitude Fixed Base Aerial Photography for Monitoring Riparian and Channel Habitat Conditions, Report to Sealaska Corporation.

Konecny, G., 1964, Application of the Wild C12 stereometric camera to structural engineering, American Society of Photogrammetry Proceedings, March.

Lyons, E.H. 1967, Forest Sampling with 70-mm Fixed Air-base Photography from Helicopters, Photogrammetria,, Vol. 22, pp. 213-231.

Paine, D.P. and McCadden, R.J. 1988. Simplified forest inventory using large-scale 70mm aerial photography and tarif tables, Photogrammetric Engineering and Remote Sensing 54(10), pp. 1423-1427.

Rhody, B. 1977, A new, versatile stereo-camera system for large-scale helicopter photography of forest resources in central Europe, Photogrammetria, 32, pp. 183-197.

Rhody, B. 1981, A combined inventory design with aerial strip-sampling, large scale photography and ground plots for assessment of natural resources in the Sahel/Upper Volta. Allgemeine Forst und Jagdzeitung, 152, No.10, pp. 195-200.

Riser, J. 1991. Photogrammetry uses of a new generation analytical stereoplotter in forestry, M.S. Thesis. Oregon State Univ., Corvallis.

Sayn-Wittgenstein, L. and Aldred, A.H., 1972, Tree size from large-scale photos, Photogrammetric Engineering, Vol. 30, No. 10, pp.971-973.

Setzer, Theodore S.; Mead, Bert R. 1988. Verification of aerial photo stand volume tables for southeast Alaska, Res. Pap. PNW-RP-396. Portland,OR: U.S. Department of Agriculture, Forest Service, Pacific Northwest Research Station. 13 p.

Withrow, D. and Angliss, R., 1992. Length frequency of bowhead whales from spring aerial photogrammetric surveys in 1985, 1986, 1989, and 1990. Report to International Whaling Commission. 43: 463-467.

Withrow, D.E. and Angliss, R.P., 1994. Length frequency of the bowhead whale population from 1991 and 1992 spring aerial photogrammetric surveys. Report to International Whaling Commission 44: 343-346.

Veress, S.A. and Tiwari, R. S., 1976, "Fixed-frame multiple-camera systems for close-range photogrammetry, Photogrammetric Engineering and Remote Sensing; September.

Young, H. E., 1954, 1955, Photogrammetric determination of huge pulpwood piles, Photogrammetric Engineering.

USE OF HISTORICAL AERIAL PHOTOS TO EVALUATE STREAM CHANNEL MIGRATION

Jeff Barry
University of Washington, College of Forest Resources
Seattle, Washington 98195

Stephen E. Reutebuch
USDA Forest Service
Pacific Northwest Research Station
Seattle, Washington 98195

Thomas Robison
USDA Forest Service
Wenatchee, Washington 98807

ABSTRACT

Channel migration is a key link in the creation and maintenance of the riparian forest. It is important to understand that channel migration is a continuous physical process which very often leaves highly visible remnants on the landscape. Migration of these remnants was measured from a series of historical aerial photographs along five river systems on the Wenathcee National Forest.

From each photo set, measurements of the channel location were taken using an analytical stereoplotter which allowed registration of each photo set to a common ground coordinate system. Channel form and location from each time period were input into a GIS so that changes in channel parameters could be easily displayed and analyzed. These parameters, both local and upstream channel curvature, and related rates of upstream channel migration, were then used in a channel migration model to predict future channel position and form. This paper focuses on the photo measurement techniques that were used and an evaluation of the channel migration model that was applied on one reach in one time period.

INTRODUCTION

Riparian zones are the interfaces between terrestrial and aquatic ecosystems (Gregory et al., 1991). This terrestrial-aquatic ecotone plays a key role in the regulation and maintenance of biodiversity in the landscape as connection lines for, and mediator of, fluxes of material between the terrestrial and aquatic ecosystems (Piney et al., 1990; Gregory et al., 1991; Dynesius and Nilsson, 1994). Riparian zones contain species belonging to each distinct environment, as well as others suited to this unique setting (Piney et al., 1990). A main feature of the ecotone, however, is its spatial and temporal instability which can create management problems (Piney et al., 1990; Gregory et al., 1991). Understanding this instability, and the disturbance regimes that drive it, are necessary if we are to effectively manage watersheds for the future health of the riparian zone.

A number of authors have shown that the structure, composition, and spatial distribution of riparian vegetation within drainage networks is driven by both hillslope and fluvial processes (Harris, 1987; Fetherston et al., in press 1995; Oliver et al., 1985). In particular, the creation and maintenance of the riparian vegetation within lowland fluvial landscapes is intricately related to channel migration (Leopold et al., 1964; Everitt, 1968; Hupp and Osterkamp, 1984; Gregory et al., 1991; Fetherston et al., in press 1995).

As a laterally migrating channel sweeps across the valley floor, it cuts into older riparian plant communities along the outer edge of a meander and creates depositional surfaces for development of younger stands along the inner margin of a meander. As a result, the vegetation pattern and habitat characteristics of the riparian zones are continually changing. To understand and manage the distribution of vegetation and habitats along a migrating river, it is necessary to understand the rates at which the channel is likely to migrate at points of interest. Until recently, however, river meandering was a poorly understood physical process (LaPointe and Carson, 1986).

A study was undertaken to apply a Bend Migration Model (Furbish, 1991) to six stream reaches over a 10 to 50 year time period (depending upon aerial photograph coverage) in the Wenatchee National Forest, covering a range of landtype subsections and disturbance patterns (Barry, 1996). This paper only describes the application of a Bend Migration Model to one of the study reaches, the White River, over a single time period, 1962 - 1979. Details on the application of a Bend Migration Model to other time periods and reaches and an evaluation of the model's accuracy and applicability are found in Barry (1996).

To study both the spatial and temporal variability in the meandering process, the Bend Migration Model was applied to each reach over the available time periods for that reach. The questions of interest are: how accurately does the Bend Migration Model predict meander train evolution for a given channel; and, are there similarities in migration models within and between landtype subsections? Although not demonstrated here, in instances where this migration model can be accurately applied, the ability to predict future migration rates and patterns will assist in the management of the riparian zone and the channel itself.

This paper is separated into three sections. First is a discussion on the aerial photograph analysis and map development process. The second section presents the analysis of the Bend Migration Model applied to the White River reach. The last section, the conclusion, presents a discussion on the overall accuracy and applicability of the Bend Migration Models developed for this research.

AERIAL PHOTOGRAPH ANALYSIS AND MAP DEVELOPMENT

In order to apply the Bend Migration Model it is first necessary to know the initial spatial conditions of the reach. Using this initial spatial information, the migration model can be used to predict channel displacement. This predicted displacement can than be compared to historic channel migration measured from aerial photography that corresponds to the simulated time period. Both initial and final channel positions were digitized from historical aerial photos taken 17 years apart using an analytical stereoplotter (Carto Instruments AP190[1]). Each photo set was oriented in the stereoplotter to a common ground coordinate system so that relative channel changes could be calculated. For this research, the ground coordinate system used was the North American Datum 1927. The process of controlling multiple sets of photography over the same area from different time periods is explained in more detail in Reutebuch and Gall (1990).

Images were viewed under 8X magnification and channel coordinates were digitized using the stereoplotter software. The primary advantage of using the stereoplotter is that the spatial data (i.e. bankfull channel location) from different years and different scales of photography can be properly registered to a common coordinate system and easily entered into a GIS or CAD package for subsequent analysis and map production (Ballerini and Reutebuch, 1994). Use of the stereoplotter allows more accurate mapping from resource scale (1:12,000 - 1:24,000) photography. Table 1 presents the planimetric root mean square error (RMSE) for each stereopair used in the White River channel mapping (only mapping from the 1962 and 1979 photos is discussed in this paper). The planimetric RMSE presented below is an indicator of the registration

---

[1] Use of trade or firm names in this publication is for reader information and does not imply endorsement by the U. S. Department of Agriculture of any product or service.

error associated with the location of each channel relative to one another on the maps produced, and from which migration data were measured. Therefore, changes in channel centerline less than the planimetric RMSE are not significant.

Table 1
Planimetric RMSE of each stereopair for the White River reach.

Year	Planimetric RMSE (m)
1949	+/- 1.88
1962	+/- 1.49
1979	+/- 1.75
1992	+/- 1.29

Once the ground coordinates have been imported into a GIS, it is a simple matter to create maps from which the necessary channel measurements and calculations can be taken (described in the following section). From these maps, the channel centerline coordinates were calculated and input into a spreadsheet. The channel centerline coordinates were determined by defining the farthest downstream cross section as some arbitrary point (e.g. 100 m, 100 m). By measuring the distance and angle to the channel centerline at the next cross section, and repeating these measurements to the upstream end of each reach, it is possible to calculate the planimetric coordinates of each point along the channel centerline for this arbitrary plane. Both channel form and centerline distance, however, are still correct. Once in a spreadsheet, a series of equations can be applied to each reach/year combination allowing the Bend Migration Model to iterate, on a yearly basis, and predict the final location of each reach's channel centerline.

RESULTS

White River (1962 to 1979)
The Bend Migration Model for this time period was applied to the 1962 channel form. Figure 1 shows the location of the initial 1962 White River reach in relation to the actual 1979 channel form and location, as well as the predicted 1979 channel form and location.

From Figure 1 it appears that the Bend Migration Model for the 1962 to 1979 White River reach accurately predicted both the direction and magnitude of the actual channel migration. Similar to the previous time period, the most upstream bend experienced large lateral migration directed toward the outer-bank, and again, just downstream of this tight bend, the White River migrated toward the inner-bank. Along the farthest upstream and subsequent downstream bend, the Bend Migration Model accurately predicted the direction and magnitude of channel migration.

Channel migration along the straight middle segment was predicted with increased accuracy, however, the predicted direction of channel migration was still incorrect along the upstream half of this segment. Along the final third of this reach, channel migration was consistently directed toward the outside-bank through each of the final two bends - as expected from channel migration theory. The Bend Migration Model predicted the migration pattern with reasonable accuracy, both in direction and magnitude, along this lower third of the White River reach.

Figure 2 compares actual channel migration to predicted channel migration at each point along the channel centerline. The predicted channel migration line appears to follow a pattern similar to the actual migration line, which may be expected from the data presented in Figure 1. The Bend Migration Model accurately predicted both the magnitude and direction of channel migration throughout the entire reach except for a 100 m stretch near the upstream end of the White River reach. From Figure 1, it appears that this incorrect prediction occurs along the upper half of the straight segment. The Bend Migration Model predicts negative channel migration along this part of the reach, however, as Figure 1 and 2 indicate, the actual migration was in the positive direction.

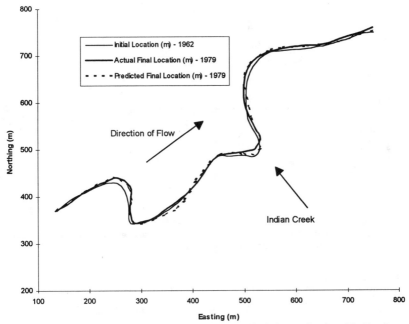

Fig. 1. Map showing location of 1962 White River, and both the actual and predicted location of the 1979 White River.

One possible explanation for the positive migration (toward the left-bank) along this 100 m stretch shown in Figure 2 may be that the channel migration was still influenced by the large amount of channel migration along the upstream bend, which drove the high velocity flow against the outside-bank. A second explanation may be that this segment of the White River pushes up against the valley wall, and therefore, unlike the rest of the channel banks, the material making up the valley walls has not been worked and reworked by the river as the channel migrates across the valley floor. An important assumption in the Bend Migration Model is that the channel banks are composed of similar material throughout the entire length of the reach. Therefore, where the channel is confined by the valley walls this assumption may not hold, and the predicted results would not be expected to accurately portray the actual situation. In this situation, the Bend Migration Model predicts the direction and magnitude of channel migration based upon an incorrect assumption: that the channel is migrating through and against material that is similar to what the channel was migrating through where the Bend Migration Model was developed (i.e. unconfined flood plain).

Downstream of the 600 m point, the Bend Migration Model correctly predicts the location of the large spike of positive migration between the 400 m and 550 m points. The predicted magnitude, however, is close to three times less than the actual magnitude. The next spike of negative migration is correctly predicted, as is the sustained positive migration over the next 200 m.

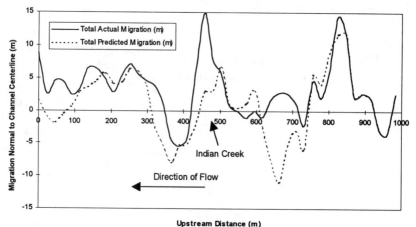

Fig. 2. Actual and Predicted Migration (m) Normal to Channel Centerline versus Upstream Distance (m) for the White River (1962 - 1979).

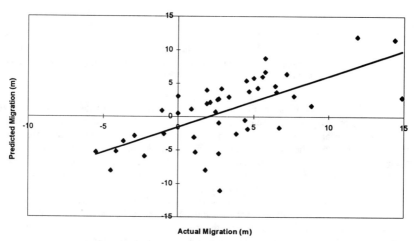

Fig. 3. Predicted Migration (m) versus Actual Migration (m) for the White River (1962 - 1979). Correlation Coefficient = 0.6675

Figure 3 presents a scattergram of actual migration versus predicted migration values at each cross-section (located one channel width apart). This plot illustrates the degree of common variation between predicted total migration and total actual migration. In addition, the correlation coefficient is presented as a means of quantifying the accuracy of the Bend Migration Model (McCuen, 1993). The calculated correlation coefficient indicates a goodness-of-fit value of 0.667.

No significant floods occurred over this 17 year period. The largest flood occurred in 1974, and had a return period of 18.2 years. The White River was able to migrate as expected because neither the White River nor Indian Creek experienced major floods during the time period examined. Barry (1996) found that extreme floods during other time periods and along other reaches included in this study influenced the migration pattern, thus, model accuracy declined.

CONCLUSIONS

Many inaccurate predictions of channel migration are related to the unpredicted and unrecorded effects of natural factors such as varying bank erodibility, random occurrences of large woody debris (LWD), and perhaps the random influences of forest animals in influencing bank erodibility. In addition, the complex interdependence (mechanically, temporally, and spatially) of the many geomorphic, geologic, climatic, and ecological variables that influence the development of a Bend Migration Model make it difficult to determine exactly what is considered similar between reaches for which similar models are to develop. However, three factors, decided upon prior to beginning this research, greatly decrease the complexity, or standardize the complexity between reaches.

First, the method of site stratification greatly decreases the natural variability between reaches within a landtype subsection. Complex interactions between geology, ecology, and climate still persist, however, channels within a single landtype subsection have evolved under similar geological, ecological, and climatic conditions. In addition, the primary disturbance patterns and disturbance regimes are assumed similar throughout the entire area bounded by that landtype subsection.

Second, the selection of similarly classified reaches (i.e. response reaches) with similar degrees of channel confinement within landtype subsections further standardizes the complexity between the selected reaches. Montgomery and Buffington (1993) state that "...assessing channel condition and predicting channel response requires identification of functionally similar portions of the channel network." By further stratifying the sites selected to represent "functionally similar portions of the channel network," it is assumed that the processes which formed each reach, and the predicted response of each reach are similar.

The final factor which standardizes environmental complexity is related to physical parameters embedded in the Bend Migration Model coefficients. Physical parameters include many standard channel geometry measurements (e.g. bankfull width, bankfull depth, channel slope, cross-channel slope, channel curvature). However, these physical parameters used to develop a Bend Migration Model were manipulated into dimensionless quantities. As a result, comparisons of Bend Migration Models were not complicated by basic differences in the size of the rivers of interest.

Although the Bend Migration Model predicted channel migration reasonably well for the reach and time period discussed above, often, Bend Migration Models applied to other reaches and time periods did not perform as well (Barry, 1996). He found that many inaccurate predictions were readily explained by unexpected variables (e.g. LWD or fire) clearly visible on the aerial photography which further indicates the need for coupling migration models with a probabilistic treatment of natural variability. Due to the unexpected variability found in nature, Furbish (1991) states: "Thus, there is decided value in coupling mechanical treatements ofmeander behaviour with probabilistic ones to formally treat uncertainty in meaner-train evolution."

The aerial photo measurement techniques work well when multiple photo series at appropriate scales are available over long periods of time. This is often the case in forested areas in this country. Channel visibility is an additional requirement for the success of aerial photo measurements. In several of the photo series examined by Barry (1996), heavy vegetative cover, shadowing, and poor quality of some older photos made it impossible to clearly see channel characteristics of interest.

A stereoplotter is essential for the accurate measurement of channel location over time. Its optics greatly enhance the detection and measurement of small channel features and allow their accurate measurement. Use of a stereoplotter allows establishment of a common coordinate system between photo series and allows measurement from photos of different scales taken with cameras of different focal lengths at different times. In addition, the stereoplotter also allows for digital transfer of data directly into spatial analysis programs.

REFERENCES

Ballerini, M. and S. Reutebuch. 1994. Measuring Landslide Dynamics from Historical Aerial Photos. In: Greer, Jerry D., ed. Remote Sensing and Ecosystem Management: Proceedings of the fifth Forest Service Remote Sensing Applications Conference; 1994 April 11 - 15; Portland, OR. Fall Church, VA: ASPRS; pp. 211 - 220.

Barry, J. 1996. Application and Evaluation of a Channel Migration Model. M. S. Thesis. University of Washington; Seattle, Washington: 159p.

Everitt, B. L. 1968. Use of the Cottonwood in an Investigation of the Recent history of a Floodplain. American J. of Science 266: 417 - 439.

Fetherson, K. L., R. J. Naiman, and R. E. Bilby. Large Wood Debris, Physical Process and Riparian Forest Development in Montane River Networks of the Pacific Northwest. Geomorphology. (submitted).

Furbish, D. J. 1991. Spatial Autoregressive Structure in Meander Evolution. Geological Society of America Bulletin 103: 1576 - 1589.

Gregory, S. V., F. J. Swanson, W. A. McKee, and K. W. Cummins. 1991. An Ecosystem Perspective of Riparian Zones. BioScience 41: 540 - 551.

Hupp, C. R. and W. R. Osterkamp. 1985. Bottomland Vegetation Distribution Along Passage Creek, Virginia, in Relation to Fluvial Landforms. Ecology 66: 670 - 681.

Gregory, S. V., F. J. Swanson, W. A. McKee, and K. W. Cummins. 1991. An Ecosystem Perspective of Riparian Zones. BioScience 41: 540 - 551.

LaPointe, M. F. and M. A. Carson. 1986. Migration Patterns of an Asymmetric Meandering River: The Rouge River, Quebec. Wat. Res. Res. 22: 731 - 743.

Leopold, L. B., M. G. Wolman, and J. P. Miller. 1964. Fluvial Processes in Geomorphology. San Francisco: W. H. Freeman.

McCuen, R. H. 1993. Microcomputer Applications in Statistical Hydrology. Englewood Cliffs, New Jersey: Prentice Hall.

Montgomery, D. R. and J. M. Buffington. 1993. Channel Classification, Prediction of Channel Response, and Assessment of Channel Condition. Department of Natural Resources. Olympia, Washington. Report TFW-SH10-93-002.

Oliver, C. D., A. B. Adams, and R. J. Zasoski. 1985. Disturbance Patterns and Forest Development in a Recently Deglaicated Valley in the Northwestern Cascade Range of Washington, U. S. A. Canadian J. of Forest Research 15: 221 - 232.

Piney, F., H. Decamps, E. Chauvet, and E. Fustec. 1990. Functions of Ecotones in Fluvial Systems. Pages 141 - 169, in R. J. Naiman and H. Decamps (editors). The Ecology and Management of Aquatic-Terrestrial Ecotones. UNESCO, Paris, and Parthenon Publishing Group, Carnforth.

Reutebuch, S. E. and B. F. Gall. 1990. Using Historical Aerial Photos to Identify Long-Term Impacts of Forestry Operations in Sensitive Watersheds. In: ASPRS technical paper: 1990 ASPRS-ACSM fall convention; 1990 November 5 - 9; Anaheim, CA. Fall Church, VA: ASPRS.

1995 SCANNED AERIAL PHOTOGRAPHY OF THE KISATCHIE NATIONAL FOREST IN LOUISIANA

Calvin "Pat" O'Neil
U. S. Geological Survey
National Wetlands Research Center
700 Cajundome Blvd.
Lafayette, La 70506

Lawrence R. Handley,
Steve Hartley,
James B. Johnston,
National Biological Service
National Wetlands Research Center
700 Cajundome Blvd.
Lafayette, La 70506

Bruce Coffland
ATAC
National Aeronautics and Space Administration
Ames Research Center
Moffett Field, Ca. 94035

Lynn Schoelerman
U.S. Forest Service
Kisatchie National Forest
2500 Shreveport Highway
Pineville, La 71360

ABSTRACT

In January and February 1995, the National Aeronautics and Space Administration (NASA) Ames Research Center collected approximately 3,500 frames of 1:65,000 scale and 3,200 frames of 1:32,500 scale color infrared (CIR) aerial photography covering the state of Louisiana. The project was funded by Coastal Wetlands Protection, Planning and Restoration Act (CWPPRA) National Biological Service, the New Orleans District of the U. S. Army Corps of Engineers (USACE) and the Louisiana Department of Environmental Quality (LDEQ). A concerted effort to satisfy natural resource imagery requirements for the National Biological Service's Louisiana GAP Project, these state and federal agencies and a university (University of Southwestern Louisiana), are in the process of scanning the 1:65,000 aerial photography for the entire state of Louisiana and will place the digital data on CD-ROM for distribution. As a much smaller project, the National Wetlands Research Center (NWRC) of the National Biological Service (NBS), with cooperation of the University of Southwestern Louisiana (USL), have scanned frames of 1:32,500 CIR aerial photography for a portion of the Kisatchie National Forest. The aerial photography has been digitally mosaicked and rectified to provide a rectified coverage of the Kisatchie National Forest, Vernon Ranger District. The mosaic was rectified using Global Positioning System (GPS) and other reference data points collected within portions of the Kisatchie National Forest. Geographic data sets have been added to the digital mosaic. The mosaicked data provides a demonstration of the potential of digitally scanned aerial photography in conjunction with other geographic information systems (GIS) data sets for general natural resource planning purposes.

INTRODUCTION

Through partnership of resources and personnel of the LDEQ, New Orleans District USACE, Coastal Wetlands Protection, Planning and Restoration Act and the NBS, the entire state of Louisiana was photographed by the NASA AMES/ER-2 aircraft in January and February of 1995. Several very highly detailed and time constrained objectives of each contributing agency were satisfied by the overflight. If each agency were resolving their remote sensing needs independently the mission objectives could not have been achieved. The rolls of aircraft CIR film (3,500-9 x 9 inch frames and 3,200-9 x 18 inch frames) provided by the NASA/ER-2 platform are available to the public at the United States Geological Survey (USGS), Earth Resource Observation System (EROS), Sioux Falls, South Dakota.

Several east-west flight lines of NASA- ER-2, 9 by 18 inch (9 x 18), 24 inch focal length, approximately 1:32,500 scale, CIR aerial photography cover significant portions of central and southern Louisiana. The ER-2 Hycon HR-732 camera, CIR (9 x 18) aerial photography is not as familiar or as a standard format as the 9 inch by 9 inch (9 x 9), 12 inch focal length, RC-10 mapping camera, CIR aerial photography of the ER-2. However, the large scale (1:32,500 or approximately 1"=2708'), of the 9 x 18 coverage (approximately 4.6 miles x 9.2 miles) provides highly detailed and descriptive representations of surface features such as trees and shrubs. Except for more extensive use by the United States Forest Service (USFS), limited use of 9 x 18 coverage has been attempted on a general application level.

NWRC is involved in a program entitled GAP, which is a national program of partnerships designed to identify areas of ecological significance that could be managed to conserve biological diversity; it seeks to identify "gaps" in our system of conservation lands. As part of a state wide effort to classify the land cover and land use of the state of Louisiana into 20 different categories, the NWRC is using Landsat Thematic Mapper digital imagery. However, to guide the Louisiana GAP project's classification of satellite imagery, the Center was interested in using the state wide CIR aerial photography and particularly the 9 x 18 CIR aerial photography. The recently acquired 9 x 18 coverage provided an opportunity to develop rectified 9 x 18 digital photo mosaics.

The Kisatchie National Forest, near Alexandria, Louisiana, provided an opportunity to test some of the applications of the 9 x 18 coverage for the GAP project. The USFS manages the Kisatchie National Forest ranger districts and has a complete set of digital geographic information system (GIS) data layers. The 9 x 18 CIR aerial photography covers all of the Vernon Ranger District and the southern half of the Evangeline Ranger District (See Figures 1, 2 and 3). The compliment of available USFS Arc/Info GIS data and the CIR photography of the Vernon and portions of the Evangeline Ranger Districts were ideally suited for imagery exploitation.

PROCEDURES

CIR 9 x 18 transparencies were acquired, reviewed, and prepared for field identification of TM GAP classification categories and collection of (GPS) registration points. Two days were spent in the field checking the GAP classification and collecting GPS points in accessible areas. A Trimble Pathfinder Pro XL GPS receiver was used to record a vehicle movement sampling rate of five seconds and a point collection rate every second. The USFS Kisatchie National Forest, Alexandria, Louisiana, provided the GPS base station data support for differential correction computations.

After photo acquisition and GPS data collection, a series of procedures were employed using computer-assisted techniques and methodologies to produce a rectified digital image mosaic of the Vernon Ranger District. The procedure was as follows:

Scanning (26 -9 x 18 color infrared transparencies)
        Hardware/Software = Sharp JX610 color scanner; Gateway 2000 Pentium IBM-
                compatible computer with 64 megabyte RAM, 2 gigabyte hard disk;
                Adobe Photoshop and Easy CD Pro software.

        Product = 26 digital image files at 300 dpi (pixel size 9 feet), 8 bit image, and
                6.53 megabyte image per TIFF file/170 megabyte TIFF file.

Image Preparation (enhancing images and producing CD-ROM)
        Hardware/Software = Adobe Photoshop software (contrast stretching), Gateway 2000
                and CD-ROM drive CD Writer Yamaha optical disk.

        Product = 26 contrast stretched TIFF image files on CD-ROM disk.

Image Processing (mosaicking and rectifiying of Vernon Ranger District)
        Hardware/Software = UNIX Data General Aviion 530, 8 bit display and 2 gigabyte
                hard disk workstation with PCI version 5.2 software (image
                processing).

        Product = 14 PCI image files mosaicked and UTM georegistered into one
                digital image mosaic 120 megabyte ERDAS LAN file.

Final Product (inkjet plot of digital image mosaic of Vernon Ranger District)
        Hardware/Software = UNIX Silicon Graphics Indy, 24 bit display, 20 gigabyte hard
                disk workstation. Laser Master inkjet plotter. ArcView (UNIX),
                version 2.1a GIS software.

        Product = One UTM digital image mosaic inkjet plot and one 75 megabyte
                postscript file copied to an 8mm tape.

A more detailed explanation of the computer-assisted processing and methodologies are briefly described below:

Twenty six 9 x 18 inch color infrared transparencies covering the Ranger Districts were scanned at the facilities of the Universtiy of Southwestern Louisiana, in Lafayette, La., by a Sharp JX610 color scanner which was attached to a Gateway 2000 Pentium IBM compatible computer with 64 megabyte RAM and a 2 gigabyte hard disk. Adobe Photoshop software was used to manipulate the photos which were scanned at 300 dots per inch (dpi) as an 8 bit image. The selection of the 300 dpi scanning rate facilitated ease of handling image files while providing good image detail (pixel size represents approximately 9 feet). Each image was enhanced to minimize photo vignetting (darkening) and saved as a 6.53 megabyte image file. The resulting 26 tagged interchange file format (TIFF) files consumed approximately 170 megabytes of computer disk space. The Easy CD Pro software was used to write the scanned and enhanced image TIFF format files to an optical disk with a CD-ROM drive (CD Writer Yamaha).

A Data General, Aviion 530, 8 bit display, UNIX workstation with PCI Version 5.2 image processing software was used to prepare the CD-ROM image files for georegistration and to mosaic the Vernon Ranger Districts digital photos. The digital photos were mosaicked by locating suitable common points on each photo and "stitching" them together. GPS and other reference points were used to georegister the digital mosaic photos and produce a rectified mosaic image. The mosaic file of the Vernon Ranger District, covering the 14 original photos, 123 megabyte file was imported to a Silicon Graphics, INDY 24 bit display workstation with ArcView (Unix), version 2.1a GIS software. The final digital image product was saved as a 75 megabyte postscript file.

Several Arc/Info GIS data layers were overlaid on the Vernon Ranger District mosaic for inspection and review. The combined mosaic and selected GIS data layers presented a good display for general planning purposes for large areas. The final product was a color paper plot of the Kisatchie National Forest, Vernon Ranger District from a LaserMaster inkjet plotter and an 8mm cartridge tape of scanned 9 x 18 photography of the Vernon and Evangeline Ranger Districts.

CONCLUSION

Through the joint interests and resources of federal, state and university entities, many natural resource objectives were achieved by obtaining NASA ER-2 CIR aerial photography for the State of Louisiana. Different formats and scales of high altitude photography can be manipulated by available computer scanning systems, image processing software, and presented as various displays to satisfy specific agency requirements. The products have been put to a specific use by the U. S. Forest Service as visual aids in looking at various alternatives for the U. S. Army, Fort Polk, La., military reservation expansion designs and as GIS backdrops for resource managers.

ACKNOWLEDGMENTS

Mitchell Barton (USFS), Kelly Mouton (NBS/Johnson Control World Services), Robert Greco (NBS/Johnson Control World Services), Helena Schaefer (NBS) and Nathan Handley (University of Southwestern Louisiana) provided techinical assistance that would have made the project possible to achieve and are hereby recognized for the efforts.

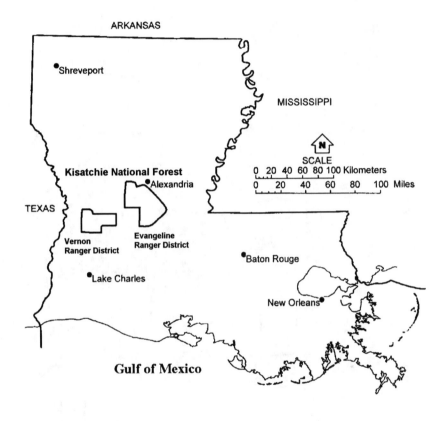

Figure 1. Location of the Vernon and Evangeline Ranger Districts, Kisatchie National Forest, Louisiana.

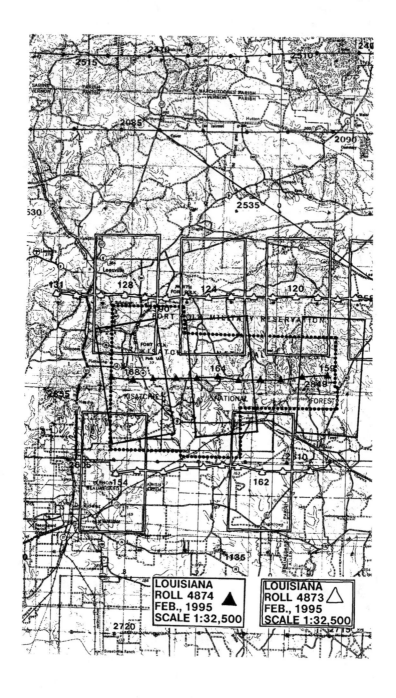

Figure 2. Plot coverage index of NASA ER-2, 9" X 18" color infrared aerial photography of Vernon Ranger District.

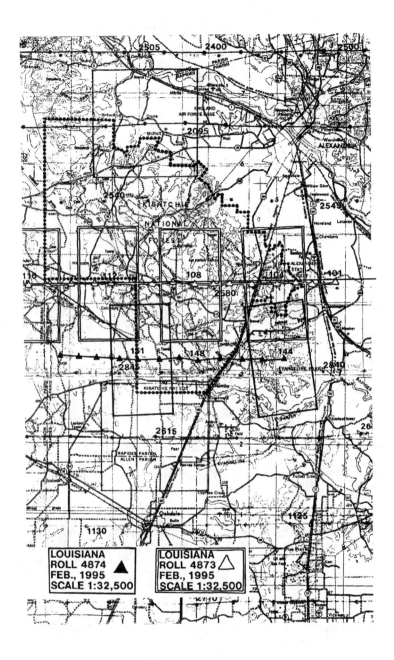

Figure 3. Plot coverage index of NASA ER-2, 9" X 18" color infrared aerial photography of Evangeline Ranger District.

USING "PSEUDO" DIGITAL ORTHO PHOTOGRAPHY
or
THE GOOD, THE BAD AND THE UGLY

Ken Winterberger
Pacific Resource Inventory, Monitoring, and Evaluation
PNW Research Station, Forestry Sciences Lab
Anchorage, Alaska 99503

ABSTRACT

To conduct an inventory of coastal Alaska there is a need for current high resolution imagery to document sample plots and aid in the location of these plots. Imagery of coastal Alaska is difficult to obtain primarily due to weather constraints. While there is some recent cloud-free SPOT panchromatic coverage of parts of coastal Alaska, acquisition costs are high and image resolution is not high enough to be useful for sample plot location in the field. Available hard copy orthophotomaps are being scanned and rectified to produce "pseudo" digital ortho photography for field and office use. Real digital orthophotos have higher resolution but cost more than "pseudo" digital orthophotos, take up much more disk space and take longer to process. Advantages and disadvantages of "pseudo" digital orthophotos are illustrated and discussed. Image rectification accuracy is examined in relation to image use in geographic information systems.

INTRODUCTION

In Alaska, the acronym T.A.D.I.A. (Things are Different in Alaska) is sometimes used to acknowledge the difficulties that must be dealt with, peculiarities that are regularly observed, and the sometimes convoluted methodologies that are used to accomplish what might be considered simple tasks elsewhere. Pacific Resource Inventory, Monitoring, and Evaluation (PRIME) for Alaska takes pride in being able to use convoluted methodologies to accomplish apparently simple tasks when necessary. This is the story behind one of those convoluted, but successful, methodologies.

The inventory of vegetation resources can be done more easily using current high resolution imagery for sample documentation and planning. It has always been difficult to obtain this sort of image data for Alaska. Resource scale photography has been flown for certain projects and for certain management entities. However, the high costs associated with logistical support of aircraft make regular image acquisition over large areas of Alaska prohibitive. A short growing season (time from snow melt to snow fall) and cloud cover during the growing season make aerial photo acquisition in certain parts of Alaska difficult (Southeast and Southcentral Coast) or nearly impossible (Alaska Peninsula and Aleutian Islands). Because of these difficulties, we have always been in the mode of trying to make do with available data and images.

In the late 1970's and early 1980's aerial photography was acquired for most of Alaska at scales of 1:60,000 (color infrared) and 1:120,000 (panchromatic) through a cooperative venture between several federal land management agencies, the State of Alaska, and NASA. In many parts of Alaska, this is still the best program are 15 years old. The color infrared photos have been used to do land cover mapping through photo interpretation, to verify land cover maps produced through satellite image analysis, and they have been used to document the location of field samples. The panchromatic photos have been used to update existing USGS map sheets, develop new map sheets, and produce orthophoto maps for certain "high use" areas of the state (see Figure 1). PRIME continues to use this raw image data where more current or higher resolution images do not exist.

Figure 1. An example of orthophoto map, Seward (D-6), Alaska prepared from 1:120,000 scale panchromatic aerial photographs taken August 24, 1978.

One of the major problems associated with the use of this raw data is that it is difficult to register to existing ground control. In gentle terrain, information has been transferred to and from these photos using zoom-transfer scopes and simple light-table rubber-sheeting. In the typically rough terrain found in many parts of Alaska these techniques have worked less well, but have been used just the same. Before the days of geographic information systems (GIS) these methodologies worked fine. There was little need for spatial accuracy because the spatial analysis that was being done was accomplished using techniques like dot grid and planimeter area measurement, and mylar overlay and map sheet spatial correlation.

As spatial accuracy has become more important with advances in digital image processing and GIS, image rectification has become much more important. It was recognized by regular users of GIS that extracting useful information from aerial photography and entering it into GIS databases was not an easy matter.. Especially in rough terrain, the photos didn't match the maps. This led to a situation where existing aerial photos were almost being ignored as viable GIS data sources.

One of several methodologies that was developed to allow data extraction from photography involved the use of orthophotography. First, and importantly, hard copy orthophoto maps were made. Early orthophoto maps were something of a novelty but quite useful. The orthophoto maps could be used as an intermediate step in the process of extracting spatially accurate data from aerial photos for entry into GIS. One scenario for photo data extraction and GIS data entry might have included the following steps. First, land cover features would be delineated on the original photographs (used because of their higher resolution and the ability to examine them under a stereoscope). Next, the information would be transferred from the original photographs to an orthophoto base (usually by hand, or with the aid of a zoom-transfer scope). Finally, the orthophoto base would be digitized, thus allowing "photogrammetrically correct" line work to be entered into a GIS database from a photo source. The steps could be time consuming and tedious. In these early methods, digital processes and data storage were kept to a minimum. Computer processing power, and digital data storage capabilities were limiting factors.

As computer power, and data storage capabilities have improved over the last several years, data extraction methods have changed to take advantage of the newest technology. Only a few years ago 500 megabytes of raster data might be considered nearly unmanageable. Today, a desktop computer with less than a gigabyte of storage is uncommon.

Recent interest in the use of digital orthophoto maps has come about mainly due to improved computer technology. However, the cost of new digital orthophoto maps can be relatively high at $700-$800 per quarter quad. The data can consume prodigious amounts of disk space, and is often resampled to reduce image size. To make use of existing hard copy orthophoto maps and avoid the cost of building new digital orthophotos from old imagery a methodology has been developed to build a set of "pseudo" digital orthophotos for the National Forests in Alaska.

METHODOLOGY

There are a number of steps and considerations that must be dealt with in the process of making consistent "pseudo" digital orthophotos.

<u>Orthophoto Map Preparation</u>

The original hard copy orthophotos can be scanned "as is" at their original scale, or reduced prints can be made from the original to scan on a desktop scanner; we opted for scanning reduced prints. Scanning from the original orthophoto map scale would have eliminated the step of producing reduced copies, but the large scanners needed for the job are not common and the scanning process would likely have to be outsourced. The files resulting from large image scans can be difficult to manage unless the scanning is done at lower resolutions.

Figure 2. Original reduced orthophoto copies too wide for desktop scanner resulted in having to scan in pieces. Port Alexander A-3 NE.

The orthophoto originals were reduced so that they fit on a desk top scanner with a 22 cm. by 36 cm. scanning surface.. The percent reduction varied by latitude and map sheet. At the southern end of the Tongass National Forest, the orthophotos were reduced to 60 percent of their original size. At higher latitudes the originals were reduced to 66 percent of their original size. Even after a fair amount of planning we had to deal with orthophotos that were too wide to be scanned in a single pass on the desktop scanner. In these cases, the orthophotos were scanned in two passes (see Figure 2) with the results being stitched together using standard desktop image processing software (see Figure 3). This stitching process reduced the number of image files that needed to be dealt with and reduced the total amount of disk space needed by eliminating overlap . In hindsight it would have been better to make sure that all reductions were done to accommodate desktop scanner dimensions.

Scanning

The process of scanning was begun "in the blind". The desktop computer we started with had an older video board with limited capabilities. We were not able to display the scan preview at high resolution with a large number of greyscales. Midway through the process we traded up to a desktop computer with better video capability. This made a great deal of difference in helping to adjust critical highlight and shadow settings. The goal in making these prescan adjustments was to capture the widest range of greyscales in the land cover types of most importance. Without adjustment, too many greyscales might be captured in ice and snow or water. While automatic adjustment worked for orthophotos with neutral density, high contrast images can be a problem.

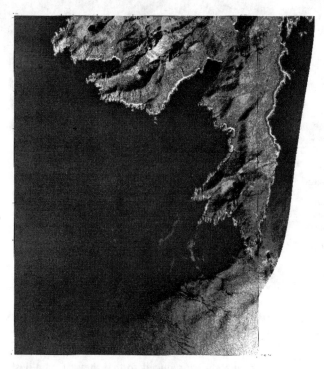

Figure 3. An example of stitching separate scanned parts together. The resulting image file is smaller than the sum of the original parts by several megabytes. Port Alexander A-3 NE.

The scan settings took advantage of the maximum scanner resolution without interpolation (400 X 400 dots per inch), maximum sharpening, and maximum greyscales. The scanned image data were saved in tagged image file format (TIF) with file sizes averaging 16 megabytes.

One of the reasons we opted for scanning on in-house equipment was the availability of a local area network and several large disk drives attached to that network. Rather than having to save the images to tape or some other removable media to transfer between sites, we were able to store many image data sets online, allowing us to deal with large areas easily.

<u>Image Re-rectification</u>
After scanning the orthophoto maps came the image re-rectification process. The graphic TIF images were converted to standard band interleaved by line (BIL) format for rectification, storage and further distribution. BIL format was selected because of its standardization and the ability to open or import BIL format using most image processing software. Our ultimate goal is to provide "pseudo" digital orthophoto maps (vintage 1978) on compact disk.

After conversion to BIL format, the images were "warped" using the known corner coordinates as control (essentially, a simple rotation). The images were rectified using a polynomial transformation to an obverse mercator projection (Alaska State Plane, Zone 1) with units in feet. The North American Datum, 1927 was used to match existing maps. The data was resampled to produce 10 ft. X 10 ft. pixels using a cubic convolution resampling to get the best looking images. This resampling was appropriate because pixel digital numbers are not being used for critical processing.

Root mean square (RMS) error of the polynomial transformations have averaged less than one pixel, as expected. There have been a few cases in which RMS errors have been relatively high (approaching 3 pixels). We have yet to determine the reason for these discrepancies. There are three things that might contribute to these discrepancies: 1) the original orthophoto map corners are wrong, 2) the scanning process has distorted the image (unlikely, as several images have been re-scanned and re-rectified several times with similar RMS errors), 3) identifying the exact map corners in some raster images is difficult.

<u>Image Quality Improvement</u>
After the orthophoto maps have been re-rectified, they can be easily integrated with other digital data, e.g., other raster data, vector data, and databases containing spatial information. However, before final image products can be prepared, image quality often needs to be improved.

The most useful "image quality improvement " that can be made involves image histogram adjustment. To complicate matters, no single histogram adjustment technique is best for all images. In fact, depending on the required results, different histogram adjustments can be made to accommodate different uses. For the most part histogram adjustment is a trial and error proposition. In general histogram adjustment should be made based on the image area of interest. Unless the entire image area is covered by a single land cover type, using the entire image as a basis for histogram adjustment rarely works well.

If the orthophotos are to be printed for field use, histogram adjustment is further complicated. Because of the difference in greyscale reproduction capability and gamma between the computer monitor and most moderately priced printers, what you see is not necessarily what you get! In fact, images that print well can look awful on screen.

Image Mosaics

While individual hardcopy orthophoto maps are useful, digital orthophoto maps are more useful when they can be combined to cover a larger area. Once the orthophoto maps have been re-rectified, the individual images can easily be combined, with one major drawback to be overcome. In the scanning process it is impossible to avoid inclusion of non-image portions of the orthophoto map (see Figure 4). These non-image areas can be reduced in the scanning process through careful cropping, but complete elimination must be dealt with digitally. Different software and different techniques produce markedly different results.

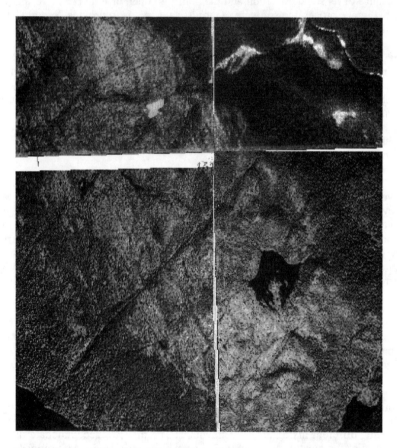

Figure 4. Example of digital pseudo orthophoto maps mosaiced together before dealing with non-image borders.

The quickest and easiest technique for removing non-image areas involves turning off image pixels with values above a certain level, e.g., an algorithm like, **if image 1 is less than 250 then image 1 else null**, works well. However, this method only works well if non-image area pixel values are well above image area pixel values. Where some image area pixels are the same as non-image area pixels, null areas within the image occur (see Figure 5).

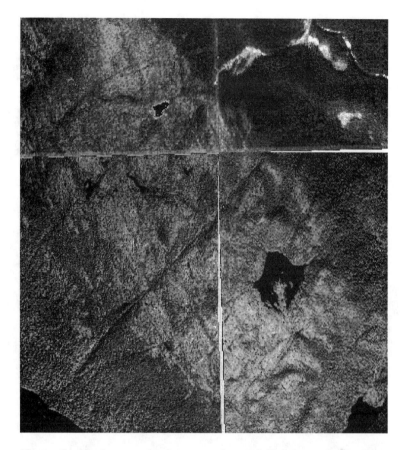

Figure 5. Northwest orthophoto map (upper left) has non-image edge removed using pixel value (threshold) mask. Note that this method can cause anomolies within image area, with null pixel being introduced where image area pixels have values equal to non-image area pixels.

The best technique for eliminating non-image areas from a mosaic involve the use of masking regions. The actual procedure used to produce and implement masking regions varies by software. Once a masking region is produced, an algorithm like, **if in "masking region" then image 1 else null** works well (see Figure 6).

Frame 6. An example of image masking using regions based on orthophoto map corner coordinates to eliminate non-image pixels.

Discussion and Conclusions

Pseudo digital orthophoto maps can be very useful. Their production takes some planning and care but is well worth the effort if alternatives are limited.

Pseudo digital orthophotos can be used to produce inexpensive hardcopy orthophoto maps at scales useful for field use. Resulting orthophoto maps can have accurate scale and angle references. While these field maps may not have the resolution and image quality of "real" photographs, their orthophotographic nature make them very useful, especially when they are used along with "real" photos.

Pseudo digital orthophotos can be used in there digital form as backdrops for "heads up" digitizing. Even if newer standard digital orthophotography is available, the production of pseudo digital orthophotos may be warranted for change detection mapping purposes.

The digital orthophotos can also be merged with land cover classification data produced using lower resolution image sources. The resulting images can be manually interpreted more easily given the texture information provided by the orthophoto. The orthophoto can also be used to produce a texture channel for use in digital classification of lower resolution data.

ACCURACY OF USGS DIGITAL ELEVATION MODELS IN FORESTED AREAS OF OREGON AND WASHINGTON

Stephen E. Reutebuch
USDA Forest Service
Pacific Northwest Research Station
Seattle, Washington

Ward W. Carson
College of Forestry
Oregon State University
Corvallis, Oregon

ABSTRACT

USGS digital elevation models (DEMs) are being increasingly used in a wide variety of terrain analysis and land-use planning systems. The accuracy of the output from these systems is dependent on the accuracy of the associated DEMs. The authors have developed a method for performing a low-intensity test of the accuracy of DEMs using ground coordinate data collected as a by-product of aerotriangulation projects. The accuracy of a sample set of 39 DEMs covering national forests in Oregon and Washington was evaluated using this method. Accuracy varied considerably between eastern and western parts of Oregon and Washington. The method used to produce a DEM also influenced its accuracy. The elevational root mean square error (RMSE) published by the USGS for DEMs produced using the line trace method were on average over 50% lower than the RMSE computed from independent ground data. The published elevational RMSE of DEMs produced using the photogrammetric technique agreed on average with the RMSE computed from independent ground data. A suite of programs was developed that allows the accuracy test to be applied in any area where aerotriangulation data, or other independent ground coordinate data such as GPS points, and DEMs are available.

INTRODUCTION

A digital elevation model (DEM) is an array of numbers that contains the elevation of the ground surface at a series of sample points. The sample points can be regularly spaced horizontally, or they can be randomly spaced. The U. S. Geological Survey (USGS) produces and distributes 4 types of regularly spaced DEMs that correspond to standard USGS map series: 7.5-minute; 15-minute; 30-minute; and 1-degree DEMs. In this study, we are examining the 7.5-minute DEMs and all further references are to this series.

Planners use DEMs to compute a variety of terrain attributes that are used in the development and assessment of management alternatives (Twito et al. 1987; Moore et al. 1991). Most geographic information systems and remote sensing image analysis systems have utilities that use DEMs to produce slope, aspect, and elevation classifications for use in further analyses. Although the accuracy of such analyses is directly dependent on the accuracy of the DEM data, surprisingly few studies of DEM accuracy have been reported. Many studies have been conducted to investigate the effects of DEM characteristics and hypothetical error levels on derived terrain values (Vieux 1993; Vieux and Needham

1993), but most have not actually measured individual DEM accuracy against independent ground coordinate measurements. Bolstad and Stowe (1994) used 42 ground elevation measurements to compare the accuracy of a single USGS DEM of Blackburg, West Virginia with a DEM produced from satellite images.

In our study, independent ground elevation measurements were used to assess the accuracy of 39 USGS DEMs from forested areas of Oregon and Washington. Eighteen of these DEMs were produced using the line-trace method where: contour lines from USGS 7.5-minute quadrangle topographic maps are digitized to produce the gridded elevation data that make up the DEM. The remaining 21 DEMs were produced via a photogrammetric profiling method (USDI 1990). Nineteen of the DEMs were of areas west of the crest of the Cascade Mountains. The remaining 20 DEMs were from east of the crest.

In the study, the accuracy of line-trace DEMs was compared to the accuracy of photogrammetric DEMs to see if the DEM production method might influence DEM accuracy. Additionally, the accuracy of DEMs from the areas east of the Cascades was compared to the accuracy of DEMs from western areas of Oregon and Washington to see if differences in vegetation cover and general topographic roughness might influence accuracy. Finally, the root mean square error (RMSE) of each DEM computed in this study from independent ground coordinates was compared to the RMSE reported by the USGS for each DEM to see if the reported USGS RMSEs compare favorably with the RMSEs computed from independent ground data.

USGS DEM accuracy assessment technique
It is important to understand how the USGS computes the RMSE of a DEM and the source of check point coordinates used in the accuracy calculations. RMSE is defined as:

(Eq. 1) $\mathbf{RMSE} = [\Sigma(e_c - e_d)^2/n]^{0.5}$

where: RMSE is the elevational or vertical root mean square error;
           e_c is the elevation of the check point;
           e_d is the DEM elevation for the check point; and,
           n is the number of check points.

In computing RMSEs, the USGS uses check points whose coordinates are measured from 7.5-minute topographic maps at well defined points such as bench marks, spot elevations, and points located directly on contour lines. In other words, the RMSE computed and published by the USGS is an indication of how accurately a DEM matches its associated topographic map--it is not a direct estimate of how well the DEM represents the actual ground surface.

Aerotriangulation Ground Coordinates used as independent check points
In many instances, a photogrammetric technique known as aerotriangulation (AT) is used to fit a block of aerial photos to a ground coordinate system (Wolf 1983). After the AT process is completed, photos can be used for map or orthophoto production. In the AT process a list of ground coordinates are generated for several points that are clearly visible on the photos. There are usually around 10-20 AT points per photo that are well distributed around the photo area. In selecting these AT points, the photogrammetrist

311

attempts to locate points in open, relatively flat areas with minimal shadows; therefore, the AT point locations are biased toward positions where it is easier to see the ground surface on the aerial photos.

In the late 1980s and early 1990s, the National Forests of Oregon and Washington undertook a project to generate new 1:24,000 orthophotos for each 7.5-minute quadrangle that included National Forest lands. A set of 1:40,000 aerial photos were flown and aerotriangulated for each forest. Global positioning system (GPS) survey points were used to establish ground control for the aerotriangulation. The typical horizontal and vertical RMSE for each photo block was less than 1 m. Several thousand AT points were produced for each forest as a by-product of this orthophoto project. These AT ground coordinates were considered to be independent "true" ground coordinates and were used to compute an independent RMSE for 39 DEMs in this study.

METHODOLOGY AND RESULTS

A total of 39 DEMs (Table 1) were randomly selected from the following 4 categories of DEMs: 1) line-trace DEMs for areas west of the Cascade Mountains; 2) line-trace DEMs for areas east of the Cascade Mountains; 3) photogrammetric DEMs for areas west of the Cascade Mountains; and, 4) photogrammetric DEMs for areas east of the Cascade Mountains.

The X-, Y-,and Z-coordinates (cast on the Universal Transverse Mercator projection system) of each AT check point was matched with its 7.5-minute quadrangle. The horizontal coordinates of the check point were used to determine where the point was located within its corresponding DEM. The DEM elevation at the check point was then computed by a three-way linear interpolation between the 4 DEM elevations that define the DEM grid cell within which the check point is located (Lemkow 1977). In addition to elevation, local slope, aspect, and roughness at each check point were computed so that possible correlations between these variables and elevation error could be investigated. No strong correlations were found between elevation error and any of these variables.

The elevation error of the DEM at each check point was computed by subtracting the interpolated DEM elevation at the check point from the aerotriangulated elevation of the check point. The arithmetic mean of the DEM elevation errors was computed for each DEM and for each DEM class. This mean error provided an indication of elevation bias.

The RMSE for each DEM was then computed using Eq. 1. The summary results of these error calculations are given in Table 2, along with the RMSEs published by the USGS for each of the 4 sets of DEMs used in the study. The published RMSEs and the elevations contained within the study DEMs are rounded to the nearest meter by the USGS; therefore, all study results have been rounded to the nearest meter. In Figure 1, the RMSE computed using the AT check points is displayed beside the RMSE published by the USGS for each individual DEM in the study. Figure 2 displays the average AT check point RMSE and the USGS RMSE for each of the 4 classes of DEMs.

Table 1. List of DEMs and the number of aerotriangulation check points used to compute independent RMSEs.

Western DEMs

DEM Name	No. of Check Points
Line-Trace DEMs (total=9)	
Matheny Ridge, WA	53
Eden Valley, OR	25
Tidewater, OR	34
North Fork, OR	30
Hebo, OR	28
Snider Peak, WA	54
Ophir Mountain, OR	23
Sixes, OR	23
Mt. Jupiter, WA	34
Photogrammetric DEMs (total=10)	
Green Mt., WA	20
Wren, OR	29
Stevens Creek, WA	36
Mt. Jefferson SW, OR	26
Bernier Creek, WA	17
Randle, WA	20
Hamilton Buttes, WA	17
Beaver, OR	36
McCoy Peak, WA	21
Smith River Falls, OR	30

Eastern DEMs

DEM Name	No. of Check Points
Line-Trace DEMs (total=9)	
Miller Lake, OR	12
Mount Lago, WA	32
La Fleur Lake, WA	31
Aladdin Mt., WA	23
Cliff Ridge, WA	24
Midnight Mtn., WA	15
Martin Peak, WA	22
Copper Butte, WA	32
Coleman Peak, WA	20
Photogrammetric DEMs (total=11)	
Sycan Marsh East, OR	18
Little Squaw Back, OR	8
Campbell Reservoir, OR	20
Foster Butte, OR	19
Alsup Mt. OR	28
Five Hundred Flat, OR	24
Trout Creek Butte,OR	9
Johnson Saddle, OR	22
Myrtle Park Meadows, OR	27
Graylock Butte, OR	27
Wickiup Spring, OR	21

Table 2. Results of the elevation error comparisons by DEM class. All values have been rounded to the nearest meter.

DEM Class	No. of Check Points	Max. Elv. Error (m) (-)	Max. Elv. Error (m) (+)	Mean Elv. Error (m)	RMSE (m) Check Pts.	RMSE (m) USGS
West Photog.	252	-11	41	2	7	6
West LT	304	-27	42	1	8	3
East Photog.	223	-11	11	-1	4	4
East LT	211	-24	14	1	5	2
All Photog.	475	-11	41	0	6	5
All LT	515	-27	42	1	7	3
All West	556	-27	42	1	7	5
All East	434	-24	14	0	5	3

313

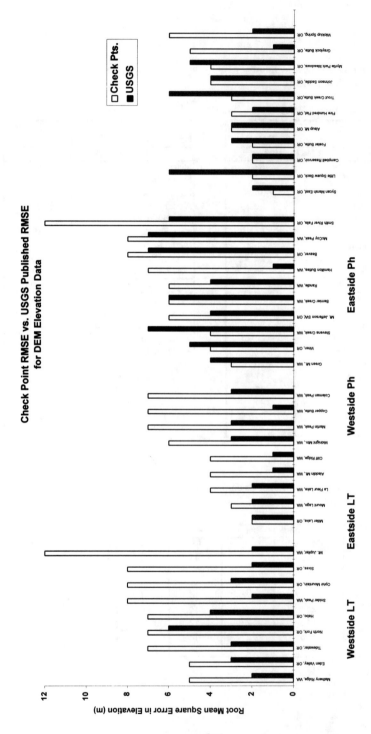

Figure 1. The RMSE computed from the aerotriangulation check points is displayed beside the RMSE published by the USGS for each DEM used in the study.

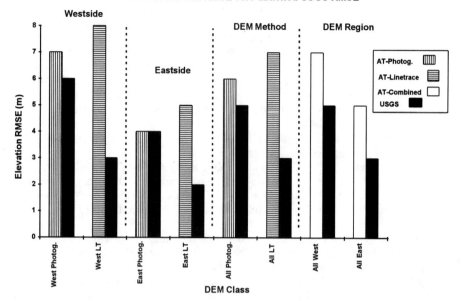

Figure 2. Average RMSE of each DEM class: computed from AT check points compared to the published USGS values.

DISCUSSION

Before discussing the results of this study, it is important to note that the RMSE published by the USGS for each DEM is computed with check points measured from its corresponding 7.5-minute topographic map, not with check points generated from independently measured terrain elevations. In other words, the published RMSE indicates how well the DEM matches its associated map sheet, not the actual terrain surface that the map sheet depicts. Map generated check points are used to check both photogrammetric and line-trace DEMs.

The AT check points used in this study were measured independently of the 7.5-minute topographic maps. These independent points were used to compute an RMSE that indicates how well a DEM actually models the terrain surface, not simply the surface depicted on the map.

DEM RMSEs published by the USGS
The published RMSEs for all DEMs in the study were 7 m or less (Fig. 1). The eastside line-trace DEMs had the lowest average USGS RMSE (2 m), followed by the westside line-trace DEMs (3 m), then the eastside photogrammetric DEMs (4 m), and finally the westside photogrammetric DEMs (6 m).

Because the line-trace DEM production technique uses map contours as its source of elevation data, one would expect that line-trace DEMs would fit their corresponding maps better than the photogrammetric DEMs which utilize a three-dimensional stereomodel as their source of elevation data. As shown in Figure 2, this was the case for the DEMs used in this study. The average RMSE published by the USGS for the line-trace DEMs was 3 m; whereas, the average for the photogrammetric DEMs was 5 m.

Additionally, due to heavy vegetative cover and highly dissected terrain on the westside of the Cascades, one would expect that the published RMSEs for westside DEMs would generally be higher than those published for eastside DEMs. Again, this was the case (Fig. 2). The average published RMSE for westside DEMs was 5 m; whereas, the average for eastside DEMs was 3 m.

DEM RMSEs computed using AT check points
The check point-derived RMSEs for all DEMs in the study were 12 m or less, with most values being 8 m or less (Fig. 1). The eastside photogrammetric DEMs had the lowest average RMSE (4 m), followed by the eastside line-trace DEMs (5 m), then the westside photogrammetric DEMs (7 m), and finally the westside line-trace DEMs (8 m).

Because the photogrammetric DEM production technique uses a three-dimensional stereomodel of the terrain (stereo aerial photos) as its source of elevation data, one would expect that these photogrammetric DEMs would fit the ground better than the line-trace DEMs which use map contour lines as their data source. As shown in Figure 2, this was the case. The average RMSE (computed from the AT check points) for the photogrammetric DEMs was 6 m; whereas the average for the line-trace DEMs was 7 m.

As was found for the published USGS RMSEs, the check point-derived RMSEs also differed due to geographic region. The average RMSE for westside DEMs was 7 m; whereas, the average for eastside DEMs was 5 m (Fig. 2).

It should be noted that the AT check points used in this study probably do not provide a random check of DEM accuracy because they are intentionally located in flatter, more open areas by the photogrammetrist. Therefore, one would expect the computed RMSE to be greater if a truly random sample of check points were used.

Comparison of the USGS published RMSEs to the AT check point-derived RMSEs
Considering either the published RMSEs or the check point-derived values, all of the DEMs in the study meet the USGS DEM Level 1 accuracy standard which states that an RMSE of 7 meters or less is desired, but that up to 15 m is acceptable, and that no single point shall have be in error over 50 m (Fig. 1 and Table 2). However, it is obvious from Figures 1 and 2 that the published RMSEs differ considerably from the RMSEs computed using independent AT check points.

On average, the USGS reported a higher RMSE for photogrammetric DEMs (5m) when compared to line-trace DEMs (3m). When independent AT ground coordinates were used, the average RMSE of photogrammetric DEMs (6m) was slightly lower than the RMSE of line-trace DEMs (7m).

The average published RMSE for photogrammetric DEMs (5m) was very close to the average RMSE computed using independent ground coordinates (6m). The published RMSE of an individual photogrammetric DEMs was off by up to 6 m (Fig. 1).

316

The average published USGS RMSE value for line-trace DEMs (3 m) was over 50% lower than the RMSE value computed using independent ground coordinates (7 m). The published RMSE of individual line-trace DEMs were off by up to 10 m (Fig. 1). In 18 of the 19 line-trace DEMs, the published RMSE was lower than the RMSE computed using independent ground coordinates. In the remaining line-trace DEM, the published RMSE was equal to the RMSE computed from independent ground coordinates.

Range of elevation errors by DEM class
The range of elevation errors computed at the AT check points is given in Table 2. The line-trace DEMs had a wider range of errors (-27 m to +42 m) than the photogrammetric DEMs (-11 m to +41 m). A wider range of elevation errors were also encountered in the westside DEMs (-27 m to +42 m) than in the eastside DEMs (-24 m to +14 m).

CONCLUSIONS

The authors have developed a technique for assessing the accuracy of USGS DEMs using independent ground coordinates. The technique is easy to apply and can be used to rapidly check large numbers of DEMs in a single run if an accurate set of ground data is available. In the study reported here, ground coordinates generated during aerotriangulation of a block of photos were used as check points; however, other data sets, such as GPS points and benchmark coordinates, can also be used.

In this study, the RMSE values published by the USGS indicate that line-trace DEMs matched the USGS 7.5-minute topographic maps better than the photogrammetric DEMs. However, the AT check point-derived RMSEs indicated that the photogrammetric DEMs matched the actual ground surface better than the line-trace DEMs.

In addition, the RMSEs published by the USGS for photogrammetric DEMs were generally close to the RMSEs computed from the independent check points. However, the RMSEs published for line-trace DEMs were generally 50% lower than the RMSEs computed from independent check points.

ACKNOWLEDGMENTS

The authors wish to thank the geometronics staff of the Forest Service Regional Office in Portland, Oregon. Particular thanks goes to Rod Dawson for many helpful suggestions, Bob Race for providing AT data and DEMs, and Roger Crystal for his enthusiastic support of the study.

REFERENCES

Bolstad, P. V.; T. Stowe. 1994. An evaluation of DEM accuracy: elevation, slope, and aspect. *Photogrammetric Engineering and Remote Sensing*, 60(11): 1327-1332.

Lemkow, D. Z. 1977. Development of a digital terrain simulator for short-term forest resource planning. M.S. Thesis. Vancouver, B.C.: University of British Columbia: 207p.

Moore, I. D.; R. B. Grayson; A. R. Ladson. 1991. Digital terrain modeling: a review of hydrological, geomorphological and biological applications. Hydrological Processes, 5(1): 3-30.

Twito, R. H.; S. E. Reutebuch; R. J. McGaughey; C. N. Mann. 1987. Preliminary logging analysis systems (PLANS): overview. Gen. Tech. Rep. PNW-GTR-199. Portland, OR: USDA Forest Service, Pacific Northwest Research Station. 24p.

USDI, U.S. Geological Survey. 1990. Digital elevation models: Data users guide 5. USDI Geological Survey, Reston, VA: 51p.

Wolf, P. R. 1983. *Elements of Photogrammetry*, 2nd Ed., New York, McGraw-Hill: 628p.

UTOOLS AND UVIEW: ANALYSIS AND VISUALIZATION SOFTWARE

Robert J. McGaughey
USDA Forest Service
Pacific Northwest Research Station
University of Washington, Bloedel 361
P.O. Box 352100
Seattle, WA 98195-2100

and

Alan A. Ager
USDA Forest Service
Umatilla National Forest
2517 SW Hailey Ave.
Pendleton, OR 97801

ABSTRACT

UTOOLS is a collection of programs designed to integrate a variety of spatial data in a way that allows versatile spatial analysis and visualization. UTOOLS software combine raster, vector and attribute data into "spatial databases" where each record represents a square pixel of fixed area, and each field in the database represents a map layer, theme, or attribute. UTOOLS includes a number of common GIS functions, including procedures for calculating buffers, slope, aspect, patch size, convexity, and a view index. UTOOLS also includes UVIEW, a visualization program designed to produce two- and three-dimensional images of digital elevation models (DEM), attribute data, and vegetation patterns at watershed and landscape scales. UTOOLS programs address the daily analysis needs of resource professionals charged with managing large land areas to provide a variety of commodity and non-commodity outputs.

INTRODUCTION

Resource analysis on federal forest lands in the Pacific Northwest has become increasing complex during the last decade Analysis has shifted from procedures that emphasize single species or specific sites to more holistic approaches that concentrate on entire ecosystems and large landscapes. Recent years have seen a shortage of suitable analysis tools for use on individual ranger districts within the U.S. Forest Service. To address this shortage, the Forest Service has begun procurement and implementation of Geographic Information System (GIS) technology. However, delays in procurement and implementation have left many district offices without sufficient analysis capability to answer the questions being asked of them. Resource specialists are being asked to resolve difficult management issues but often do not have adequate analysis capabilities to support their decisions.

The assessment and study of ecosystems and their dependent species is a complex task demanding a variety of analysis tools, including geographic information and decision support systems, simulation and deterministic models, and appropriate data linkages. As part of an effort to create more appropriate spatial analysis tools to address the needs of aquatic, wildlife, and other biologists and specialists within the Forest Service, the authors

319

have developed of a set of software programs called UTOOLS[*]. UTOOLS software is designed for use by field biologists and resource specialists working in district or forest offices within the U.S. Forest Service or in equivalent positions in other land management agencies. The programs have been widely applied within the Forest Service (McKinney 1995, Mellon et al 1995, Ager and Hitchcock 1992) and outside the Forest Service (Oliver and McCarter 1996) for analysis of a variety of resource issues since their first release in 1992. In this paper, we describe the capabilities of UTOOLS and discuss its application to natural resource analysis and planning.

OVERVIEW

UTOOLS is a collection of programs designed to integrate a variety of spatial data in a way that allows versatile spatial analysis and visualization (figure 1). One of the primary goals of UTOOLS is to provide analysis tools that can capture a "snapshot" of existing databases, allow the incorporation of a variety of data types, and provide analysis capabilities to address watershed-scale management issues. UTOOLS software combine raster, vector and attribute GIS data into "spatial databases" where each record represents a square pixel of fixed area, and each field represents a map layer, theme, or attribute. This raster data structure, maintained in PARADOX[**] format, lends itself to a wide range of analysis procedures and simplifies linkages between spatial map themes and their attributes. PARADOX queries can be used to perform many types of map operations such as intersects, joins, unions, and subject recoding. Rule bases can be applied as PARADOX queries to synthesize new layers from combinations of other layers and their attributes. The analysis capability accessed using PARADOX queries and scripts is augmented by UTOOLS programs that perform specific, raster-based, spatial analysis operations commonly used in wildlife and other resource analyses, e.g., buffering and patch identification. UTOOLS capabilites include a number of common GIS functions, including procedures for calculating buffers, slope, aspect, patch size, convexity, and a view index.

In operation, users invoke specific programs to carry out desired operations. UTOOLS programs include:

UCELL5	Creates PARADOX spatial databases using vector-based GIS coverages and performs common GIS functions.
UVIEW	Visualization program used to compose planimetric maps and perspective images depicting data and analysis results.
FLY	Command script generation utility designed to produce animation path scripts for UVIEW.
UMAP	Display utility used to compose planimetric maps depicting data and analysis results.
HEIWEST	Habitat effectiveness indicator calculator for Western Oregon.
HEICALC	Habitat effectiveness indicator calculator for the Blue Mountains region of Eastern Oregon.

---

[*] UTOOLS/UVIEW programs are in the public domain. Recipients and users may not assert any proprietary rights over the programs, documentation or example data files distributed with UTOOLS.

[**] The use of trade or firm names in this publication is for reader information and does not imply endorsement by the U.S. Department of Agriculture of any product or service.

DISPLAY   Landscape diversity calculator (McNeil and Flather 1992).
OVERLAY   Utility to perform intersections using vector-based map coverages.
MAP2MOS   Utility to convert raster-based spatial databases into vector-based map
            coverages.

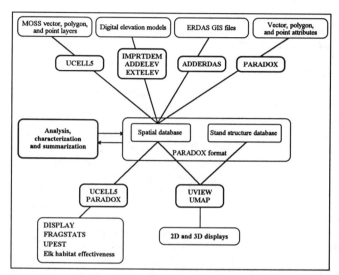

Figure 1.   Schematic of the analysis and data conversion capabilities of UTOOLS.

UTOOLS includes utilities to convert and manipulate data from a variety of sources to augment the spatial databases created by UCELL5. These utilities allow users to integrate USGS digital elevation models, raster-based GIS coverages, ERDAS GIS data coverages, and ASCII text tables containing map attributes.

UTOOLS provides data linkages to several public domain programs developed to address specific analysis problems, e.g., calculation of landscape pattern metrics (FRAGSTATS, McGarigal and Marks 1993; DISPLAY, McNeil and Flather 1992), and elk habitat assessment (HEICALC, Hitchcock and Ager 1992; HEIWEST, Ager and Hitchcock 1992).

Data in spatial databases can also be exported to ASCII raster files readable by MOSS and ARC-INFO. It is also possible to convert raster data in a PARADOX spatial database into MOSS import/export vector format. These data linkages extend the functionality of UTOOLS programs and the associated PARADOX spatial databases to encompass a broad spectrum of spatial analyses.

UTOOLS operates on IBM-compatible personal computers using the MS-DOS, Windows95, or Windows NT operating systems. Effective use of UTOOLS requires microcomputer configurations with a 66 Mhz 80486 (or faster), 4Mb of RAM, at least 300 MB of available space on the hard drive, and an SVGA display adapter. Lesser configurations are suitable for browsing the example data included with UTOOLS, but will not provide satisfactory performance for most analysis problems. UTOOLS programs can be used without PARADOX (version 3.5 and later), however, their usefulness is limited in its absence. UTOOLS programs will run from within a DOS window in

321

Windows 3.X and Windows95. UTOOLS programs will run from within a DOS window in Windows NT but the high resolution, 256-color graphics modes will not be available in UVIEW.

BUILDING SPATIAL DATABASES FOR USE IN UTOOLS

The fundamental data used by UTOOLS programs are PARADOX "spatial databases" built by converting data layers from their native formats to the PARADOX spatial database format using the UCELL5 program. A typical database construction sequence proceeds as follows:

- Assemble various project data, including vector and raster map layers, attribute data from GIS-linked databases, and USGS digital elevation models.
- Convert the vector-based layers to raster-based format and import them into PARADOX using the UCELL5 program. UCELL5 accepts polygon, line and point data in MOSS import/export format using either the State Plane or Universal Transverse Mercator (UTM) coordinate system.
- Import attribute data into PARADOX format and join the resulting database with the spatial database using a PARADOX query with the MOSS subject code as the key.
- Add elevation data to the spatial database by processing USGS digital elevation data using the IMPRTDEM and ADDELEV programs. IMPRTDEM imports the data to a PLANS (McGaughey 1991) format. In this format, the elevation data can be read by UVIEW as a terrain model or it can be added to PARADOX with the ADDELEV program.
- Create a visualization terrain model for UVIEW using elevation data contained in the spatial database using the EXTELEV utility. This operation is necessary to interpolate elevations for cell corners using the elevations for cell centers contained in the spatial database.
- Use the ADDERDAS utility to add ERDAS data layers, such as themes derived from landsat scenes obtained by the Forest Service for Region 6.
- Compute additional fields representing buffers, slope, or aspect using the UCELL5 program. PARADOX queries can be used to compute new fields representing combinations of existing fields or complex calculations using existing fields in the spatial database.

The pixel size of a spatial database can be varied according to the needs of the project. Smaller pixel sizes result in less spatial error in the vector to raster conversion process, but result in large databases that are slow to query and map. Most UTOOLS applications for watershed-scale projects use a pixel size of 30-100 meters per side resulting in databases of 30,000-250,000 records and 10-50 Mb in size.

SPATIAL ANALYSIS WITH UTOOLS

Given the UTOOLS spatial database, operations such as overlays and unions are accomplished using simple, one-step PARADOX queries. In contrast, typical vector-based GIS systems store layers and attributes separately, and thus every layer must be overlaid and associated attribute tables must be linked to complete similar operations. While the latter is acceptable when questions are straightforward, most analyses involve repetitive querying of multiple layers and attributes. Analyses generally involve comparing and cross validating data assembled from many sources, simulating effects of management activities, or designing the spatial layout of treatment areas. Seldom are the queries and map operations needed to answer these questions perfected on the first attempt. Analysis processes generally evolve as more is learned about the quality of the resource data and the effect of the data quality on the final outcome. This evolution process is hindered when data are stored as fragments of the whole ecosystem, e.g., data cataloged according to management or mapping boundaries rather than watershed boundaries. UTOOLS programs provide the ability to integrate all of the necessary information into a single database simplifying the overall analysis process and providing a means to easily accomplish multi-resource analyses. In this way, UTOOLS provides an excellent environment for developing prototype analysis and modeling schemes for later integration into corporate level GIS environments.

LANDSCAPE VISUALIZATION CAPABILITIES IN UTOOLS

UVIEW is a display program designed to produce two- and three-dimensional images of digital elevation models (DEM), attribute data stored in PARADOX spatial databases, and vegetation patterns at watershed and landscape scales. UVIEW evolved from terrain viewing software developed for the PLANS system (McGaughey 1991), the Vantage Point visualization system (Fridley et al. 1991), and the stereoplotter visualization system (McGaughey 1992). The images produced by UVIEW provide a readily understood visualization depicting spatial analysis results and existing or desired landscape conditions.

UVIEW provides a flexible system for viewing a digital elevation model. UVIEW allows users to specify exact coordinates for the head and focus locations or interactively select a head and focus location while viewing a simple perspective representation of the DEM. UVIEW simulates camera lenses with focal lengths ranging from 15mm to 400mm and vertical exaggeration values ranging from 0.1 to 4.0. Users can also "fly" over and around a low resolution image of a DEM using a mouse controlled "virtual trackball".

UVIEW can render a DEM using a variety of methods and resolutions (figure 2). Users typically use the wire frame representations, profiles and grid, for positioning and exploration of the terrain surface. They use the solid surface representations to display attribute data from the PARADOX spatial database and simulated vegetation. Computer systems equipped with a VESA compatible graphics adapter capable of displaying 256 colors in at least 640 by 480 pixel resolution can display lighted, solid surface representations. These rendered images add realism to a scene by simulating a light source and computing various shades of color depending on the orientation of the ground surface as defined by the DEM. UVIEW identifies and eliminates hidden areas in solid surface representations by sorting and drawing DEM cells starting with the cells farthest from the head location. UVIEW displays a planimetric view of the DEM, represented as contour lines using a user-specified contour interval, along with attribute data from PARADOX databases and vector and polygon data files.

UVIEW supports interactive query development allowing the user to specify combinations of database attributes to control the color of individual cells on solid surface representations. The query system supports a variety of query operators including string pattern matching, numeric comparisons, and boolean operators. UVIEW displays query results in both perspective and planimetric views.

UVIEW provides a script language to facilitate generating sequences of images. Sequences can represent a landscape from different viewpoints or changing landscape conditions from a single viewpoint. UVIEW scripts can generate images with no user interaction providing a simple method of reproducing a standard set of images depicting alternative management scenarios. The UV360FLT and FLY utilities assist with the creation of scripts by creating animation path scripts to generate "fly-through" sequences.

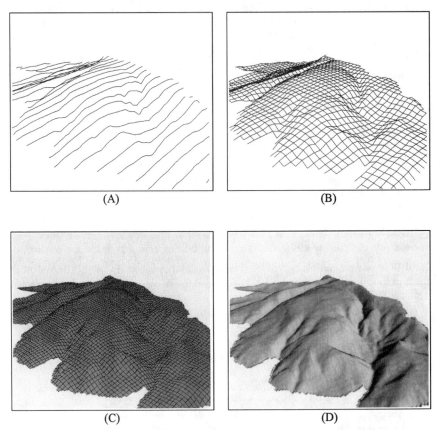

Figure 2.   UVIEW represents digital elevation models using (A) coarse and fine resolution profiles, (B) coarse and fine resolution grid, (C) solid surface, and (D) lighted, shaded surface.

UVIEW can model vegetation patterns to simulate existing or desired landscape conditions. The primary goal in vegetation modeling is to simulate overall landscape texture and pattern rather than specific, detailed vegetation structure. UVIEW provides two methods to model vegetation patterns:

- use estimates of canopy closure contained in the PARADOX spatial database to generate tree cover for each pixel,
- use vegetation codes contained in the PARADOX spatial database and a second PARADOX database containing structure definitions for each possible vegetation code.

Canopy closure based vegetation modeling represents vegetation patterns over an entire landscape. This method represents differences in stand densities well but does not represent differences in stand composition and structure (figure 3). UVIEW represents all values of canopy closure using the same type and size of tree, only the density of plants varies.

Vegetation modeling based on structure definitions represents both stand density and stand composition. Stand structure definitions consist of layer descriptions with each layer in a vegetation type described by the type of plant, height, crown diameter, crown ratio, a factor describing the variability of the size parameters, and the number of plants per unit area (acres or hectares). UVIEW represents a variety of plant types ranging from grass to mature, healthy conifer and hardwood trees. Vegetation structure descriptions can consist of up to 36 layers. Practical descriptions contain two or three layers.

EXAMPLE APPLICATIONS OF UTOOLS

UTOOLS and UVIEW have been applied to a wide variety of analysis projects. Some of the applications are specific to geographic areas but most are applicable across a broad range of conditions. All applications share a common theme: resource specialists using the analysis capabilities of UTOOLS to address specific problems that were previously difficult or impossible to solve due to technical, implementation, or access limitations with existing GIS software.

Wildlife Habitat Analysis

HABSCAPES (Mellon et al 1995), developed on the Mount Hood National Forest, provides analysis capabilities for wildlife populations in a community context on large landscapes. HABSCAPES uses spatial databases created using UTOOLS and maps its results using UVIEW. The goal of the system is to predict the occurrence of all terrestrial vertebrate and aquatic amphibian species relative to landscape pattern over large geographic areas. HABSCAPES links UTOOLS spatial databases to databases containing wildlife habitat relationships and life history characteristics using custom FORTRAN programs and PARADOX scripts. Although HABSCAPES was developed for forest-level analysis on the Mount Hood National Forest, it is finding wide application for watershed analyses and adaptive management areas throughout Region 6.

Spatial Analysis of SMART Stream Survey Data

McKinney (1995) describes the application of UTOOLS to multi-scale analysis of Forest Service stream survey data stored in the stream management, analysis, reporting, and tracking (SMART) system used in Region 6. The goal of this analysis is to better understand cause and effect relationships of both in-channel and upland interactions. Prior to the adoption of UTOOLS, there was no usable methodology to analyze and view these data. McKinney (1995) developed a procedure for linking the SMART data to stream segment maps for visualization using UVIEW and methods to analyze the data using PARADOX queries and scripts. These procedures were used to assess aquatic habitat relationships within the Columbia River basin as part of the Eastside Ecosystem Management Project.

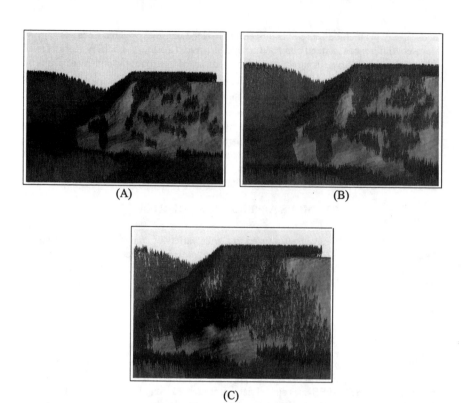

Figure 3.   UVIEW can depict vegetation patterns using canopy closure estimates for each pixel in the spatial database*** . Image (A) represents a heavy harvest with aggregated retention, image (B) represents a light harvest with aggregated retention, and image (C) represents a harvest with dispersed retention.

Starkey Ungulate Research Project

The Starkey ungulate research project is a long-term study of cattle, deer and elk interactions on the Starkey Forest and Range Experiment Station near LaGrande, Oregon (Johnson et al 1991). The project uses a Loran C animal tracking system to record animal movements throughout the grazing season. A major component of this investigation is to refine existing habitat use models for cattle, deer, and elk. As part of this effort, a wide array of habitat data for the project area have been assembled in UTOOLS databases. Animal telemetry data (locations) have also been converted to spatial database format and merged with habitat data based on their coordinates. The association of habitat data with animal use in a single data structure allows for rapid exploration of the relationship between animal locations and habitat characteristics.

*** Data and images were produced by the students of FE 444, Winter quarter 1995, College of Forest Resources, University of Washington.

With the UVIEW scripting language, movies of the telemetry data are created depicting cattle, deer, and elk movements over the grazing season on a landscape image created with a UVIEW terrain model (figure 4). Vegetation can be added to these images to create a realistic image of the landscape with animal locations. This technique provides a dynamic depiction of animal behavior over time that yields valuable insights into animal-habitat relationships and inter specific interactions.

Figure 4.   Telemetry data depicting elk, deer, and cattle locations over a one month period on the Starkey Forest and Range Experimental Forest.

Landscape Management System

The Landscape Management System (LMS), developed at the University of Washington, is designed to assist in landscape level analysis and planning of forest ecosystems by automating the tasks of stand projection, graphical and tabular summarization, stand visualization, and landscape visualization (Oliver and McCarter 1995). LMS includes many separate programs that simulate stand growth and development, produce graphical or tabular displays, and store or process inventory information. The primary function of the LMS program is to connect these diverse programs into a cohesive system. LMS uses UTOOLS and UVIEW to produce landscape visualizations of forest conditions over time using a variety of alternative management strategies.

SOFTWARE DISTRIBUTION

The UTOOLS distribution consists of three files. UTOOLS1.EXE and UTOOLS2.EXE contain the UTOOLS programs and documentation and UTOOLS3.EXE contains an example data set for use with UTOOLS and the tutorial exercise described in the manual (UMANUAL.DOC in the UTOOLS1.EXE archive).

UTOOLS and UVIEW can be obtained via anonymous FTP using the INTERNET at forsys.cfr.washington.edu. Use "anonymous" as your login name and your internet email address as a password. The UTOOLS1.EXE, UTOOLS2.EXE, and UTOOLS3.EXE files are stored in the /pub/software/utools directory.

UTOOLS and UVIEW home pages and links to the FTP site mentioned above are maintained on the world wide web at:

http://forsys.cfr.washington.edu/utools.html
http://forsys.cfr.washington.edu/uview.html.

U.S. Forest Service users can retrieve UTOOLS and UVIEW from the following RIS site:

Host:	R06F14A
Staff:	PUBLIC
Drawer:	UTOOLS
Folder:	UTOOLS
Files:	UTOOLS1.EXE
	UTOOLS2.EXE
	UTOOLS3.EXE

UTOOLS and UVIEW can also be obtained, by written request, from the authors.

SUMMARY

UTOOLS meets a critical need for simple and efficient spatial analysis tools by removing many of the barriers commonly found in large GIS systems. UTOOLS programs address the daily analysis needs of resource professionals charged with managing large land areas to provide a variety of commodity and non-commodity outputs. Because UTOOLS is easy to learn, runs on commonly available computer hardware, and utilizes existing data; it provides an excellent learning platform for resource specialists. Using UTOOLS programs, they can explore the capabilities of GIS while conducting real-world analyses to support and enhance their resource management decisions.

ACKNOWLEDGMENTS

We are indebted to Mark Hitchcock, David Hatfield and Bill Connelly for their contributions to UTOOLS software. Ed Pugh, Jim Merzenich, Ken Tu, Shaun McKinney, Bill Connelly, Bob Clements, Dave Kendrick, Alison Reger, Jim McCarter, Jeremy Wilson, and many others provided valuable feedback on early versions of this software.

LITERATURE CITED

Ager, A.A., and M.E. Hitchcock. 1992. Microcomputer software for calculating the Western Oregon elk habitat effectiveness index. USDA For. Serv., Gen. Tech. Rep. PNW-GTR-303. Portland, OR: U.S. Department of Agriculture, Forest Service, Pacific Northwest Research Station.

Ager, A.A. and R. McGaughey. 1994. Operations manual for UTOOLS. Pendleton, OR: U.S. Department of Agriculture, Forest Service, Umatilla National Forest.

Fridley, J.L., R.J. McGaughey, and F.E. Lee. 1991. Visualizing engineering design alternatives on forest landscapes. American Society of Agricultural Engineers paper #917523. Presented at the 1991 International Winter Meeting. Chicago, IL. December 17-20.

Hitchcock, M.E. and A.A Ager. 1992. Microcomputer software for calculating elk habitat effectiveness index on Blue Mountain winter range. Gen. Tech. Rep. PNW-GTR-301. Portland, OR: U.S. Department of Agriculture, Forest Service, Pacific Northwest Research Station.

Johnson, B.K., J. Noyes, J.W. Thomas, and L. Bryant. 1991. Overview of the starkey project: current measures of elk vulnerability. In: Elk Vulnerability - A Symposium; April 10-12. Bozeman, MT. Bozeman, MT: Montana State University: 225-228.

McGaughey, R.J. 1991. Timber harvest planning goes digital: PLANS–preliminary logging analysis system. Compiler 9(3):10-17.

McGaughey, R.J. 1992. Three-dimensional terrain visualization with a real-time interface to a photographic stereomodel. In: Proceedings of the fourth Forest Service remote sensing application conference; April 6-11, 1992; Orlando, FL. Bethesda, MD: American Society of Photogrammetry and Remote Sensing. 204-211.

McGarigal, K. and B.J. Marks. 1995. FRAGSTATS: spatial pattern analysis program for quantifying landscape structure. Gen. Tech. Rep. PNW-GTR-351. Portland, OR: U.S. Department of Agriculture, Forest Service, Pacific Northwest Research Station.

McKinney, S.P. 1995. Spatial analysis of SMART stream survey data with UTOOLS. AquaTalk: Region 6 Fish Habitat relationship Technical Bulletin 10 (September). Portland, OR: U.S. Department of Agriculture, Forest Service, Region 6 Regional Office.

McNeil, B., and C. Flather. 1992. Operations manual for DISPLAY. Internal publication. Fort Collins, CO: U.S. Department of Agriculture, Forest Service, Rocky Mountain Forest and Range Experiment Station.

Mellon, K., M. Huff, and R. Hagestedt. 1995. HABSCAPES interpreting landscape patterns: a vertebrate habitat relationships approach. In: Analysis in support of ecosystem management, Analysis workshop III. Fort Collins, CO. Apr. 10-13. USDA Forest Service. Ecosystem Management Analysis Center, Washington, DC. 135-145.

Oliver, C.D. and J.B. McCarter. 1996. Developments in decision support for landscape management. In: M. Heit, H.D. Parker, and A. Shortreid (ed), GIS Applications in Natural Resources 2. GIS World Books. 501-509.

AN ECOLOGICAL APPROACH TO ASSESS VEGETATION CHANGE AFTER LARGE SCALE FIRES ON THE PAYETTE NATIONAL FOREST

Susan L. Boudreau
Payette National Forest
Ecology Program
McCall, Idaho
208-634-0700 ext.745

and

Paul Maus
Remote Sensing Application Center
Salt Lake City, Utah
801-975-3660

ABSTRACT

The summer of 1994 proved to be the most active fire season in the Payette National Forest's recorded history. Three fires ignited in early August: the Corral, Blackwell, and Chicken Complexes burning over 290,000 acres. These three fires covered a broad range of diverse vegetative communities, associated landtypes, and hydrologic features. Analysis of the ecological changes caused by fire and the ensuing management opportunities, was critical.

The ability to quantitatively document burn mosaics is integral to the Payette National Forest's effort to implement an ecological approach to management. Landsat TM satellite imagery and high altitude photography became an integral part of a Forestwide Geographical Information System (GIS) and allowed for processing and analyses of resource data. With support from the Remote Sensing Applications Center (RSAC) a pre- and post-burn vegetation classification was developed for the Forest's GIS describing the vegetation that existed before the fire and the severity of the burn within those vegetation types after the fire.

INTRODUCTION

Traditionally most inventories conducted by the Forest Service have found each functional unit inventorying and mapping landscapes according to its specific needs and purpose. Such a fragmented approach is incapable of addressing the many complex processes shaping today's ecosystems. It is imperative that efficient and consistent methods are utilized in sampling abiotic and biotic attributes that provide the characterization of ecosystem composition, structure, pattern, and function (Spies 1994). The key message coming from various national meetings, is the need to move rapidly toward integrated resource management, develop needed tools and methodology, and have consistent approaches to ecosystem management. As stated by Jack Ward Thomas in mid-November 1993, "...the Forest Service has reached an historical crossroads for natural resource management. We face the immediate challenge of managing resources in an integrated and coordinated manner, that is scientifically sound and ecologically-based, linking all elements of the

330

landscape to meet the ever-changing demands of human beings, under ever-changing natural conditions." There can no longer be any misunderstanding regarding the expressed paradigm shift within the Forest Service. Implementation however, is proving to be much more difficult than philosophical expression.

The Payette National Forest, like many other Forests within the National system, is attempting to shift it's stewardship focus from site-specific functionalism to that of ecologically-driven landscape management. During the summer of 1994 the Payette experienced three major concurrent wildland fires: the Blackwell, Corral and Chicken Fire Complexes. The combined perimeters exceeded 290,000 acres. The fires burnt in cover types ranging from low elevation Ponderosa pine/savannah to high elevation spruce-fir. Although fire is but a single change agent associated with broadscale ecological processes, these fires and the associated post-fire management opportunities are significantly influencing present management direction.

The Ecology Program on the Payette National Forest was charged with providing, within a Geographical Information System environment, an informational databases for biotic (i.e., existing vegetation) and abiotic (i.e., landtypes and landtype associations) resources. If indeed broadscale ecological assessment is to be achieved in a timely and economical manner, available technologies which support such assessment must be used. The aftermath of these massive fires provided an ideal opportunity for the Payette N. F. in cooperation with the Remote Sensing Applications Center (RSAC), Salt Lake City, Utah, to use satellite imagery to develop a pre- and post-fire vegetation classification. Similarly, TM imagery coupled with high elevation NASA photography acquired during the summer's fires, is being used to assess and map both fire intensity and severity.

The 1994 fires had given the Payette a rare opportunity to evaluate a "change of condition" over a broad landscape and monitor plant succession within the Englemann spruce (*Picea engelmannii*) and subalpine fir (*Abies lasiocarpo*) zone. This paper will discuss the outcomes of the process and examine the role of Remote Sensing and GIS as an important tool for taking a broadscale ecological approach to management of landscapes.

CASE STUDY

Broadscale Area

The 1994 fires encompassed 290,000 acres within the Idaho Batholithic Section (Bailey 1994) in southcentral Idaho. This includes the entire South Fork Salmon River drainage, the North Fork Payette River drainage above Payette Lake, and those watersheds draining directly into the Salmon River on the New Meadows and McCall Ranger Districts of the Payette National Forest (Figure 1).

331

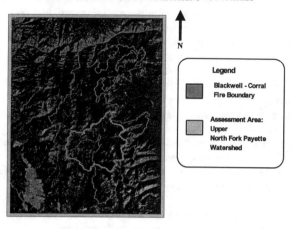

Figure 1. Landsat TM Image - October 7, 1994

Idaho Batholithic Section is characterized by mountains with alpine ridges and cirques at higher elevations (McNab and Avers 1994). The large u-shaped valleys with broad bottoms indicate that the area has been strongly glaciated. Mature surfaces are dissected with major drainages deeply incised, resulting in steep breaklands. Soils are generally shallow to moderately deep, with loamy to sandy textures. Volcanic ash accumulations in some soils causes them to be very productive. Precipitation ranges from 20 to 80 inches (510 to 2,030 mm) mostly occurring during fall, winter, and spring as snow. Climate is maritime-influenced, cool temperate with dry summers. Mean air temperature ranges from 35 to 46 F (-4 C) in the high mountains. Growing season ranges from 45 to 100 days.

The landscape is composed of a dynamic array of forest, shrubland, grasslands, and aquatic ecosystems. The forests are dominated within the Englemann spruce (*Picea engelmannii*) and subalpine fir (*Abies lasiocarpo*) zone with a mixture of Ponderosa pine (*Pinus ponderosa*), lodgepole pine *(Pinus contorta)*, and white bark pine (*Pinus albicaulis*), western larch (*Larix occidentalis*). This highly diverse landscape has been influenced by a variety of ecological processes including fire, insects, disease, and variations in climate and weather patterns. Fire one of the primary ecological change agents, has been significantly altered by three factors: 1. unregulated grazing by cattle and sheep in the late 1800's removed large proportions of grass and other light fuels needed for lightning-ignition and initial spread of fire (Jones 1989); 2. A reduction in the numbers of Indian ignited fires after 1900 (Barret and Arno 1982; Arno 1983; Gruell 1983); 3. the advent of effective, organized fire suppression beginning in 1934 (Steele et al. 1986; Barrett 1988). Driven by economic considerations, these changes in forest resource management and utilization have resulted in altered fire regimes throughout the landscape.

Assessment Area

The assessment area chosen for this study was the Upper North Fork Payette River Watershed located within the Blackwell and Corral Fire Complex (Map 1).

METHODS

Satellite Images

The classification was derived from two Landsat TM satellite images acquired pre-fire September 17, 1993 and post-fire October 7, from the Earth Observation Satellite Company (EOSAT).

Reference Data

Ground reference data was used in the satellite image classification process as will as for the assessing accuracy of both the photo interpreted map and satellite image classification (Table 1).

Level 1: Cover Class	Level 2: Crown Closure	Level 3: Types	Level 4: Size Class	Level 5: Structure	Burn Intensity
Non-vegetation (<10% veg cover) 1 - rock 2 - water 3 - snow 4 - duff 5 - bare ground 6 - other non-veg Vegetated 7 - forested (>= 10% tree crown closure) 8 - Coniferous 9 - Deciduous 10 - Non-forested 11 - Shrub (>=20% shrub) 12 - Grass/forb (<20% shrub)	Non-forested (<10% tree crown closure) 1 - Non-forested/sparse Forested 2 - Open (10-33% closure) 3 - Moderate (34-66% closure) 4 - Closed (>66% closure)	1 - Non-vegetated 2 - Wet grass/forb 3 - Dry grass/forb 4 - Wet Shrub 5 - Dry shrub 6 - Deciduous/ Cottowood-Aspen 7 - Conifer spp.1 8 - Conifer spp. 2 9 - Conifer spp.	1 - non-vegetated 2 - grass/forb 3 - low shrub (>50% <=2.5 ft tall) 4 - tall shrub (>50% >2.5 ft. tall) 5 - seedling/sapling (<5 in. d.b.h.) 6 - seedling/sapling (>seedling/sapling >=20%and pole+ medium+large trees are also >= 20%) 7 - pole: 5-8.9 in. d.b.h. 8 - medium: 9-15.9 in. d.b.h. 9 - large: 16-24 in. d.b.h. 10 - very large: >24 in. d.b.h.	Decision for structure classes has not been finalized.	Unburned: patches were areas which did not burn in 1994. Low intensity: burn patches were characterized by occasional scorched tree canopy areas (red needles), and a mosaic pattern of lightly burned ground litter. Moderate intensity: burn patches had red needles or foliage on 50% to 100% of the tree crowns or shrubs. Ground char was variable. High intensity: burn patches exhibited both continuous ground and crown fires. The canopy was completely consumed leaving only black needleless trees and leafless shrub stems or root collars.

Table 1. Satellite Imagery Classification Scheme for Existing Vegetation: Pre & Post-Fire Assessment (compiled by S.Boudreau).

One hundred eighty-six ground locations were visited in the Corral-Blackwell fire complex between early September and late October, 1994. Ground locations were georeferenced using Global Positioning Satellites (GPS) and these locations were differentially corrected to an accuracy of five meters. All site information was input into the Payette National Forest's geographic information system (GIS) and linked to it's GPS determined ground location. Data collected at the ground site locations were intended to represent a homogeneous area of one acre.

Pre-Fire Classification

An unsupervised classification was initially produced by the ISODATA clustering routine using ERDAS Inc. Software to produce 60 spectral classes. To improve this clustering, a supervised classification using a minimum of three training sites per spectral class were identified on the TM image covering the range of elevation, aspect, and geographic area for that spectral class. A minimum size of 10 pixels (2.5 acres) was used to adequately locate each training site. Approximately 200 training sites were simultaneously located on the TM image and 1:15,840 aerial resource photographs. Training sites were photo interpreted by Payette Forest personnel familiar with the landscape using the same classification criteria defined in Table 1. This data was used to supervise the pre-burn TM classification. The same procedure was duplicated for the post-burn TM classification. Similarities between spectral statistics for each classification were compared using a complete linkage cluster analysis with Statistical Analysis Software (SAS).

To date, Level 1: Cover Classes, Level 2: Crown Closure, and Level 3: Types have been completed. Level 4: Size Class and Level 5: Structure have not been completed (Figure 2).

Landsat Thematic Mapper Burn Classification

An unsupervised classification was completed using ISODATA program from ERDAS image processing software (ERDAS, Inc., Atlanta, GA). Sixty spectral classes were produced by the unsupervised classification. These were subsequently recorded to one of four possible burn intensity class values (Table 1) based on visual analysis, input from the Payette National Forest ecologists who collected the ground data, and selected ground data points with associated burn intensity information. Figure 3 displays the map of burn intensity resulting from the satellite image classification. Pixels were classed for fire intensity only within the perimeter of the fire complexes. Those ground sites used in the classification process (a total of 15 sites) were not used during the accuracy assessment phase.

GIS Analysis

The pre-burn vegetation was classified by density and eight cover type classes (Figure 2) and then summarized within each burn intensity type (Figure 3). A final image of Vegetation Loss (Figure 4) was produced by overlaying Figure 2 and Figure 3.

Level 1: Cover Class / Level 2: Crown Closure	Density
Water	446
Rock	1192
Grass/Forb	2742
Shrub	11305
Hardwoods	300
Open Conifer 10 - 33% closure	5563
Moderate Conifer 34 - 66% closure	15069
Closed Conifer 67 - 100% closure	18650
Total	54967

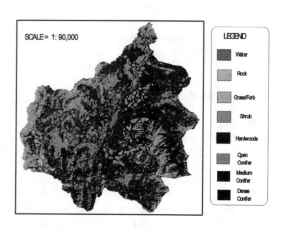

Figure 2. Vegetation Cover Density Layer Within the Assessment Area: Upper North Fork Payette River Watershed Pre-fire.

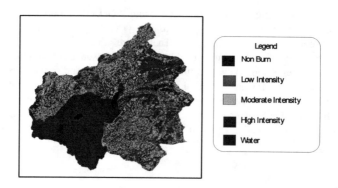

Figure 3. Burn Intensity Layer Within The Assessment Area: Upper North Fork Payette River Watershed (Scale = 1:90,000).

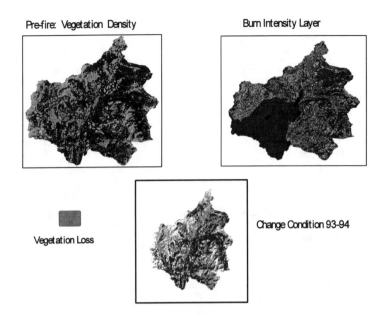

Pre-fire: Vegetation Density

Burn Intensity Layer

Vegetation Loss

Change Condition 93-94

Figure 4. Vegetation Changes Within The Assessment Area: Upper North Fork Payette River Watershed (Scale = 1:90,000).

POTENTIAL APPLICATIONS

The Payette National Forest faces the immediate challenge of managing forest and grassland resources according to defensible broadscale ecological principles. Ecosystem health themes of biodiveristy, watershed stability, water quality and long-term site productivity are replacing multiple-use economic themes. Management actions supporting *Ecological Stewardship* are directed at maintaining or restoring ecosystems and the associated ecological structure, composition and function (Thomas 1995).

A combination of automated image classification and visual image interpretation can be used to detect changes in vegetation after disturbance. The aftermath of 1994 fires resulted in significant changes throughout the landscape. Analysis of these changes relative to the original vegetation type, allowed a better understanding of spatial landscape qualities and prove to be an effective tool for further analysis. For example we are able to:

1. Describe large-scale patterns in ecological systems and provide information to support guidelines for assessing the impacts of the landscape distribution of disturbances (i.e., management activities, natural processes) on biodiversity (i.e., coarse filter), as projected through time.

2. Have easy access to long-term, continuous data over a broadscale area than has previously been practical thru traditional means.

3. Detect changes in successional patterns, phenology, and production.

4. Locate permanent vegetation plots for long-term monitoring of successional pathways and TES plant species.

5. Track the historical range of variation of vegetative communities necessary for management planning.

6. Identify tree mortality (i.e., ecosystem health).

7. Describe structural features of mature and old-growth forest ecosystems.

8. Assess wildlife habitat diversity resulting from changing landscape mosaics created by disturbance.

9. Model and simulate landslide prone areas over time.

10. Identify of severely burned areas for appropriately designing salvage logging opportunities.

ACKNOWLEDGMENTS

A program will only succeed if the people who believe in it strive to make it happen. This was apparent, by the overwhelming commitment and professionalism of the folks who were involve in this study. They were all highly trained and intensely motivated professionals who obviously had a passion for the welfare of natural resources.

We thank Rob Morrow, Brad Sanders, Calvin Farris, Marilyn Olson, and Francis Husso from the Payette National Forest - Ecology Program, USDA Forest Service for their support and hard work even when the times got hard and money was limited. Special thanks to the team at the Remote Sensing Application Center, Henry Lachowski, Jay Powell, and Tim Wirth. They never said "NO" and always found away to address the bazaar requests from us.

REFERENCES & SUGGESTED READINGS

Arno. S. F. 1983. Ecological Effects and Management Implications of Indian Fires. *In* proceedings: *Wilderness Fire Symposium*. USDA Forest Service. General Technical Report INT-182. Ogden, Utah. 5 pp.

Barrett, S. W. And S. F. Arno. 1982. Indian Fires as an Ecological Influence in the Northern Rockies. Journal of Forestry. 5 pp.

Gruell, G. E. 1983. Indian Fires in the Interior West: A Widespread Influence. *In* proceedings: *Wilderness Fire Symposium*. USDA Forest Service. General Technical Report INT-182. Ogden, Utah. 6 pp.

Diaz, Nancy and Dean Apostal. 1992. Forest Landscape Analysis and Design. USDA Forest Service, PNW region. R6 Eco-TP-043-92. 62pp.

ECOMAP. 1993. National Hierarchical Framework of Ecological Units. USDA Forest Service, Washington D.C., 21pp.

ECOMAP. 1994. National Hierarchical Framework of Aquatic Ecological Units In Northern America. USDA Forest Service, Washington, D.C., 92pp.

EOSAT - Earth Observation Satellite Company. 5850 T.G. Lee Coulevard-Suite 650, Orlando, Florida 32822. 407-856-7828.

ERDAS, Inc. 2801 Buford Highway, NE - Suite 300, Atlanta, Georgia 30329-2137. 404/248-9400.

Fox III, Lawrence and John D. Stuart. 1994. Detecting Changes In Forest Condition Following Wildfire, Using Image Processing and GIS. Presented at the ASPRS/ACSM 1994 Annual Convention, Reno Nevada. April 25-28, 1994.

Jones, Melanee. 1989. History of Early Livestock Graxing in the Area of the Payette National Forest. Regional Office Historic Files.

McNab and Peter E. Avers. 1994. Ecological Subregions of the United States: Section Descriptions. U.S.D.A. Forest Service WO-WSA-5.

Spies, Thomas A. 1993. Ecological Perspective: The Nature of Mature and Old-Growth Forest Ecosystems. *In:* Remote Sensing and GIS in Ecosystem Management. Edited by V. Alaric Sample Island Press, Washington D.C.

Steele, R., R.D. Pfister, R.A. Ryker, and J.A. Kittems. 1981. Forest Habitat Types of Central Idaho. General Technical Report INT-114. Intermountain Forest and Range Experiment Station. USDA Forest Service. Ogden, Utah. 138 pp

Thomas, Jack Ward. 1995. Ecological Stewardship Workshop. Tucson, Arizona. In draft.

RESOURCES AT RISK:
THE BOISE NF FIRE-BASED HAZARD/RISK ASSESSMENT

T.A. Burton, D.M. Dether, J.R. Erickson, J.P Frost, L.Z. Morelan, W.R. Rush,
J.L. Thornton and C.A. Weiland
Boise National Forest
Boise, ID 83702

L.F. Neuenschwander
University of Idaho
Moscow, ID 83843

ABSTRACT

On the 2.6-million-acre Boise National Forest (NF) in southwestern Idaho, wildfires
have burned nearly 50 percent of the ponderosa pine forest over the last nine years.
Much of this forest has burned with uncharacteristic intensity. Ponderosa pine forests are
now among the most endangered and threatened ecosystems in the U.S. The historic fire
regime -- one marked by nonlethal surface fires that removed dense understories of
saplings or pole-sized trees and increased nutrient availability -- has changed. The
altered fire regime now results in severe, stand-replacing fires that kill large areas of
forest and return them to grass- and shrub-dominated landscapes. Preliminary analysis
shows the remaining ponderosa pine on the Boise NF could be lost within the next 20
years to severe, stand-replacing wildfire.

In partnership with the University of Idaho, the Boise NF has developed a Geographic
Information System (GIS)-based "hazard/risk assessment" model that estimates where the
forest ecosystems are most at risk to severe, large wildfires burning in conditions outside
the historical range of variability (HRV), and evaluates important resources at risk to
these fires. The hazard/risk assessment links five submodels. When the submodels are
linked, the assessment estimates where severe, large wildfires burning in conditions
outside HRV would severely deplete late-successional habitat needed by old-growth-
dependent and other wildlife species, accelerate naturally-high levels of erosion and
sedimentation, and increase the likelihood that identified fish populations will not persist.

The hazard/risk assessment is most appropriately used to approximate the relative size
and extent of the fire-based ecosystem problem on the Forest -- the result of excluding
fire from fire-adapted ponderosa pine ecosystems. It is intended to "nest" between the
large-scale analysis undertaken as part of the Upper Columbia River Basin assessment,
and the site-specific evaluation performed for landscape- and project-level analysis.

INTRODUCTION

The Boise NF has an especially acute focus on forest ecosystem health:

**Its ponderosa pine forests are among the endangered and threatened
ecosystems in the U.S. (Noss et al, 1995).**

Historically maintained by frequent, low-intensity fire, the 1.1-million acres of ponderosa
pine forests encompassed by the Boise NF have been altered by decades of fire
suppression, grazing and logging that removed fire-adapted species. In these and other

areas throughout the Interior West, ponderosa pine forests are now dominated by dense stands of Douglas-fir and other fire-sensitive species (Noss et al, 1995).

When wildfires now occur in ponderosa pine forests with altered fire regimes, they are more intense, severe and larger than traditionally experienced. The historic, nonlethal surface fires that removed dense understories of saplings or pole-sized trees and increased nutrient availability have been succeeded by stand-replacing fires that return large areas of forest to grass and shrubland (Crane and Fischer, 1986).

On the Boise NF, wildfires in ponderosa pine forest have been increasingly large and severe since 1986. Nearly 500,000 acres of National Forest land (about 50 percent of the Boise NF's ponderosa pine forest, and almost 20 percent of the land managed by the Forest) have burned. Many of these acres have burned with uncharacteristic intensity. Costs to suppress these fires and undertake emergency watershed rehabilitation exceeded $100 million dollars. In many severely burned areas, soil productivity, and aquatic, wildlife and plant habitat, have been critically damaged (USDA Forest Service, Boise NF, 1992; 1995).

Preliminary analysis shows the remaining ponderosa pine forest could be fragmented, with only isolated pockets remaining, within the next 20 years (Neuenschwander, 1995). To respond to this threat to the Forest's ponderosa pine ecosystem, a Forest interdisciplinary team, working in partnership with the University of Idaho, has developed a GIS-based "hazard/risk assessment."

The assessment estimates on a relative, Forestwide basis where forest ecosystems are most at risk to severe, large wildfires burning in conditions outside the historical range of variability (HRV), and evaluates important resources at risk to these fires. The hazard/risk assessment links five submodels -- forested vegetation outside HRV, fire ignition, wildlife habitat persistence, watershed hazard (erosion and sedimentation potential) and fisheries condition. When linked, these submodels estimate where severe, large wildfires burning in vegetation conditions outside HRV would alter the composition, structure and function of an ecosystem by:

- severely depleting late-successional habitat needed by old-growth-dependent and other wildlife species;
- accelerating naturally-high levels of erosion and sedimentation; and
- increasing the likelihood that identified fish populations will not persist.

DEVELOPMENT OF THE HAZARD/RISK ASSESSMENT

In developing the hazard/risk assessment, the team used GIS tools, state-of-the-art computer software designed to process and analyze spatial information.[*]

---

[*]The assessment was written using ARC/INFO Version 7.03 and uses automated machine language (AML) to process data in the GRID, ACRPLOT, ARCEDIT and TABLES modules. Most of the analysis was performed using rasterized data in the GRID module, ARCPLOT for graphic output, and TABLES for reports. Data was analyzed and displayed on a system that included an IBM RISC-6000 "390" server and AIX 3.2.5 operating system on a Thinwire Ethernet local area network (LAN).

The assessment was formulated through the following steps:

1.   Five GIS submodels were first created to evaluate hazards for specific resources. These submodels included forested vegetation outside HRV, fire ignition, wildlife habitat persistence, watershed hazard (erosion and sedimentation potential), and fisheries condition.

2.   For each of the five submodels, a relative hazard rating, ranging from 1 (lowest) to 5 (highest), was assigned to each subwatershed. (The 378 subwatersheds on the Boise NF are drainages averaging 6,000 acres in size.)

The submodels and sample hazard ratings include:

Forested Vegetation Outside HRV

Locates areas where ponderosa pine is or once was climax or a major seral species, and examines the density of the forested vegetation in these areas (based on June, 1992 LANDSAT satellite imagery classification). Subwatersheds with moderate to high hazard (3 or higher on the 1-5 scale) for this submodel are those where 25 percent or more of the subwatershed consists of moderate or dense Douglas fir, ponderosa pine and grand fir; Douglas-fir and ponderosa pine; or Douglas-fir.

Fire Ignition

Evaluates where fires - both lightning- and human-caused - have historically started (1956-1994), based on Boise NF fire records. Subwatersheds with moderate to high risk (rated 3 or more on the 1-5 scale) are those where 4 or more fire starts have occurred in any one section (640 acres) throughout the 39-year fire history. This submodel assumes fire starts will continue to occur where they have historically.

Wildlife Habitat Persistence

Examines where large, extensive areas of late-successional forested habitat occur outside HRV, and where they would be limited following a stand-replacing fire. Subwatersheds with moderate to high hazard (rated 3 or higher on the 1-5 scale) for wildlife habitat persistence include those where 15 percent or more of the subwatershed would remain as late-successional habitat following wildfire, with only one (or no) patch at least 350 acres in size. (Low-elevation subwatersheds which primarily consist of grass, brush and shrublands are not included in this analysis.)

Wildfire burning in an altered regime in dense, late-successional habitat could alter the successional pathway, changing the current vegetation structure to shrub/brushfields and displacing or eliminating populations dependent on the late-successional habitat for several hundred years. Large, severe wildfire could also result in ecosystem simplification, with greater landscape homogeneity, and loss of biodiversity (including genetic diversity) [Neuenschwander, 1995].

Watershed Hazard (Erosion and Sedimentation Potential)

Evaluates potential natural sediment yield, as determined from landtypes (areas with similar soils and landforms, and therefore similar hazards and capabilities). Moderate to high subwatersheds (rated 3 or higher on the 1-5 scale) for watershed hazard are those with an average potential natural sediment yield of 35 tons/square mile/year or more.

Fisheries Condition

Selects spring/summer chinook salmon and bull trout as indicator species, because in Idaho chinook have been listed as "endangered," and bull trout as "warranted but deferred," under the Endangered Species Act of 1973. For chinook salmon, moderate and high hazard subwatersheds (rated 3 or higher on the 1-5 scale) are those where spawning and rearing habitat for chinook salmon exists. For bull trout, moderate and high hazard subwatersheds are those where within strong regional populations, there is risk that local populations will not persist: those populations relatively lower in abundance, smaller in areal extent, isolated from other populations and therefore less likely to recover from uncharacteristic fire.

The fisheries condition submodel assumes that large wildfires burning in conditions outside HRV would lead to environmental disturbances (floods, etc.) that decrease the likelihood of persistence for those fish populations low in abundance (chinook salmon) or important to regional populations (local bull trout populations).

3. An overall "high risk" rating was assigned to a subwatershed if it received moderate ("3") or higher hazard ratings from ALL FIVE submodels.

4. As shown in Figure 1, a watershed was rated as "high risk" if at least ONE subwatershed within it received an overall high risk rating. (The 82 watersheds on the Boise NF are larger drainages, about 30,000 acres in size, which consist of several subwatersheds.) This assignment reflects the Forest's observation that the recent uncharacteristic wildfires are burning across vast landscapes and entire watersheds.

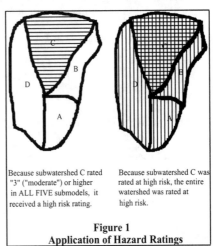

Because subwatershed C rated "3" ("moderate") or higher in ALL FIVE submodels, it received a high risk rating.

Because subwatershed C was rated at high risk, the entire watershed was rated at high risk.

Figure 1
Application of Hazard Ratings

RESULTS

The hazard/risk assessment was designed in part to answer two questions:

Where are forest ecosystems most at risk to severe, large wildfires burning outside HRV?

- Based on current information and analysis, the forest ecosystems most at risk to severe, large wildfires burning outside HRV include large areas of moderate and dense forest where ponderosa pine is or was a major seral species, and where moderate to high numbers of fires have occurred. By linking the fire ignition submodel (which can identify those subwatersheds with moderate to high levels of fire ignition), with the forested vegetation outside HRV submodel (which can identify those subwatersheds with moderate to high hazard for forested vegetation outside HRV), the assessment estimates that up to 152 subwatersheds (total of 1,196,781 acres) are those most at risk to severe, large wildfire burning in vegetation conditions outside HRV (Figure 2).

What important resources are at risk to these severe wildfires?

- To determine what important resources are at risk to these fires, the hazard/risk assessment estimated where severe, large wildfires burning in vegetation conditions outside HRV would affect specific wildlife, watershed and fisheries resources. By linking all five submodels included in the assessment, analysis indicates that in 20 watersheds (total of 610,389 acres), all of these important resources could be affected by severe, large wildfires burning in vegetation conditions outside HRV (Figure 3).

Comparative information is illustrated in Table 1:

Table 1
Forest At Risk

	Acres	% of Forest 1/
Most at Risk to Large, Uncharacteristic Fire	1,196,781 2/	40%
Important Resources Affected by Fire	610,389 3/	20%

1/ Percentage relative to Boise NF encompassed area of 3,000,000 acres, as captured in 1992 LANDSAT satellite imagery. Includes about 350,000 intermingled State, other Federal and private land. Net Boise NF is about 2,650,000 administered acres.

2/ Figure represents all acres within 152 subwatersheds, including some grass and shrublands, subalpine fir, etc.

3/ Figure represents all acres within 20 watersheds, including some grass and shrublands, subalpine fir, etc.

344

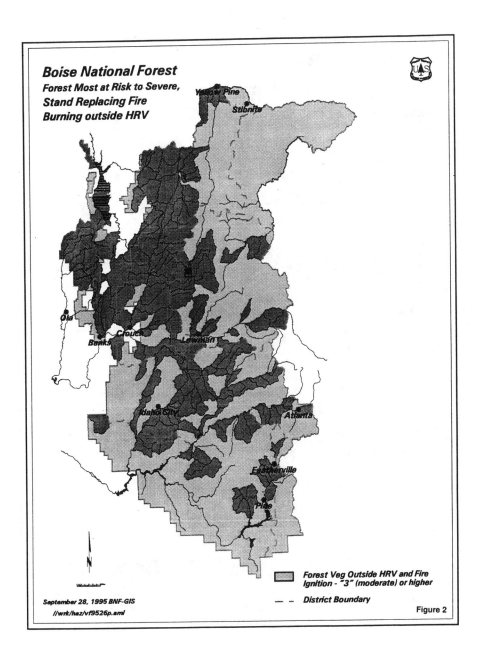

Boise National Forest
Forest Most at Risk to Severe,
Stand Replacing Fire
Burning outside HRV

Yellow Pine
Stibnits
Ola
Crouch
Banks
Lowman
Idaho City
Atlanta
Featherville
Pine

N

Forest Veg Outside HRV and Fire
Ignition - "3" (moderate) or higher

District Boundary

September 28, 1995 BNF-GIS
//wrk/haz/vf9526p.aml

Figure 2

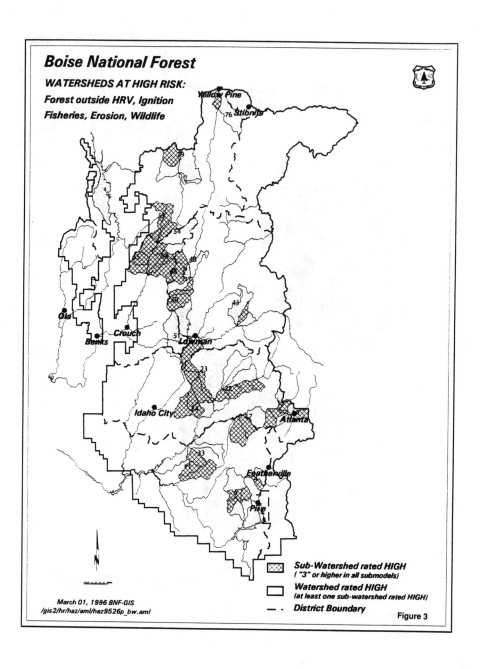

Boise National Forest

WATERSHEDS AT HIGH RISK:
Forest outside HRV, Ignition
Fisheries, Erosion, Wildlife

Sub-Watershed rated HIGH
("3" or higher in all submodels)

Watershed rated HIGH
(at least one sub-watershed rated HIGH)

District Boundary

March 01, 1996 BNF-GIS
/gis2/hr/haz/aml/haz9526p_bw.aml

Figure 3

APPLICATIONS AND CONCLUSIONS

The hazard/risk assessment is designed to evaluate the relative size and extent of the Boise NF's challenge in managing sustainable, resilient and resistant ponderosa pine ecosystems. It also tells land managers where to "go look closer" -- where to begin evaluating site-specific conditions at a finer scale, where to begin determining a "desired future condition" for a landscape at risk, and finally, where and what specific projects might be designed and undertaken, if needed, to begin restoring sustainable ecosystem conditions across the landscape.

The assessment is intended to "nest" between the large-scale analysis undertaken as part of the Upper Columbia River Basin assessment, and the more site-specific evaluation performed for watershed- and landscape- or project-level analyses. The assessment is compatible with the Forest Service National Hierarchical Framework of Ecological Units. The Forest lies in Section M332A (Idaho Batholith) of Province M332 (Middle Rocky Mountain Steppe - Coniferous Forest - Alpine Meadow) [McNab et al, 1994]. Habitat types developed as a basis for the wildlife persistence model were in turn developed based on "section" information established by the Upper Columbia River Basin assessment. Information from the hazard/risk assessment can thus be aggregated to ecological "sections" at a larger scale.

Because the assessment was developed to analyze conditions on a Forestwide basis, it should not be used for more site-specific watershed- and landscape- or project-level work without further evaluation and refinement.

The hazard/risk assessment represents an important addition to the "analysis toolbox" available to today's land managers. It recognizes the potential for large, severe wildfires burning with altered fire regimes to damage important resources, and to substantially interrupt successional pathways historically not experienced on the large scale we see today. Because it focuses on potential effects to fisheries populations, late-successional wildlife habitat, and sedimentation, the hazard/risk assessment highlights the consequences of severe, stand-replacing fire burning outside historical patterns to disturb the dynamics of an entire ecosystem.

Given the potential loss of ponderosa pine-dominated forests on the Boise NF in the next 20 years, the hazard/risk assessment can be a primary tool for prioritizing areas most at risk, for further evaluation. The model's structure is particularly well suited to examine situations like this in which time and resources for assessment and resolution are limited, because the assessment uses selected criteria to progressively narrow the area of consideration to one which is "do-able."

The assessment's use of GIS as the "modeling medium" is particularly appropriate in examining landscape conditions, because GIS can analyze large amounts of data and sophisticated relationships across extensive areas. Since GIS is a widely-used, state-of-the-art analysis tool, it lends itself especially well to sharing information among resource specialists from different agencies and organizations. It also facilitates expansion of the hazard/risk assessment to incorporate different ownerships and boundaries (if desired), since the challenges to ecosystem health cross jurisdictional boundaries and affect resources and resource users at many scales.

Forest scientists recognize that to restore the resistance and resilience of ecosystems with altered fire regimes, land managers must use several tools, including fire and timber harvest (Agee, 1995; Mutch, 1995). The Forest will need to conduct low-intensity fire under prescribed conditions, to begin restoring fire-dependent ecosystems, as well as to remove ground fuels and recycle nutrients; and thinning to remove less fire-resistant trees such as Douglas-fir and grand fir, while leaving the larger, fire-resistant ponderosa pine. (In today's altered landscapes, thinning is needed to remove trees from dense areas where prescribed fire alone could result in a lethal, stand-replacing wildfire.) By identifying the areas most at risk, the hazard/risk assessment takes land managers "to the ground" to look closer, with the possible outcome that some of these restoration treatments may be prescribed. If so, the hazard/risk assessment may then support the adjustment in the Forest's management course needed to incorporate different types of timber harvest, and more extensive use of prescribed fire, than traditionally undertaken.

References

Agee, J.K. 1995. Forest fire history and ecology of the Intermountain West. In, Inner Voice, March-April, 1995. Pp. 6-7.

Crane, M.F.; and W.C. Fischer. 1986. Fire ecology of the forest habitat types of central Idaho. General Technical Report INT-218. USDA Forest Service, Intermountain Research Station.

McNab, W.H.; and P.E. Avers. 1994. Ecological Subregions of the United States: Section Descriptions. WO-WSA-5. USDA Forest Service, Washington, D.C.

Mutch, R.W. 1995. Prescribed fire and the double standard. In, Inner Voice, March-April, 1995. Pp. 8-9.

Neuenschwander, L. 1995. Unpublished data presented to Boise NF District interdisciplinary teams; February - April, 1995.

Noss, R.F., E.T. LaRoe III, and J.M Scott. 1995. Endangered ecosystems of the United States: a preliminary assessment of loss and degradation. Biological report 28, National Biological Service, U.S. Department of the Interior. Pp. 12, Appendix A.

USDA Forest Service, Boise NF. 1992. Foothills Wildfire Timber Recovery Project. Environmental Assessment. Pp. 3, 29-31.

USDA Forest Service, Boise NF. 1995. Boise River Wildfire Recovery. Final Environmental Impact Statement. Pp. III-38-43, III-51-52, III-75-77, III-79, III-80-83.

TECHNICAL UPGRADES TO THE USDA FOREST SERVICE FIREFLY INFRARED MAPPING SYSTEM

Ronald C. Wicks
Research, Development and Engineering Center
US Army Missile Command
Redstone Arsenal, AL 35898

BIOGRAPHICAL SKETCH

The author currently serves as the Chief of the Technology Demonstration Branch of the Missile Guidance Directorate, an element of the Research, Development and Engineering Center (RDEC) at the US Army Missile Command (MICOM), Redstone Arsenal, Alabama. He has a Bachelors in Chemical Engineering from the Georgia Institute of Techology, a Master of Arts in Public Administration from the University of Oklahoma, and a Master of Business Administration from Alabama A & M University. He has been involved in various aspects of missile development work for over 30 years.

ABSTRACT

For the past year and a half, MICOM RDEC personnel have been working with the US Forest Service, National Interagency Fire Center (NIFC), Boise, Idaho, in an effort to incorporate several modifications to the Firefly system to improve the system's performance. The Firefly is an aircraft mounted computer system using data from a dual detector infrared (IR) linescanner to produce firemapping imagery. The modified Firefly system has been successfully tested in firemapping operations in the initial fires of the 1996 fire season.

INTRODUCTION

This Firefly upgrade effort resulted from meetings with NIFC personnel over a period of several months in 1994 and early 1995 as part of the MICOM RDEC's technology transfer program. Initial discussions were conducted to determine if there might be some evolving IR technologies which could be applied to the firemapping problem. Firemapping generally involves the use of aircraft equipped with IR sensors to detect fires and then produce in-flight imagery suitable for use by the fire fighting forces on the ground. The Firefly system had been developed during the late 1980's and early 1990's to produce freeze-frame images based on the IR scanner data, with the fires being displayed in red on a gray scale background. The Firefly is mounted in a twin-jet Sabreliner and has been deployed for the past few years in operational conditions with limited success. Our discussions covered many potential improvements to Firefly, but eventually centered around three problem areas in which NIFC personnel desired near-term improvements. These were system reliability, system accuracy, and data delivery by radio link to the Incident Commander (IC) at the Fire Camp. These tasks are described below.

SUPPORT TASKS

Concerning system reliability, numerous crashes of the 33 Mhz Macintosh computer being used as the Data Processing System (DPS) had been experienced over the past few years, some requiring reformatting of the hard drive and replacement of system cards, others requiring in-flight operational downtime to restart the computer. The source of the problem had not be found, and NIFC wanted to replace the Macintosh. It was recommended that it be replaced with a DOS-based computer, to take advantage of the many low-cost off-the-shelf hardware and software items available in DOS format. Replacement of the computer hardware would also require a rehosting of the software to the DOS environment. This task consisted of the procurement and checkout of a new 100 Mhz Pentium computer and the necessary peripheral cards and drives, along with a rewriting of the DPS software, and interfacing the new DPS to the Signal Processing System (SPS) computer and the Navigation (NAV) computer.

The second area of concern, system accuracy, resulted from the rather large positioning errors being experienced in the location of the fires on the maps. An error analysis had been performed by MICOM and the results showed that errors greater than 500 feet were being produced, errors too large for any practical use by the IC to determine how to control the fire. It was recommended that the Global Positioning System (GPS) being used, one with an accuracy of +/- 100 meters, be replaced with the newer Rockwell PLGR (Precision Lightweight GPS Receiver), which could provide accuracies of +/- 16 meters in the PY-code. This task involved the procurement and installation of a PLGR with its airborne qualified receiving antenna, and the writing of a special software package to reformat the PLGR data into a format which would be accepted by the NAV computer, without rewriting all the NAV software.

The third area involved the need for a radio link to transmit the firemapping imagery to the IC's ground station from the aircraft. This had been attempted a few years earlier using a packet modem transceiver, but problems were experienced and the technique was abandoned. It was our recommendation that a low-cost commercially available wireless Local Area Net (LAN) be used to transmit digital data over the link at a rate of about 2 Mbits/sec. The power level of this 915 MHz spread spectrum wireless LAN is limited to one watt by Federal regulations, so the line-of-sight range would also be limited to a few miles, probably about 3-5 miles, depending on the operational conditions. This downloading system would be sufficiently fast to negate the need for lengthy aircraft loiter times over the ground station, minimizing the time-aloft required for the relatively high-cost Sabreliner. The specific task was to design, procure, assemble, install and test a prototype datalink.

BRIEF SYSTEM DESCRIPTION

The aircraft is a Rockwell Sabreliner Model 75/A80, with a cruise speed of 420 knots, an endurance of 3.5 hours, and a ceiling of 45,000 feet. For imaging operations, the aircraft is flown at a speed of about 220 knots.

A schematic of the Firefly system is shown as Figure 1. The Infrared system is a Daedalus line scanner operating in both the 3-5 micron (um) and 8-12 um bands. It has a 2.5 milliradian (mr) Instantaneous Field of View (IFOV), an 80 degree total FOV, and produces 100 lines per second and 1024 pixels per line. It has a moving window display, a computer to digitize the data, a printer to produce a real-time continuous strip image, and a power supply system. The Daedalus strip chart is produced independently from the Firefly system and is marked with aircraft heading and attitude data gathered from the

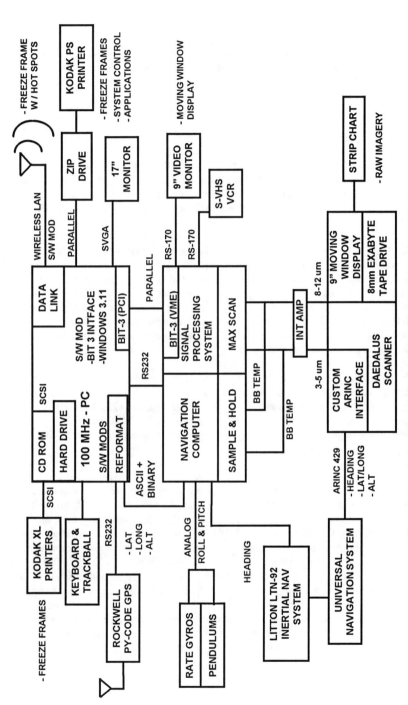

Figure 1. U. S. FOREST SERVICE "*FIREFLY*" SYSTEM

FFDIAGS.PPT

inertial navigation system. The strip chart image uses target detection marks placed on the sidebar to indicate the scanner lines on which a fire detect occured.

The 8-12 um IR detector provides the background image and the 3-5 detector senses hotter temperatures. These two analog signal streams are passed from the Daedalus line scanner to the SPS, where they are digitized, displayed on the SPS monitor, and made available to the DPS. The comparison between the two channels in the SPS provides the 600 degree Celsius threshold hot spot detection capability.

The Navigation computer processes the data from the analog rate gyros and pendulums and provides the digital results to the SPS for the heading and attitude values. These values used for Firefly are separate from those used by the Daedalus. The latitude, longitude, and altitude come from the PLGR, and are currently being reformatted in the DPS to look like the data that used to be available from the prior GPS receiver. Eventually, the Navigation computer software will be rewritten to accept the PLGR data directly.

The 100 Mhz DPS computer, with a 2.2 GB hard drive, provides the capability to process the digital data into a continually updating scrolling format to allow the operator to view the imagery. As fires are detected, as shown by red pixels, the operator can capture the current 1024 by 1024 IR image and fire points into a freeze-frame file. These freeze-frames are automatically stored on the hard drive for later analysis and printing. Each freeze-frame is about 1 MB in size, so several hundred images can be stored before downloading to the archival medium. The scrolling imagery is also recorded digitally on the hard drive as a log file for later review.

The output devices of the Firefly are a moving window display for operator viewing, VHS video recorder, and a photographic quality freeze-frame Kodak printer. Previously, an optical recorder was used for archiving of the image data, but a Zip drive has now been installed to replace the optical drive. The Zip drive has a 100 MB capability and can also be used for overnight transmittal of image data by disk for quick review at the MICOM labs of possible system problems.

The wireless datalink consists of a LAN card and an amplifier card mounted in the new DPS computer, LAN software, and a flight certified antenna mounted on the underside of the aircraft. The ground computer uses the same set of LAN cards and software, and a 13 element Yagi antenna with 14 db gain. Since the Yagi is a directional antenna, it must be oriented in the direction of the aircraft. Testing at MICOM has demonstrated the effectiveness of the system when the antenna is mounted on a vertical mast at a 45 degree angle and pointed to the center of a circular orbit of the aircraft at 5,000-7,000 feet above groung level (AGL) and offset by about 1 mile. Approximately 6 seconds is required to transmit one image in an uncompressed format. With compression, the transmission time per image should be about 2 seconds. The datalink software operates in the Windows environment, and the transfer of files is a rather simple move of freeze-frame files from the local drive (on the aircraft) to the network drive (at the ground station), which is procedurally the same as for an office desktop transfer.

SUMMARY OF ACCOMPLISHMENTS

The following is a summary of the upgrades accomplished by MICOM from April 1995 through April 1996 for both the primary system mounted on the aircraft and the secondary backup system retained in the MICOM lab for continued software effort.

Hardware Upgrades to the Primary Aircraft System:
- Replaced the Macintosh computer with a faster and more reliable DOS-based PC (100 MHz Pentium, 2 GB Hard Disk, 64 MB RAM, CD-ROM, Windows 3.11)
- Replaced the Macintosh monitor with higher resolution 17 inch PC monitor
- Replaced the previous TNL-2000 GPS with more accurate military Rockwell PLGR
- Installed an aviation certified Rockwell GPS antenna on the Sabreliner
- Developed software to reformat the PLGR data to TNL-2000 format
- Installed a wireless LAN for the Data Link, using commercial off-the-shelf hardware and software; non-license 1-watt system
- Installed an aviation certified datalink antenna on the Sabreliner

Secondary Backup/Laboratory System:
- Identified and corrected problems with one SPS processor board
- Constructed a test environment identical to the aircraft system, including SPS, NAV, GPS, PC/DPS, and power supply for hardware-in-the-loop software development and testing in the laboratory

Software Upgrades:
- Completely rewrote the original Mac-based code for the new DOS/Windows environment, including:
  - PC interface to the SPS through Bit3 interface cards which included the writing of a Bit3 device driver for Windows (the driver was available only in DOS)
  - Integration of the Prefly operation into the Firefly program, eliminating separate startup operations
  - Reliable and robust operation
- A simplified user interface with more automated operations, using icons for primary functions, in addition to pull-down menus
- Developed full freeze-frame scrolling window display on PC of IR imagery for freeze-frame capture and continuous color display of system status
- Implemented continuously updating automatic brightness and contrast enhancement of IR imagery, with manual fine-tuning controls
- Implemented freeze-frame storage on hard disk in standard picture format for later recall, review, printing, down link, and other post-mission operations
- Implemented SPS data logging to hard disk, capturing selected portions of all SPS data obtained in mission flights, for:
  - additional post-processing analysis
  - desktop, software-only simulation (no hardware-in-the-loop required)
  - classroom training
  - further system development
- Improved Velocity/Height (V/H) calculation for better imagery

Performance Tests:
Data Link Tests:
-Several antennas were tested

- No data transmission at any range with low gain omni-directional antenna
- Usable data transfer at shorter ranges with high gain omni-directional antenna
- Reliable data transfer at long ranges (8-10 miles) with manual pointing of the high-gain Yagi directional antenna; however, manual pointing is not operationally feasible for night-time operations when aircraft may not be visible
- Optimal operational transfer using Yagi at fixed 45 degree position with aircraft in an offset circular loiter pattern

Sensor Tests:
- Used four NIFC charcoal fire pots (18, 12, 9, and 6 inches square) on MICOM test range, with aircraft flying at altitudes of 4,000 to 13,000 feet AGL
- Successful detection of all 4 fire pots at altitudes up to 9,000 feet AGL
- Successful detection of 3 largest fire pots at 10,000 feet AGL
- Successful detection of 2 largest fire pots at 11,000 and 12,000 feet AGL
- Successful detection of largest fire pot at 13,000 feet AGL

ADDITIONAL EFFORTS

In addition to the three initial upgrade tasks, several other efforts have been or are in the process of being accomplished. Electronic upgrades to the Sample and Hold Box were deemed necessary after some testing proved the box to be inadequatedly designed and built. A task is now being initiated to build new integrated circuit boards for the S&H Boxes. An effort is underway to migrate the software from Windows 3.11 to Windows 95 to allow the use of the 32-bit processing capability in Win 95. Planning is underway for the design and implementation of a next-generation Firefly system. Consideration is being given to the use of a 1.25 mr line scanner, operating at 200 lines per second. The DPS would be further upgraded to a 150-200 Mhz Pentium dual-processor computer to handle the increased data rate. Also, the functions of the present SPS and NAV computers will probably be incorporated into the new DPS. The Windows NT operating system would also be utilized to provide even greater capabilities. This computer system would be able to handle the data processing requirements of a new generation Firefly.

CONCLUSION

The Firefly system, with the modifications described in this paper, was successfully operated in late April 1996 in support of several fire fighting efforts involving wildland fires in the Arizona and New Mexico areas. The data link was not utilized due to the lack of any established ground stations, but plans are underway by NIFC personnel to provide the ground station hardware for the data link operation. It is anticipated that continued technical support will be provided by MICOM in the form of further refinements and improvements to the Firefly system as the result of feedback obtained through operational use.

THE APPLICATION OF FORWARD LOOKING INFRARED IMAGING SYSTEMS IN WILDLAND FIRE SUPPRESSION

William J. Krausmann
USDA Forest Service, Southwestern Region
517 Gold Avenue, SE
Albuequerque, New Mexico 87120
and
Phillip C. Hicks
USDA Forest Service, Chetco Ranger District
555 Fifth Street
Brookings, Oregon 97415

ABSTRACT

Forward Looking Infrared (FLIR) Imaging Systems have been used intermittently within the Forest Service since the mid-1970's. The first operational FLIR systems were deployed during the 1994 fire season. The systems were intended to be used by Air Attack Group Supervisors (ATGS) as tools for monitoring the effectiveness of chemical retardant drops from aerial tankers. Results from both the 1994 and 1995 fire seasons indicate that the FLIR systems can be effective tools for analyzing retardant drops. The systems also proved useful for monitoring and analyzing current fire line conditions: In a number of instances, video tapes acquired by FLIR systems of fire perimeters were used by fire teams to aid in the development of strategies to combat wildfires.

USDA Forest Service Regions are being encouraged to acquire and operate additional FLIR systems due to the positive results demonstrated during the first two seasons of implementation. Several Forest Service Regions are presently outfitting their Agency owned aircraft with the hardware to accept FLIR units. Other Forest Service Regions utilizing contract aircraft are requiring these aircraft to be equipped with hardware and wiring to accept FLIR units. Multi-agency financing and operation of FLIR systems is being investigated, and additionally, private contractors are being encouraged to develop FLIR observation and mapping systems for hire. Within the next couple of years Forest Service leadplane aircraft with FLIR will be operationally tested for fire support purposes.

INTRODUCTION AND HISTORICAL PERSPECTIVE

The USDA Forest Service has employed thermal infrared imaging (TIR) systems since the mid-1960's (Warren and Celarier, 1991). The thermal sensitivity and smoke penetration capabilities of a range of thermal infrared sensing systems have proven useful over time for fire detection, mapping, analysis of chemical retardant drops, and numerous other applications. Thermal infrared imaging technology has, over three decades, contributed to increaseing the safety of the wildland fire environment while helping to preserve countless acres of forest and rangeland.

Development of TIR systems for wildland fire applications has proceeded along two different technological paths within the USDA Forest Service, thermal line scanning

355

systems, and forward looking infrared (FLIR) imaging systems. Thermal line scanning systems have traditionally been the primary tools used to provide thermal infrared information about wildfires to incident commanders. FLIR systems however are beginning to play an equally important role in fire management.

Thermal line scanning systems with sensitivity in the 3-5 micrometer region of the electromagnetic spectrum have been in development since 1962. The Firefly system is the most current incarnation of thermal line scanner systems. It is flown on board a Sabreliner jet aircraft at high altitude, generally at night. Firefly was designed to be a fire mapping tool and has the capability to acquire nadir imagery of the entire perimeter of large fires once or perhaps twice a night. The system is best thought of as a strategic planning tool used for large geographical area fire detection, and first-time large fire mapping (Warren, 1991).

FLIR systems have been used intermittantly within the USDA Forest Service since the mid-to-late 1970's. FLIR systems are typically flown on light aircraft or helicopters at low elevations (generally under 4000 feet above the ground, often much lower). FLIR systems have a narrow field of view compared to line scanner systems, a video format, and spectral sensitivity in the 8-14 micrometer region of the electromagnetic spectrum. These characteristics make them useful as day and night tools for gathering near real-time information about specific locations or activities along the perimeter of large fires. FLIR systems may consequently be considered to be tactical fire management tools (Warren, 1991).

In 1983 the Operational Retardant Evaluation (ORE) project was initiated using FLIR systems to evaluate fire retardant and aerial retardant delivery systems on wildfires (George, et. al., 1989). The FLIR systems enabled observers to evaluate the placement and effectiveness of retardant drops. The ORE project also demonstrated that FLIR systems could overcome a number of operational problems on wildfires that occur due to smoke obscuration, to include:

1. Discovering gaps in retardant lines.
2. Inability to see the location and distribution of retardant.
3. Placement of water or foam from helicopters.
4. Misjudgement of fire rate of spread.
5. Lack of recognition and use of natural barriers to anchor a retardant line.
6. Failure to recognize and make known how long a retardant line will hold.
7. Inefficient aerial suppression due to lack of accurate fire situation information (George, et. al., 1989).

The success of the ORE project in part lead to the development of the FIRE MOUSE TRAP (FMT)[1] in 1984. The FMT consisted of a FLIR unit, a VHS recorder, video monitor, a LORAN and later a GPS receiver, a video caption generator, and a laptop personal computer. The FLIR and navigation reciever were provided by contractors during implementation, the Forest Service contributed the remaining equipment. A ground segment used to generate maps from data collected in the air also included a

---

[1] FIRE MOUSE TRAP - Flying InfraRed Enhanced Maneuverable Operational User Simple ElecTronic Reconnaisance and Patrol - There has been considerable speculation as to the conditions under which this acronym was developed.

plotter, VCR, and monitor. Fires were mapped by flying over the fire perimeter in a helicopter or fixed wing aircraft containing the FMT system. The FLIR was used to maintain position over the fire line and to locate hot spots. The LORAN or GPS system collected positions over the fire line that were stored on a laptop and later downloaded to the plotter to produce fire maps at a user specified scale. The advantages of the FMT system included availability, flexibility, low cost, user control, and timeliness of fire line information. The system was hampered by the limited field of view of the video format, and by the fact that it is often difficult to acquire coordinates exactly over a fire perimeter. Smoke and extremely strong convection currents can make flying unsafe.

AIR TACTICAL GROUP SUPERVISOR/FLIR SYSTEM

While the general capabilities and value of FLIR systems had been demonstrated in the ORE project and in several seasons of FMT operations, their value as an air attack tool had not been fully explored. In 1993 the USDA Forest Service began planning the development of a FLIR program directed at enhancing the capabilities of Air Tactical Group Supervisor (ATGS). The system was implemented for the first time during the 1994 fire season in various configurations and on various aircraft by Regions 1,2,3,5, and 6. The system was used again during the 1995 fire season in Regions 3, 5, and 6.

The 1994 fire season was a "proof of concept" period. The intent of the national program was to determine:

1. The ability of FLIR systems to aid in "real time" evaluations of chemical retardant and water drops.
2. The utility of FLIR systems for evaluating retardant line integrity, retardant coverage level, the effectiveness of retardant drops, and to assist the ATGS in creating and maintaining an effective retardant line.
3. The ability of FLIR systems to effectively locate spot fires.
4. Any other uses for the FLIR that would aid the ATGS in performing the air attack mission more effectively (Moody, 1994).

Three FLIR systems were purchased for the evaluation. The systems were stationed in Regions 3, 5, and 6. The system in Region 3 was migrated north as the fire season progressed, seeing considerable service in Region 2, and eventually in Region 1. The ATGS/FLIR Systems were declared operational for the 1995 fire season, but saw limited service due to low fire activity.

The FLIR unit utilized by the Regions was either an Inframetrics 445G-Mk I, or Mk II. Both the Mk I and Mk II are stabilized thermal IR imaging systems with closed cycle cooling. They are light weight (41 pounds complete) and have low aerodynamic drag (the gimbal is only a 9 inch sphere). The difference between the two systems is that the Mk II has a bore sighted color video system in addition to FLIR capability.

The FLIR operator, located behind the pilot and ATGS, operated the pistol grip contol unit (joy stick) and viewed the imagery on a 7 inch black and white, aft facing monitor (Regions 1, 5 and 6), or on a 10 inch Sony Trinitron color montior (Regions 2 and 3). A 5.5 inch color, flat plate, color LCD monitor was mounted directly in front of the ATGS. This enabled both the pilot and ATGS to view the live FLIR imagery or recorded replays as appropriate.

The ATGS/FLIR System also has video record and playback capability. Both Sony 8mm VCR's and Panasonic VHS/SVHS VCR's were used for recording and playback in different Regions. Recorded FLIR (and in the case of Mk II systems, color video) imagery was used by the ATGS for real time assessment of retardant drops and fire line conditions, by fire teams for planning purposes, and by air operations personnel for training purposes. Regions 2 and 3 included a video caption generator that wrote GPS locations from the aircraft navigation equipment into the video images. All audio communications including crew intercom, VHF-AM, and VHF-FM radio communications were recorded onto the video sound track.

OPERATION PROCEDURES

Two to four hours of training were typically required for personnel to become competent with the FLIR system. Instruction in TIR fundamentals and a ground orientation with the system generally preceeded flight training. Instruction was provided by a factory technical representative when possible.

Both an ATGS and FLIR operator were included on most of the fires where FLIR was used during the 1994 and 1995 fire seasons. The FLIR operator, generally an ATGS trainee, continually scanned the fire perimeter with the FLIR system looking for hot spots in the fire line, spotting outside the fire line, and potentially unsafe conditions along the fire line. Any anomalies that were discovered were reported to the ATGS. The ATGS examined the anomaly on an available monitor and would take appropriate action if necessary. When retardant was being dropped, the FLIR operator locked onto the incoming aerial tanker and monitored the entire drop sequence. Specific attention was paid to where retardant was placed, how densely it was distributed, and in searching for any breaks in the retardant line. When a retardant drop occurred the ATGS generally monitored it and assessed its effectiveness. When the workload became extreme the ATGS would occasionally rely on the FLIR operator to assist in assessing the effectiveness of a retardant drop if the operator was an ATGS trainee.

FLIR systems were initially envisioned as a tool that would be operated by the ATGS. A key component of the FLIR operational evaluation tests during the 1994 fire season was to determine if adding FLIR operation to the ATGS workload would negatively impact performance of the standard ATGS functions. Evaluation forms were given to each ATGS and FLIR operator at the end of their missions. The evaluation forms provided the ATGS and FLIR operator with an opportunity to comment on the overall effectiveness of the FLIR system. Space was also available for the ATGS to discuss the impact of the FLIR system on job performance. System advantages and disadvantages were also highlighted, and any operational problems were described.

The FLIR systems were often used to provide direct support to the Incident Commander or the Operations Branch on the ground even though this was not its intended primary application. Video data collected by the FLIR systems was widely distributed to incident management teams on several fires where they were used for planning fire operations. In several instances in Region 3, the FLIR aircraft was removed from the ATGS role to directly support fire operations. In one instance, the Air Operations Branch Director personnally flew with the FLIR system to map spots on a contained fire and plan

358

mop up operations.

RESULTS AND DISCUSSION

Most evaluators believed that the FLIR system greatly enhances the capabilities and effectiveness of the ATGS. Acceptance of FLIR by the ATGS community however varies between Regions. The FLIR systems seem especially popular in Regions 5 and 6. Region 3 personnel were initially less enthusiastic about FLIR technology. A possible reason for the lack of interest in Region 3 is that fires tend to produce less smoke than those in the tall timber Regions and visibility on fires is generally better. ATGS personnel in Region 3 who were initally unimpressed with FLIR capability gradually began to see its utility over the 1995 fire season. Recently, two Incident Commanders in Region 3 requested that a FLIR be based in the Region during the 1996 fire season because of its many advantages.

ADVANTAGES OF FLIR SYSTEMS

A primary concern of everyone involved in a wildfire is personnel safety. The FLIR systems deployed have demonstrated the capability to aid in making the fire environment a safer place. FLIR systems were used successfully to determine and monitor the rate of fire spread relative to the location of personnel, and to detect and monitor spot fires that might have endangered personnel: In particular, spot fires were located outside fire lines on several fires. Spot fires were located in many instances before they were putting up smoke, or could otherwise be seen by personnel on the ground. In one instance this proved frustrating to an ATGS on the Western Slope fires in Colorado in 1994. The ATGS could see a spot fire on FLIR that a lead plane pilot without FLIR could not locate for a tanker run. In another instance, FLIR was used successfully to verify a blowup situation and monitor an escape route to a safety zone/helicopter pickup point.

A primary mission of the ATGS is to determine when, where, and how much chemical fire retardant should be dropped on a given fire. The ORE project demonstrated the ability of FLIR systems to aid in retardant operations, the FLIR results from the 1994 and 1995 fire seasons verify the ORE project conclusions. FLIR systems were used successfully to observe and verify the position of retardant drops, and also water drops from helicopters. They were also useful for assessing relative coverage levels and for locating weak points or gaps in retardant coverage, allowing the ATGS to strengthen line where necessary. It was possible, in the case of water drops, to determine when the water had evaporated and additional drops were necessary to protect a section of line or deal with a troublesome hot spot. FLIR operations were especially effective in low visibility situations resulting from smoke, inversions, low sun, shadow, or when using retardants that contrast minimally with vegetation.

The ATGS on many fires used the FLIR to provide information about the fire directly to fire operations personnel. Intelligence provided by the ATGS included fire size up information, verifying spread rates and fire behavior, verifying the fire perimeter, often through smoke, and verifying the deployment of personnel, and monitoring fire line integrity. On a number of fires the ATGS also verified the location of structures relative to the fire perimeter, determined their need for protection, reported the information to the incident command, and took action if necessary. FLIR has also been used to monitor and assess the effectiveness of burnout operations. On the Miller Fire on Coronado National Forest for example, the FLIR aided in protecting homes being threatened by a backfire that changed direction due to a wind shift. It was near sunset in a smokey shaded canyon with less than 100 yards between the homes and fire front. FLIR was used to monitor the change in fire behavior and aided in establishing the line for the tanker run.

An advantage of FLIR that should not be overlooked is its use as a training aid. Lead plane and tanker pilots have shown considerable interest in reviewing the video tapes acquired by the FLIR system and discussing tactics, strategy, communications, and air operations in general. The tapes are also an excellent tool when used in formal training classes for air operations personnel. Several hundred hours of video tape have been acquried in both FLIR and color video formats. These tapes contain real time examples of air operations on large fires, including communications. The tape data also serves as a record of the air operations on a fire. Several FLIR videotapes are now part of official fire records.

PROBLEMS

Two potential problems were noted relative to FLIR operations in general, and retardant operations in particular. On hot summer days in many fire environments, especially those in the southwest, ambient heat can raise the surface temperature of bare soil and rocks to the point where they are hot enough to saturate the detectors on the FLIR system. Bare rock and soil may look like spot fires on the FLIR monitor. In the southwest where fires are common on rocky mountainsides and in semi-desert environments, detector saturation can be a problem from 1000 to sunset if the ambient temperature is high. It can be difficult to find a gain setting on the FLIR system that will provide suitable contrast to adequately monitor retardant drops, and separate hot rocks from spot fires.

An additional problem effecting FLIR operations occurs at or around sunset in most environments. The thermal inertia of most natural materials is such that, at or near sunset, they have similar radiant temperatures, producing a low contrast image on the FLIR montior. The low contrast image can make the placement of retardant drops difficult using FLIR because landscape cues may be difficult to see. At or near sunset visual clues may also be difficult to use due to smoke in the environment and shadow from low sun angles. Often these problems may be solved by adjusting the gain on the FLIR system, sometimes not. Low contrast imagery acquired during cross-over does have one advantage, it becomes quite easy to locate small spot fires initiated by lightning strikes due to their bright signature relative to a low contrast background. FLIR can be a useful fire reconnaissance tool at or near sunset.

A major goal of the 1994 FLIR evaluation was to determine if the FLIR system could be effeciently opperated by the ATGS. Evaluations from the 1994 and 1995 fire seasons indicate that the ATGS mission is generally too complex to add FLIR operation to the workload. Most ATGSs surveyed indicated that on large, complex fires they did not have the time to run the FLIR system, and if they tried, it got in the way of their normal responsibilities. They did indicate that it would be a manageable job on small, simple fires with few aircraft.

A few other operational problems cropped up during the 1994 and 1995 fire seasons.

On some implementations the FLIR gimble was located in a position where it collected engine oil, or was occasionally blocked by an aircraft wheel. Installations that used an 8mm Sony tape recorder occasionally had problems distributing the video due to the non-standard tape format. Some implementations did not have color monitors for the FLIR operator to view the color video imagery on, and all installations had problems with glare on the video screens. The Region 3 aircraft went through the 1994 fire season with a bad alternator that effected video quality. Finally, many ATGSs prefer a twin engine aircraft as an ATGS platform. A twin engine plane provides a greater margin of speed and safety. Most FLIR implementations are currently on single engine aircraft. All these problems are comparatively minor. Many have already been dealt with.

CONCLUSIONS

The FLIR program initiated in 1994 has been quite successful based on the results of the returned evaluation forms and comments from ATGSs, lead plane and tanker pilots, and incident command team members. The FLIR systems fulfilled the primary assigned role of aiding the ATGS. It is an excellent tool for assessing the need for chemical retardant drops on a fire, and for evaluating the drops once they are made, especially in smokey environments. The systems also proved useful to incident command personnel on the ground providing information on hot spots, fire line position and conditions, and helping to enhance the safety of personnel involved on the fire. FLIR systems also have provided excellent training imagery, and have been used to document fires.

The FLIR systems are most effectively utilized when both and ATGS and a FLIR operator are present on the ATGS aircraft. The ATGS workload is generally too great to allow FLIR operation as well. Having a dedicated FLIR operator is especially effective if the operator is an ATGS trainee. The trainee both operates the FLIR and acquires invaluable training. The system would also be most effectively used on a twin engine, IFR (Instrument Flight Rules) equipped, high wing aircraft. This type of aircraft would allow the FLIR to be used day and night and would assure maximum system visability.

The Inframetrics 445G Mk II FLIR system was preferred over the Mk I. In many instances, the video capability of the Mk II system provided a quick confirmation of what observers were seeing in the FLIR imagery. It was especially useful as an aid in discriminating between hot spots and hot rock outcrops. This was especially important while ATGS personnel were learning to use the systems.

FUTURE OPERATIONS

USDA Forest Service Regions are being encouraged to acquire and operate additional FLIR systems due to the positive results obtained during the first two seasons of implementation.. Several FS Regions are presently outfitting their Agency owned aircraft with the hardware to accept FLIR units. Other FS Regions utilizing contract aircraft are requiring these ships to come with hardware and wiring to accept FLIR units. Multi-agency financing and operation of FLIR systems is being investigated, and additionally, private contractors are being encouraged to develop FLIR observation and mapping systems for hire. Within the next couple of years Forest Service leadplane aircraft with FLIR will be operationally tested for fire support purposes. Canadian Fire suppression personnel have already successfully tested the use of FLIR imaging systems in their Birddog leadplanes.

FLIR system technology also continues to improve. New FLIR units are expected within the next two years that will use focal plane array technology that should provide even better quality imagery. Further, Region 3 personnel and others have experimented with developing a radio downlink for the FLIR video signal, allowing personnel in fire camp to monitor the fire environment in real time.

The interest in the application of FLIR imaging systems to the ATGS role is expanding within the USDA Forest Service. Personnel who were initally skeptical about the value of the system are rapidly becoming adherents. FLIR systems are becoming and integral part of the Air Tactical Group Supervisor mission.

ACKNOWLEDGMENTS

The authors wish to thank all the ATGS officers, trainees, and incident management team personnel within both the USDA Forest Service, and the USDI Bureau of Land Management whose comments and suggestions contributed directly to the sucesss of the FLIR program.

REFERENCES

Dipert, Duane and J.R. Warren 1988, Mapping Fires with the Fire Mouse Trap: Fire Management Notes, Vol. 49, pp. 28-30.

George, Charles W. 1989, FLIR: A Promising Tool for Air-Attack Supervisors: Fire Management Notes, Vol. 50, pp. 26-29.

Inframetrics, Inc. 1994, Model IRTV-445G Mk II Operators Manual: Inframetrics, Inc., 28 pp.

Moody, William D. 1994, Air Tactical Group Supervisor/FLIR, Operational Evaluation Report: Unpublished report presented to the national FLIR program manager.

Warren, John R. 1990, Changes in Infrared Use for Fire Management: Protecting Natural Resources with Remote Sensing, Proceedings of the Third Forest Service Remote Sensing Applications Conference, The University of Arizona, April 9-13, pp 259-274.

Warren, John R., and D.N. Celarier 1991, A Salute to Infrared Systems in Fire Detection and Mapping: Fire Management Notes, Vol 52, pp. 3-18.

Warren, John R. 1991, Selecting the "Right" Infrared System for a Firefighting Job: Fire Management Notes, Vol 52, pp. 19-20.

Integrating GIS and BEHAVE
for
Forest Fire Behavior Modeling

Jeff Campbell
Pacific Meridian Resources
421 S.W. 6th Avenue, Suite 850
Portland, OR 97204
Ph: (503) 228-8708
Fax: (503) 228-8751

David Weinstein
Pacific Meridian Resources
5915 Hollis Street, Building B
Emeryville, CA 94608
Ph: (510) 654-6980
Fax: (510) 654-5774

Mark Finney
Systems for Environmental Management
PO Box 8868
Missoula, MT 59807
Ph: (406) 329-4837
Fax: (406) 329-4877

Abstract

Research scientists and forest resource managers have long sought
the ability to model wildfire behavior. Considerable effort has been
directed toward understanding the science of fire behavior and
developing computer-based modeling techniques for predicting fire
growth. Recent advances in the field of fire science, along with the
availability of high resolution remote-sensed satellite imagery,
powerful image processing software, Geographic Information
Systems (GISs), and affordable computer hardware has enabled the
development of sophisticated, yet easy to operate fire simulation
applications.

The model's user interface has been designed so that advanced
computer and/or GIS skills or experience are not required by the
user for model execution. The model puts the power of
comprehensive forest fire behavior prediction into the hands of on
the ground resource managers where it can be most effectively
applied.

Introduction

Fire plays a very significant role in management of the pine forests of the Southeastern United States. Both wildfires and controlled prescribed burning significantly affect forest management activities from timber harvest scheduling to reforestation and thinning operations. Wildfires destroy thousands of acres of prime forest land every year while controlled burning helps to maintain a manageable fuel loading for forests susceptible to destructive wildfires. With the added complexity of managing productive forests which also support active military operations and training, monitoring and controlling both wild and prescribed fires is a particularly challenging charge for the Division of Forestry at the Camp Lejeune Marine Corps Base in eastern North Carolina.

Throughout the country as well, personnel, equipment, and financial resources are tremendously strained in the struggle to contain and extinguish wildfires. Thousands of firefighters and support staff are engaged and millions of dollars spent annually to preserve and protect human lives and personal property as well as valuable timber and recreation resources. Twelve hour work days for weeks at a time in hazardous and treacherous conditions are not uncommon for wildfire fighting personnel during heavy burning seasons such as those encountered during the past several years.

The continued emphasis on safety and resource protection by the numerous state and local agencies charged with combating wildfire has increased the need for more accurate and dependable tools for wildfire management. For several years, forest fire management personnel have utilized fire behavior models such as BEHAVE (Andrews, 1986) to aid in predicting fire behavior and subsequently mapping probable scenarios of fire spread during a given time period. BEHAVE is a non-spatial fire behavior prediction tool. Utilizing inputs of fire fuel type, topography data, weather data, and initial fuel moisture data, BEHAVE calculates fire behavior and fire characteristics for a given area. While BEHAVE is very useful for predicting fire characteristics for a given land area, the output is inherently non-spatial. In other words, the spread rates, flamelengths, fireline intensities, and heat calculations generated by BEHAVE are applicable only so long as the specified fuel type, topographic, and weather related parameters do not vary. As the topography or fuel model changes or as the weather patterns shift, BEHAVE results must be recalculated in order to provide an adequate approximation of fire behavior for the new input data regime. Considering the complexity of fuels, topography, and weather over a given landscape during a period of time, efficiently and accurately predicting fire behavior characteristics for a wildfire over time utilizing such techniques is inefficient and often impractical.

However, through the incorporation of GIS technology, developing detailed fire behavior predictions for numerous scenarios becomes not only possible but incredibly efficient and effective. The forest fire behavior model developed for this project incorporates spatial fuels and topographic data, temporal weather and wind settings and initial fuel moistures into the prediction of forest fire behavior across both time and space. The model puts the power of sound, accurate, and efficient fire behavior modeling technology into the hands of forest fire management personnel charged with coordinating the containment and extinguishing of wildfires. The model can become one of the most effective tools for managing personnel, financial, and equipment resources for battling one of the most destructive and dangerous forces of nature.

Products Required

The primary data source for the forest fire fuel model classification was Landsat Thematic Mapper imagery, geocoded and terrain corrected to UTM coordinates. Landsat TM sub-scene from Path 14, Row 36, acquired August 8, 1993, was utilized for this project. This date of imagery was chosen due to its combination of minimal cloud cover, optimum sun angle, and optimum vegetation vigor and reflectance characteristics. In addition, 1:15,840 scale color infra-red aerial photography for the study area was utilized extensively. The acquisition date of the photography was March 6, 1993. Although collected during different seasons, the aerial photography and satellite imagery were collected within five months of one another, thus minimizing the amount of landcover change occurring between the two dates of data collection.

Image classification was enhanced through the use of ancillary GIS data layers. Ancillary data utilized for the forest fire fuel model classification included various ARC/INFO GIS coverages representing past and present land-use and land-cover characteristics. The ancillary GIS coverages were used to help ensure quality control during iterative and final fuel model classifications. The fuel classification was further refined through the development and application of GIS models which examine the relationship between overstory vegetation types, soil types, recent forest management activities, forest fire fuels. Digital elevation data was utilized in the fire behavior model development phase of the project due to its direct influence of fire behavior. Within the Camp Lejeune Marine Corps Base, however, the effect of topography is minimal. A total elevation change of approximately 35 feet occurs on the base.

Extensive field collected data was also incorporated into both the forest fire fuel model classification as well as the fire behavior model development and calibration. Ground data for forest fire fuel models, overstory and understory vegetation cover, and tree crown cover was collected to establish field training sites for image classification. This data was collected on field data forms in addition to field notes and delineations regarding forest fire fuel type on aerial photography and draft classification maps. All field collected data was extensively used to develop and refine the forest fire fuel model classification.

Field data regarding past prescribed burns and wild fires collected by Camp Lejeune Forestry Division personnel were extensively used in the fire behavior model calibration process. Base forestry division personnel consistently collect detailed data regarding fire behavior (spread rates, flamelengths, burn perimeter, etc.) for fires burning on the base. Global Positioning System (GPS) units are used to map fire perimeters. The fire behavior model was calibrated using this detailed data from past fires to ensure the most accurate and reliable fire behavior predictions under local conditions. The incorporation of this past fire behavior data during the fire behavior model calibration was among the most important tasks of the entire project.

Methods

Image Classification
A combined supervised/unsupervised approach was used to classify the Thematic Mapper imagery (Congalton et al., 1992). All classification work was performed using ERDAS (Atlanta, Georgia) image processing software. The imagery was classified into the thirteen models described by Anderson (1982) and two non-fuels classes. Anderson produced a similarity chart for cross referencing the 13 fuel models he developed to the 20 fuel models used in the National Fire Danger Rating System. The classifications used for this project are listed below:

Fuel Model/Class Complex   **Model Description/Typical**

Grass and Grass-Dominated Models:

1	Short Grass (1ft)
2	Timber (grass and understory)
3	Tall Grass (2.5+ ft)

Chaparral and Shrub Fields

4	High Pocosin/Chaparral (6+ ft)
5	Brush (2 ft)
6	Dormant Brush, Hardwood Slash
7	Southern Rough/Low Pocosin (2-6 ft)

Timber Litter

8	Closed Timber Litter
9	Hardwood Litter
10	Heavy Timber Litter and Understory

Slash

11	Light Logging Slash
12	Medium Logging Slash
13	Heavy Logging Slash

Non-Fuel

14	Water
15	Bare/Non-Flammable

Each fuel model listed above is described by specific information regarding fuel loading, surface area to volume ratio of each size group, fuel depth, fuel particle density, heat content of fuel, and moisture of extinction values. This description provides the necessary information for each fuel model to allow for the automated modeling and calculation of fire spread rates and other fire behavior characteristics provided in the GIS model described below.

An initial unsupervised classification was completed for the entire study area. From this classification, obvious "Water" and "Bare/Non-Flammable" cover types were identified and masked from the satellite imagery. Portions of many of the 13 fuel models were also easily identified and labeled from the unsupervised classification. For the remaining spectrally confused classes, extensive field data was collected by Pacific Meridian image processors/foresters along with Camp Lejeune Division of Forestry personnel. The field data was used to describe supervised training sites and aid with further spectral class labeling of unsupervised classes as appropriate.

Draft forest fire fuel type classification maps were produced and plotted at 1:15,840 scale. The draft maps were reviewed, evaluated, and edited on the ground by both Pacific Meridian and Camp

Lejeune personnel for accuracy and consistency. The information obtained through the draft map review process was incorporated into the final manual editing and GIS modeling phases of the classification process to refine and enhance the forest fire fuels classification. The final raster classification of the forest fire fuel models was converted from ERDAS format to ARC GRID format for incorporation into the development and implementation of the fire behavior model.

A raster forest crown cover layer was also developed through classification of the Landsat TM imagery. Initially, "Water", "Bare/Non-Flammable", and other non-forest fuel classes previously identified in the mapping of the forest fire fuels were masked from the imagery. These areas were assigned a crown cover class of 0%. For the remaining areas, an series of unsupervised classifications was completed and labeled with one of the following crown cover classes:

1 -20% Tree Crown Cover
21 - 50% Tree Crown Cover
51 - 80% Tree Crown Cover
81 - 100% Tree Crown Cover

The resulting raster tree crown cover classification was also converted from ERDAS format to ARC GRID for incorporation into the development and implementation of the fire behavior model. The tree crown cover data layer influences the effect of wind direction as well as incoming solar radiation in the fire behavior model.

Upon completion of the raster fire fuels classification, polygon creation algorithms developed by Pacific Meridian were utilized to covert the raster GIS coverage into an ARC/INFO polygon coverage. In the coverage, each polygon was assigned a label depicting its forest fuel model class. The polygon coverage was produced to provide base forest managers with a vector coverage of fuel type which they can utilize alone or in conjunction with other vector and raster data layers for future landscape analysis projects. The polygon coverage of forest fire fuel type was not directly utilized in the fire behavior model development or implementation.

GIS Model Development

At the completion of the fire fuels classification, Pacific Meridian developed a new fire simulation application, known as *FIRE!*, which brings fire modeling capabilities to the Arc/Info GIS environment. *FIRE!* is an ArcTools-based application that allows a user to interactively specify all the required spatial and non-spatial parameters and specify the time, duration, and locations of a multiple ignition fire simulation. Vector representations of fire perimeters are graphically displayed as the simulated fire advances. Spotting potential is computed and displayed for each perimeter. At the conclusion of the simulated burn, *FIRE!* also displays raster representations of time of arrival, heat, fireline intensity, rate of spread, and flame length for the burned area. Plots of the simulation results may be generated using built-in plotting templates or customized by the by the user with the full suite of plotting tools available in Arc/Info. All of the output data is preserved as Arc/Info coverages and grids, which may be further analyzed by the user with the full range of capabilities of the Arc/Info GIS environment.

The engine of the *FIRE!* application, responsible for all the complex computations necessary for simulating fire behavior is *FARSITE* (Fire Area Simulator), a C++ program developed by Systems for Environmental Management. *FARSITE* interacts seamlessly within the Arc/Info environment as a component of *FIRE!* enhancing the spatial display and query capabilities of the GIS for fire modeling and analysis. Figure 1 outlines a flow diagram for *FIRE!*.

FIRE! allows a user to model fire behavior by defining a fire "scenario". The scenario is comprised of three sets of input parameters: landscape files, run parameters, and ignition locations. A Geographical User Interface (GUI) has been designed to allow the user to easily specify and edit the data and parameters necessary to execute each simulation scenario set. The first set establishes the spatial and temporal data to be utilized by the model. The user specifies the appropriate fire fuels, canopy cover, slope, elevation, and aspect layers required for the simulation. In addition, non-spatial data sets including weather, wind, initial fuel moistures, and fuel model adjustment factors can be created, specified, and edited.

Run parameters include the burn simulation start and end dates and times the spatial and temporal resolution of calculations performed during the simulation. For instance, a spatial resolution of greater than the 25 meters may be specified for scenarios covering very large areas in which only a gross estimation of fire behavior over a long time period is desired. Specifying a greater spatial resolution reduces the computational requirements of the model resulting in a faster simulation. However, scenarios requiring detailed information regarding fire behavior throughout a simulation area should utilize a spatial resolution at least as small as the input data sets provide.

Finally, a data set identifying the location and configuration of a fire ignition must be specified. Fire ignitions may be established as points, lines, and/or polygons and are entered interactively by clicking on the screen at the desired ignition locations.

After defining the burn scenario, the model simulation can be executed. As the model performs the necessary fire behavior calculations, vectors are displayed indicating the fire's perimeter at a user-specified time interval. The vectors may be displayed over the fuels raster data layer or the original Landsat TM imagery. At the completion of the simulation, raster data layers are produced providing the flamelength, fireline intensity, time of arrival, heat per unit area, and rate of spread of the fire for every pixel within the burned perimeter.

FIRE! has been designed and developed to overcome many of the shortcomings encountered with past fire behavior models. Fire modeling in the past suffered from two important limitations: the non-spatial qualities of early methods based solely on BEHAVE, and the limitations of raster-based models that followed. *FIRE!* employs the most recent developments in wave-based fire modeling. Wave-based models have been shown to yield the most accurate modeling results available to date.

Early modeling efforts relied on BEHAVE, a non-spatial model for estimating fire characteristics, such as flame length, rate of spread, etc., for a homogeneous surface. Fire managers had to perform three labor intensive tasks manually: delineating regions of homogeneous fuel characteristics, computing BEHAVE statistics for each of these regions, and propagating the fire front based on expert knowledge of fire behavior under local terrain and wind conditions. After each iteration of manual propagation of the fire, the steps would then have to be repeated. Modern spatial techniques still use BEHAVE algorithms for computing fire characteristics while providing automated methods for propagating the fire across a heterogeneous landscape.

Raster-based spatial models rely on cellular propagation methods. Cellular models use the constant spatial arrangement of a cell or raster landscape to solve for time of ignition. The fire is propagated through the raster in checkerboard steps, with cells igniting based on the characteristics of other cells in their neighborhood. Studies have shown this technique to yield distorted representations of fire shapes owing to the necessity of growing the fire from each burning cell in discreet steps in one of the eight cardinal directions available to the grid.

Wave models recognize the inherent wave-like behavior of wildfire; that is, that the fire front propagates as a wave, shifting and moving continuously in time and space. Wave models solve for the position of the fire front at specified times. *FARSITE* uses a technique for wave propagation, known as Huygen's principle, to expand surface fire fronts in two dimensions. While rasters are still used to represent the underlying landscape, and to record fire characteristics during the simulation, the fire perimeters are processed and stored as continuous vectors.

FIRE! is the first model that integrates advanced wave-based fire simulation methods with a widely utilized GIS platform in an easy to use, fully graphical environment. Some potential enhancements to be developed and incorporated into future versions of *FIRE!* include: improved handling of torching and spotting, with predictive techniques for simulating spot fires; real-time updates of developing weather and wind patterns recorded in the field during an actual burn; interactive simulation of containment efforts by allowing the user to create firelines and backfires during a burn; and refinements to the propagation model to include important fire behavior characteristics such as localized convection effects and pre-heating of fuels.

Results and Discussion

Three primary products resulted from this project: 1) a raster forest fire fuel model classification; 2) a GIS-based fire behavior prediction model: *FIRE!*.; and 3) a fire fuel coverage updating program. The forest fire fuel model classification is a raster data layer that can be continually updated based on forest management activity, wildfires, etc. utilizing the fuel classification update program also developed for the project. The raster data layer provides detailed information regarding the variation of forest fire fuel types across the entire landscape of the Camp Lejeune Marine Corps Base making possible the site specific forest fire behavior modeling provided in the GIS-based model. The model allows for more comprehensive, efficient, and effective planning and resource coordination by forest fire fighting managers and fire behavior specialists. The forest fire fuels map upadating program allows for the continuous, automated updating of the fuels data coverage utilized by the fire behavior model based on identified fuel changes due to fire or harvesting activitiy. Four specific characteristics of the project results that distinguish this model from other forest fire fuel classifications and fire behavior models previously developed also warrant further discussion.

<u>More Realistic Depiction of Landcover Characteristics.</u>
The fire fuels and tree crown cover classifications developed using digital satellite imagery render raster data layers depicting the continuous variation of fire fuels and tree crown cover present across the landscape. This provides a more realistic portrayal of the complexity and composition of the mapped land cover

371

characteristics. The spatial detail provided by the pixel classification of satellite imagery translates into a more realistic prediction of fire behavior by the GIS-based model as the variation of spatial factors influencing fire behavior (fuels, topography, tree crown cover, etc.) are more realistically represented. Often, photo-interpreted delineations of fuels, tree crown cover, etc. portray a deceptively homogeneous pattern of fuel and crown cover variation across the landscape. This often unrealistic homogeneity can result in an unrealistic prediction of fire behavior.

State-of-the-art fire behavior modeling.
FIRE! is based on sound state-of-the-art fire spread simulation technology. As outlined in the above description of the methodology utilized in the development of the GIS-based fire prediction model, it is evident that utilizing wave propogation techniques for fire growth simulation provides a much more realistic and defensible fire behavior prediction that those provided by cellular propogation models. *FIRE!* utilizes wave propogation technology representing the state-of-the-art methodology for modeling fire behavior.

Advanced model user interface.
The Geographical User Interface (GUI) developed for the execution of the GIS-based model allows natural resource managers not intimately familiar with computers or GIS to effectively utilize the fire behavior model. By developing and incorporating a user friendly user interface for model execution, this model puts the power of fully utilizing the functionality of the fire behavior model into the hands of a much wider range of resource managers and technicians. With minimal training and experience, nearly anyone can identify and edit the necessary input data, develop a burn scenario, execute the model, analyze the results, and produce hardcopy results output without ever being required to specify command line instructions for direct computer interfacing. The potential user-base of the model is significantly expanded due to the development of the GUI.

Open Architecture Design
The open architecture of *FIRE!* facilitates the incorporation of technological updates in fire behavior modeling as well as addition of additional analysis tools associated with fire behavior modeling. Corollary modules for smoke plume management and fire suppression techniques are scheduled to be added to FIRE's fire behavior prediction capabilities. The design of the model is such that as these additional tools are developed, they can be easily added to the current structure and executed utilizing the same GUI. In addition, tremendous flexibility has been built into the model. For instance, customized fuel models may be utilized in the fire behavior

372

prediction by the model. *FIRE!* also allows for tree crown cover and topography data sets to be represented via a number different classification schemes. Also, a great deal of flexibility is provided for identifying burning scenarios to model. The user has the ability to define the exact circumstances and parameters defining the burning scenario. This allows the user to perform sensitivity analysis on fire fuels management alternatives or to mimic the exact environmental conditions and burning times of past fires, if desired.

Other future applications include integrating change detection analysis using digital satellite imagery with fire fuels mapping and forest fire behavior modeling. For instance, a fire fuel model classification can be developed from Landsat TM imagery for vast forested areas simultaneously. Then, using change detection analysis techniques, regions of bug killed forests may be identified and mapped again using Landsat TM imagery. By comparing the original fuels classification with the mapped areas of change due to bug kill, updating of the fuels classification may be accomplished. With the pre- and post-bug kill data sets, fire behavior analysis can be completed using *FIRE!* for both fuel type scenarios. This type of analysis may be extremely beneficial for evaluating the potential effects of forest management activities related to forest pest management to potential forest fire behavior. In this type of application, the GIS-based fire behavior model assists resource managers in being pro-active in their fire management strategies as opposed to being reactive.

The application of a fire behavior model such as *FIRE!* is limited only by one's imagination. With the continued emphasis on accomplishing effective and more efficient forest fire management throughout the United States and the world, a sound GIS-based fire behavior model such as *FIRE!* provides an incredibly valuable tool for resource managers. Conceivably, with the application *FIRE!* to wild and prescribed fires, fire fighting activities from initial attack to mop-up can be better coordinated resulting in the savings of money and, much more importantly, even human life.

References

Anderson, H.E. 1982. Aids to Determining Fuel Models for Estimating Fire Behavior. USDA Forest Service General Technical Report INT-122.

Andrews, Patricia L. 1986. BEHAVE: Fire Behavior Prediction and Fuel Modeling System-Burn Subsystem, Part 1. USDA Forest Service General Technical Report INT-194.

Congalton, R.G., Green, K., and Tepley, J. 1993. "Mapping Old Growth Forests on National Forest and Park Lands in the Pacific Northwest from Remotely Sensed Data". *Photogrammetric Engineering and Remote Sensing*, Vol. 59, No. 4, April 1993, pp. 529-535.

Figure 1
FIRE! Simulation Flow Diagram

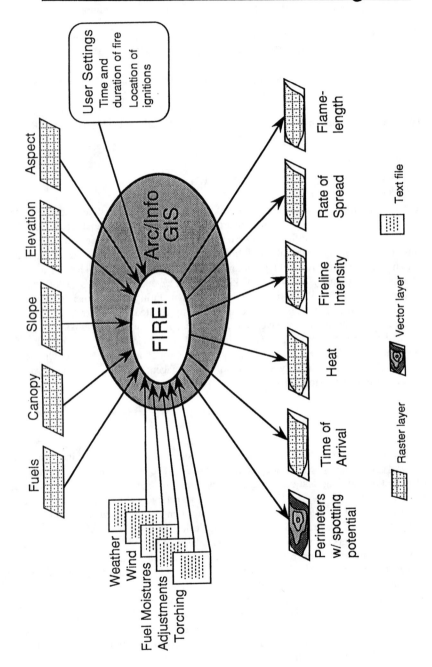

Application of Thermal Infrared (FLIR) and Visible Videography to the Monitoring and
Restoration of Salmonid Habitat in the Pacific Northwest

N. Poage[1], C. Torgersen[1], D. Norton[2], M. Flood[3], B. McIntosh[1]

[1]Oregon State University / USDA Forest Service, Forestry Sciences Laboratory, 3200
SW Jefferson Way, Corvallis, OR 97331

[2]Office of Water (4503F), U.S. Environmental Protection Agency, 401 M Street, SW,
Washington, DC 20460

[3]Topographic Engineering Center, U.S. Army Corps of Engineers, 7701 Telegraph Road,
Alexandria, VA 22315

ABSTRACT

The focus of this project is the application of thermal infrared and visible videography to
the monitoring and restoration of riparian areas important for rare and endangered species
of salmonids in the Pacific Northwest. Basin and reach-level patterns of water
temperatures within ranges critical to salmonids can be described using imagery acquired
with an airborne Forward-Looking Infrared (FLIR) thermal scanner. Thermal infrared
and visible videography of over 1500 kilometers of rivers and streams in the Pacific
Northwest was collected from a low-flying helicopter during late-summer in 1995. A
longitudinal profile of average stream temperature is presented for a 60 km section of the
Middle Fork of the John Day River, Oregon. Through the use of a global positioning
system, data derived from both thermal infrared and visible videography can be linked to
other spatially-explicit data layers in a geographic information system (GIS). Once in a
GIS, these data become readily available to land managers and policy makers involved
with monitoring and restoration efforts.

INTRODUCTION

Water temperature is of critical concern for the survival of spring-run chinook salmon
(Oncorhynchus tshawytscha) in the Pacific Northwest. The water temperature optima for
the migration, holding, and spawning life history phases of spring-run chinook salmon
are estimated to be 7-16 °C, 8-13 °C, and 6-12 °C, respectively (McCullough 1993). The
migration of adults is effectively blocked by water temperatures greater than 21 °C, and
the upper limit of water temperature for suitable habitat is 24 °C (McCullough 1993).
Summer water temperatures in eastern Oregon, Washington, and Idaho frequently exceed
thermal tolerance zones for spring-run chinook and other salmonids.

Recent work using a forward-looking infrared (FLIR) scanner to assess water
temperature patterns on tributaries of the John Day River in Oregon suggests that spring-
run chinook respond to local differences in stream temperature (Torgersen 1996,
Torgersen et al. 1995). This pattern of behavior was particularly pronounced in areas
where summer water temperatures exceeded critical thresholds. Describing how water
temperatures vary over multiple scales is, therefore, central to monitoring and restoring
salmonid habitat in the Pacific Northwest.

Thermal infrared and visible videography of over 1500 kilometers of eight Pacific Northwest river systems important for salmonids was collected during late-summer in 1995 from a low-flying helicopter. Sections of the following river systems were sampled: Asotin Creek (WA), Grande Ronde River (OR), Imnaha River (OR), John Day River (OR), Lolo Creek (ID), Tucannon River (WA), Umpqua River (OR), and Yakima River (WA). Results from the upper 60 km of the Middle Fork of the John Day River (OR) are presented in this paper.

METHODS

Thermal infrared and visible videography were collected for the upper Middle Fork of the John Day River between approximately 13:40 and 14:20 on August 25, 1995. An Agema 1000 FLIR vertically mounted on the underside of a helicopter was used to acquire thermal infrared data within a 5-55°C temperature range as grayscale imagery. A color video camera (visible spectrum) bore-sighted with the FLIR was used to record simultaneous visible spectrum data. The videography was collected from an altitude of 425 meters above the ground which provided an image swath width of 150 m.

The thermal infrared and visible data were recorded in S-Video format on separate Hi-8 video tapes. Individual frames in the thermal infrared and visible video tapes were labeled in-flight with identical SMPTE time code stamps. Additionally, the longitude and latitude of the helicopter calculated by an on-board global positioning system (GPS) were recorded at one second intervals on the sound track of both the thermal and visible video tapes using a Horita GPS3 SMPTE time code generator/reader. These coordinates and the corresponding SMPTE time code were extracted post-flight and written to an ASCII file using the Horita system. This enabled digitally captured frames from both the thermal and visible video tapes to be integrated with spatially-explicit data layers in the Arc/Info and ArcView geographic information systems (GIS).

Digital images with ground resolutions (or pixel size) of 25-30 cm were captured post-flight from the analog video tapes using a TARGA+ digitizing board and a DiaQuest video animation controller. In the captured grayscale thermal images, 256 shades of gray (0-255) corresponded linearly to the temperature range of 5-55°C. These digital images were processed by using custom Arc/Info Arc Macro Language programs to 1) convert the thermal images to Arc/Info GRIDs containing the corresponding temperature values in degrees Celsius, and 2) color-code the thermal GRIDs to display and visually enhance temperature differences, enabling the user to extract stream temperature data. The primary uses of the visible videography in this study were to identify channel characteristics associated with cold-water areas and differentiate between land and water surfaces in the thermal imagery.

In order to describe the overall pattern of water temperature for the 60 km study section of the Middle Fork of the John Day River, a longitudinal profile of average stream temperature was generated by classifying and examining thermal image frames extracted at 200 m intervals.

RESULTS AND DISCUSSION

The average stream temperature profile for the Middle Fork of the John Day River indicates that the pattern of water temperature was highly variable (Figure 1). Interestingly, the highest average water temperatures were observed at the upstream end of the 60 km study section (right side of Figure 1). Certain downstream temperature decreases can be explained by cold-water inputs from cooler tributaries (e.g., the confluence of Clear Creek at river kilometer 50). The rapid increase in average stream temperature from river kilometer 50 to river kilometer 47 is, presumably, due to the rapid mixing of the cooler tributary water with the warmer water of the main stream channel.

Other downstream temperature decreases may be the result of relatively cool groundwater seeping into the main stream channel. For example, the gradual downstream decrease in stream temperature from river kilometer 45 to river kilometer 40 occurs in a relatively wide and unshaded valley. Although a downstream increase in stream temperature is normally associated with a lack of vegetative shading, we hypothesize that the observed downstream temperature decrease is due to progressively increasing amounts of relatively cooler groundwater flowing into the main stream channel.

Figure 1. Middle Fork John Day River Average Stream Temperature
(August 25, 1995)

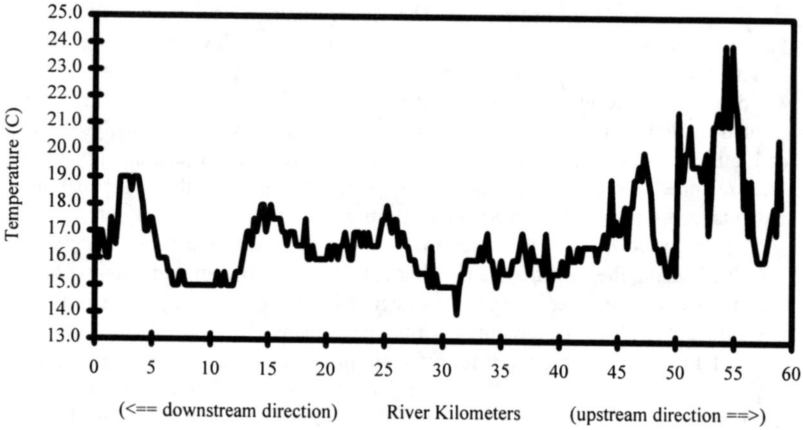

(<== downstream direction)   River Kilometers   (upstream direction ==>)

CONCLUSION

Describing stream temperature patterns is of critical importance in the monitoring and restoration of salmonid habitat in the Pacific Northwest. Aerial thermal infrared videography is an effective way to describe the spatial variability of stream temperature patterns. By providing a continuous spatial record of stream temperature, thermal infrared videography can complement more traditional in-stream temperature data-loggers which provide a continuous temporal record of stream temperature. Through the use of a global positioning system, data derived from both thermal infrared and visible videography can be linked to other spatially-explicit data layers in a geographic information system. Once in a GIS, these data become readily available to land managers and policy makers involved with monitoring and restoration efforts.

ACKNOWLEDGMENTS

The authors would like to acknowledge the invaluable assistance of Kathryn Ronnenberg in the preparation of Figure 1.

REFERENCES

McCullough, D.A. 1993. Stream Temperature Criteria for Salmon. Unpublished review for the Columbia River Inter-Tribal Fish Commission, Portland, Oregon. 18 p.

Torgersen, C.E. 1996. Multi-scalar Assessment of Thermal Patterns and the Distribution of Chinook Salmon in the John Day River Basin, Oregon. M.S. Thesis, Department of Fisheries and Wildlife, Oregon State University, Corvallis, Oregon. 96 p.

Torgersen, C.E., D.M. Price, B.A. McIntosh, and H.W. Li. 1995. Thermal refugia and chinook salmon habitat in Oregon: Applications of airborne thermal videography. Pages 167-171 in P.W. Mausel (ed.) Proceedings of the 15th Biennial Workshop on Color Photography and Videography in Resource Assessment. American Society for Photogrammetry and Remote Sensing. Terre Haute, Indiana, May 1-3, 1995.

NEW TECHNOLOGIES FOR INFRARED REMOTE SENSING AND FIRE MAPPING

James W. Hoffman
Ronald C. Grush
Space Instruments, Inc.
Encinitas, California, 92024, U.S.A.

1.0 INTRODUCTION

New technologies have recently been developed which can be applied to infrared fire detection and mapping. These technologies can be used to solve some of the problems encountered with the current generation of cryogenic infrared line scanners. They can be grouped into three primary categories as shown below. The first technology is the recent development of uncooled infrared detector arrays operating in the thermal infrared region from 7-13μ. The elimination of cryogenic systems can greatly reduce the size, weight, complexity, and cost of airborne and ground-based sensors. The second new technology is the development of high precision, high-temperature in-flight calibration systems. These systems are required to accurately map flame temperatures and distinguish burning areas from smoldering areas. The third new technology is the recent development of a method to obtain multispectral images from a moving platform with a simple filter wheel. This eliminates the need to have separate detector arrays with beam splitters or dichroic mirrors for each spectral band.

In addition to these new technologies, several existing technologies can be utilized to reduce the size and complexity of the total system. These existing technologies are pushbroom scanning with TDI (Time Delay and Integration) and selective dynamic ranges.

New technologies for infrared fire detection and mapping:
1) Uncooled thermal infrared imaging
   (To eliminate cryogenics and reduce size, weight, complexity and cost)
2) High temperature, inflight calibration
   (To provide accurate flame temperatures)
3) Multispectral imaging with a filter wheel (from a moving platform)
   (To eliminate multiple detector arrays and complex optics)

Existing technologies to be utilized:
1) Pushbroom scanning with TDI (Time Delay and Integration)
   (To eliminate mechanical scanning and reduce weight, complexity and cost)
2) Selective dynamic ranges (To prevent saturating when viewing fires)

2.0 UNCOOLED THERMAL INFRARED DETECTOR ARRAYS

Uncooled microbolometer infrared detector arrays were developed by Honeywell and have been licensed by several other companies for manufacture. Figure 1 shows a Scanning Electron Microscope photograph of one of the detector elements from an array manufactured by Loral Infrared and Imaging Systems of Lexington, Massachusetts. The array contains 245 rows by 327 columns. The pixel pitch is 46.25μ. The sensing surface is made of vanadium oxide. Because the detector array is a bolometer, it operates at room temperature. The incoming infrared radiation is absorbed by the vanadium oxide and increases the temperature of that pixel. This increase in temperature results in a change in resistance, which is read out by measuring the voltage across the pixel when current is applied. The long contact legs shown in the photograph provide thermal isolation from the neighboring pixels and result in minimal crosstalk between the pixels.

Figure 1. Uncooled microbolometer detector array

3.0 HIGH TEMPERATURE INFLIGHT CALIBRATION

One of the problems with today's commercial infrared systems is the limited calibration available. These systems are generally designed to image normal Earth scenes at temperatures from 300°K to about 350°K. For imaging forest fires at temperatures of 1100°K or more, scenes need to be imaged without saturating. As shown in Figure 2, a flame temperature of 1100°K produces a radiance value 50 times higher than a nominal 300°K Earth temperature. As most commercial systems only calibrate at temperatures less than 350°K a large error can be introduced by extrapolating out to flame temperature radiances. It is thus important to obtain a calibration point at least in the middle of this dynamic range and to also insure that the dynamic range of the system can cover flame temperatures without saturating. The system to be described later in the paper utilizes a calibration system that can calibrate anywhere from ambient up to about 750°K.

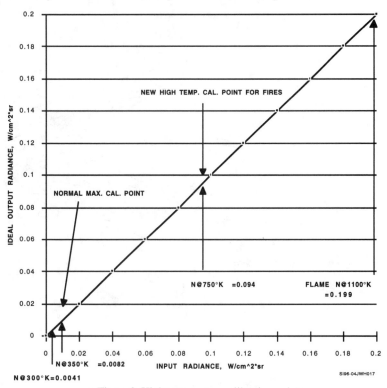

Figure 2. High temperature calibration points

Infrared sensors utilized in fire fighting applications will undergo a dust and dirt environment which will change the transmission of the optical system. A useful feature for such a system is a through-the-optics calibration system which calibrates not just the detector array but the entire instrument. Figure 3 shows the ray trace for a calibration optics that does just this. The calibration blackbody is located in the upper part of the figure and the calibration mirror is the last element on the right. The energy from the blackbody is reflected from the calibration mirror and passes through the optics and the spectral filter, onto the detector array which is located at the left. The calibration system mixes the energy from the blackbody such that it provides a uniform radiance across the entire detector array. The energy from the blackbody is spread such that even if there were a hot point on the blackbody, the energy from that point is spread uniformly across the entire detector array. This is illustrated in Figure 4 which shows the uniform energy distribution produced by five points along the length of the blackbody.

By utilizing high temperature materials and a high temperature thermocouple measuring device, a high temperature blackbody has been developed for calibration. Figure 5 shows a test run on the new blackbody in which the blackbody was heated to a temperature of approximately 480°C. This corresponds to the 750°K calibration point shown in Figure 2.

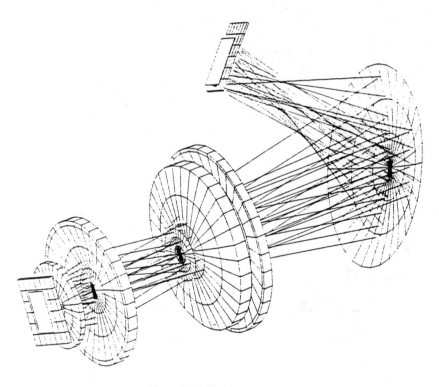

Figure 3. Calibration optics

382

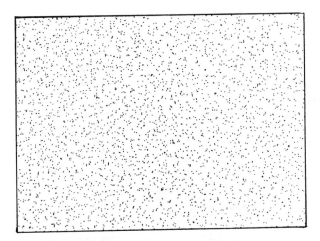

Figure 4. BB energy distribution across the detector array

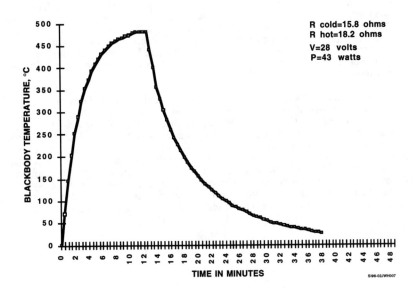

Figure 5. High temperature blackbody test

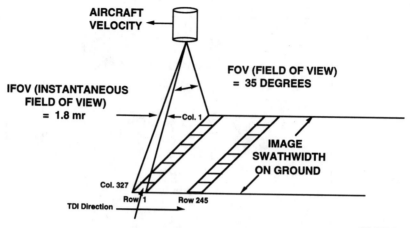

Figure 6. MIR spatial resolution and field of view

4.0 SYSTEM OPERATION

To produce as small and simple a system as possible, pushbroom scanning can be utilized. Figure 6 illustrates how this works. Each row of 327 detector elements is scanned over the ground by the motion of an aircraft (or other moving platform). A 25 mm focal length optical system will produce an IFOV (Instantaneous Field Of View) of 1.8 mr when used with the detector array described previously. 327 pixels will cover a total Field Of View of 35°. Figure 7 shows the ground resolution that will be produced at any aircraft altitude from either a 25 mm or a 50 mm focal length optics. For instance, a 25 mm focal length optics will produce a ground footprint of less than 4 meters from an aircraft altitude of 2000 meters.

Uncooled detector arrays are less sensitive than cryogenically cooled detector arrays. To make up for this loss of sensitivity the well known technique of TDI (Time Delay and Integration) can be used. In this technique the outputs from the row 2 detectors are summed to the row 1 detectors when row 2 is over the same ground location. The same is done for all the following rows. The theoretical improvement in SNR (Signal to Noise Ratio) equals the square root of the number of samples integrated. The theoretical SNR improvement for TDI by 36 rows is therefore equal to 6. In practice, platform motion and 1/f noise limit the actual performance improvement that can be obtained. In the system to be described the amount of TDI for an airborne system has been limited to 40 while the amount for a ground based system can be selected to be any value up to 200.

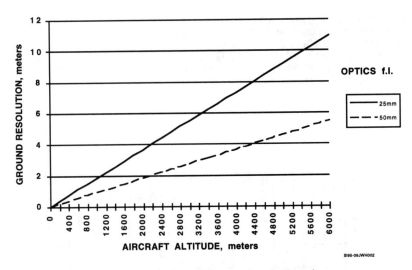

Figure 7. Ground resolution vs. aircraft alitutude

5.0 SYSTEM PERFORMANCE

By utilizing a filter wheel approach, spectral bands can be selected depending on the application. A filter wheel that is removable in the field can allow different missions to be flown on different days. This is advantageous in order to utilize the system for other applications such as crop and forestry surveys and environmental monitoring in between forest fires. Figure 8 shows a set of four spectral bands that have been selected for the nominal MIR (Multispectral Imaging Radiometer) design. The first band is centered at 8.55µ and lies approximately at the same location as the MODIS (Moderate Resolution Imaging Spectrometer) band 29 and the TIMS (Thermal Infrared Multispectral Spectrometer) bands 1 and 2. Bands 2 and 3 were selected to duplicate the AVHRR (Advanced Very High Resolution Radiometer) bands 3 and 5 which are widely used around the world. These bands also lie in the approximate location of MODIS bands 31 and 32 and TIMS bands 5 and 6. These bands can be optimized in the future for fire fighting applications. Band 4 is a wide open thermal imaging band to obtain maximum sensitivity and optimum temperature resolution.

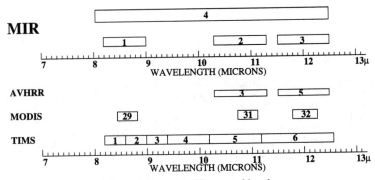

Figure 8. Sample MIR spectral bands

Most thermal imaging systems try to maintain the NEDT (Noise Equivalent Differential Temperature) at less than about 0.1° K. These systems generally utilize HgCdTe detector arrays operating at about the liquid nitrogen boiling point of 77°K. By utilizing fast refractive optics with an f/# of 0.80 and TDI of 40, the MIR system can achieve an NEDT less than 0.1°K in all of the four spectral bands shown in Figure 9. For example the NEDT in the 8.5μ band is approximately 0.04°K. The design is thus able to obtain superior sensitivity even with an uncooled detector array.

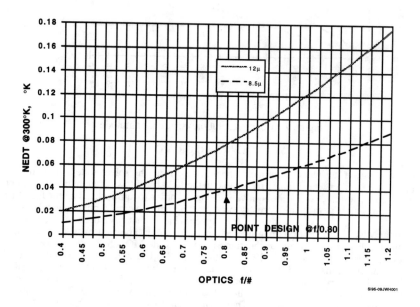

Figure 9. NEDT vs. optics f/#

6.0 AIRBORNE SYSTEMS

For fire fighting and fire mapping applications an airborne system has been developed and fabricated. Figure 10 shows the side view of the MIR system. The lower portion of the instrument contains the calibration system with its drive motor for moving the calibration mirror in and out of the field of view. The center portion of the system contains the optical bench with its filter wheel and detector array. The upper portion contains the digital signal processor. The Earth facing view is shown in Figure 11. The calibration mirror is shown in the stow position. The 25 mm optical system is seen on the right side

Figure 10. Side view of MIR

Figure 11. Earth facing view of MIR

SI96-01JWH009

7.0 TOWER BASED SYSTEM

Figure 12 shows the tower based TIR (Thermal Imaging Radiometer). The TIR is enclosed in a weather proof housing and images through a Germanium window.

Figure 13 shows the TIR with the cover removed and the door open. The base contains the pan motion drive motor and gear box and the digital electronics. The upper portion is the rotating optical head which can rotate through 360° continuously. Power and signals are transmitted through a 24 channel slip ring assembly to allow continuous rotary motion. The elevation drive shown on the right side of the opened instrument allows the optics to be pointed at the correct depression angle based on the height of the tower above the scene. The elevation drive also allows the optical system to be rotated upwards to view the calibration blackbody.

An encoder located next to the drive assembly provides knowledge of the pan angle at all times. When a fire is automatically detected by crossing a given threshold, the exact location of the fire with respect to the center of the detector array is known by the row and column numbers of the central pixel that is illuminated. This location within the detector array is then translated to an exact map latitude and longitude through the knowledge of the pan angle at the time of detection and the elevation angle of the line of sight combined with the local terrain geometry. After the fire has been detected and located, the TIR can be put into the stare mode to continuously image the fire area at a 60 Hz frame rate.

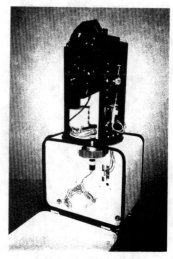

Figure 12. TIR in weatherproof housing   Figure 13. TIR opened

8.0 CALIBRATION SYSTEMS

The inflight calibration components are shown in Figures 14 and 15. The high temperature blackbody employs a hexagonal surface to produce an emissivity of approximately 95% in the 8 to 12μ region. Figure 15 shows the calibration mirror drive assembly with its d.c. torque motor and gearbox and encoder leads. The drive features a Geneva mechanism to precisely position the calibration mirror along the optical line of sight.

Figure 14. High temperature blackbody

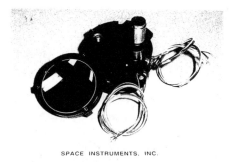

Figure 15. Calibration mirror drive assembly

SI96-01JWH012

9.0 SUMMARY

By incorporating the new technologies discussed above, an improved airborne multispectral fire mapping system has been designed and fabricated. The same new technologies have also been incorporated into the tower mounted fire detection system where the motion is provided by a 360 degree rotating platform. Both systems employ the uncooled microbolometer detector arrays, a through the optics real time calibration system, and Time Delay and Integration to improve system performance.

10.0 ACKNOWLEDGEMENTS

This project was supported by ARPA TRP Agreement Number MDA972-95-3-0023. The authors are solely responsible for the contents of the publication.

DEVELOPMENT AND UTILITY OF A FOUR-CHANNEL SCANNER
FOR WILDLAND FIRE RESEARCH AND APPLICATIONS*

Vincent G. Ambrosia[1], James A. Brass[2], Robert Higgins[3] and Edward Hildum[4]

[1]Johnson Controls World Services, Inc.
M.S. 242-4, NASA-Ames Research Center,
Moffett Field, CA. 94035-1000
[2]NASA-Ames Research Center,
Moffett Field, CA 94035-1000
[3]SIMCO Electronics, NASA-Ames Research Center,
Moffett Field, CA 94035-1000
[4]Sverdrup Inc., NASA-Ames Research Center,
Moffett Field, CA 94035-1000

ABSTRACT

The Airborne Infrared Disaster Assessment System (AIRDAS) is a four-channel scanner designed and built at NASA-Ames for the specific task of supporting research and applications of fire impacts on terrestrial and atmospheric processes and also of serving as a vital instrument in the assessment of natural and man-induced disasters. The system has been flown or integrated on numerous airframes including the Navajo, King-Air, C-130, various Lear Jet models, a Cessna 310 and a Cessna 206. The system includes a configuration composed of a 386 PC computer workstation, a non-linear detector amplifier, a sixteen-bit digitizer, dichroic filters, an Exabyte 8500 5Gb Tape output, VHS tape output, a Trimble GPS and a 2-axis gyro. The AIRDAS system collects digital data in four wavelength regions, which can be filtered: band 1 (.61-.68µm), band 2 (1.57-1.7µm), band 3 (3.6-5.5µm), and band 4 (5.5-13.0µm). The optical system has a FOV of 108 degrees, an IFOV of 2.62 mrads, and a digitized swath width of 720 pixels. The inclusion of the non-linear detector amplifier allows for the accurate measurement of emitted temperature from fires and hot spots. Lab testing of the scanner has indicated temperature assessments of over 600°C. without detector saturation. This has advantages over previous systems which were designed for thermal measurement of earth background temperatures, and were ill-equipped for accurate determination of high intensity conditions. The scanner has been flown successfully on data collection missions since 1992 in the western US as well as Brazil. These and other research and applications responses are presented along with an assessment of the design considerations for the AIRDAS system.

INTRODUCTION

The Airborne Infrared Disaster Assessment System (AIRDAS) scanning instrument was designed and built at NASA-Ames Research Center, Moffett Field, California. The initial elements of the instrument were funded through the Ames Basic Research Council's,

---

* Presented at the Sixth Biennial U.S. Forest Service Remote Sensing Applications Conference, Denver, Colorado, April 29 - May 3, 1996.

Director's Discretionary Fund and the U.S. Forest Service's Riverside Fires Laboratory. In designing the system, Ames scientists and engineers focused on specific design criteria which would enable the scanner to be a valuable research instrument as well as a cost-effective applications support scanner. These criteria included: 1) Make the total system self-contained and portable, incorporate position and altitude sensing, data display and recording capability, and full scanner controls in a single station. 2) Be able to rapidly integrate and deploy on small aircraft in order to reduce operating costs. 3) Develop a telemetry interface allowing for timely information flow to ground-based fire and disaster managers. 4) Incorporate detectors that are specifically configured to support fire research and management, fuel condition assessment, vegetation stress management, and accurate fire temperature measurements. 5) Utilize off-the-shelf components to reduce costs and allow duplicity of system construction. These criteria have led to the development and deployment of a unique instrument capable of overcoming some of the design limitations of other previous scanning and photographic systems. The AIRDAS instrument is capable of resolving high temperature profiles within intense wildfires, allowing for accurate mensuration of fire intensities.

The AIRDAS scanner was conceived, designed and constructed at NASA-Ames Research Center's Ecosystem Science and Technology Branch, High Altitude Aircraft and Medium Altitude Aircraft Branches, and the Engineering Branch (Figure 1). During the past twenty years, the development of scanner technology has allowed the simplification of many system features, allowing the science community to overcome many technological limitations (Ambrosia and Brass, 1988; Brass et al., 1987). The limitations of previous remote sensing scanner systems, including the ability to spectrally penetrate smoke plumes and define minute temperature variations in intense wildfires, have been overcome in the design characteristics of the AIRDAS (Riggan et al., 1993).

Figure 1. The AIRDAS digital scanning system. The optics head is located on the left, with the control rack and console to the right.

This paper will explore the fire science background information necessary to the understanding of the design criteria for airborne instrumentation of wildland fires. The design limitations of various satellite and airborne-based systems will be discussed and the system specifications, platform incorporation, and data analysis of the AIRDAS scanner will be presented.

BACKGROUND

Fires, natural or otherwise, play an important role in global atmospheric change and as an altering agent in landscapes and land use patterns. Fire research has become an important driving force in the development of new remote sensing systems. Systems, like the one described, are needed to characterize the components and movement of wildland conflagrations in forested environments and, more recently at the urban/wildland fringe. Due to the increased spread of urbanization and second home development into more rural, wildland environments, fire and fire monitoring have become an important consideration to the wildland manager. Monitoring and characterization are demanded, in order to mitigate fire effects in various inhabited ecosystems. Fire assessment is therefore integral to discussions of the application of remote sensing technology.

Meyers (1989) and Brustet *et al*. (1991) acknowledged the importance of satellite and airborne remote sensing technology in providing better estimates of burn area and fire intensity. Brass *et al*. (1995) summarized the measurable fire parameters from remote sensing as three different signals: the fire line itself produces thermal and visible signals; smoke and aerosols from volatilization produce an optical signal; and char produces a characteristic spectral signature in the visible and thermal infrared. These factors of biomass combustion must be characterized remotely if estimates of fire emissions and, ultimately, fire impacts are to be made.

Monitoring fires necessitates the analysis of high temperature emissivity from both the flaming front, the ash and soil background, and preheating conditions at the fore of the moving flame front. These temperature differences can be significant and can range from earth ambient temperatures to over 1200°C. The predictability of the influence of fire on an ecosystem is dependent on the burn intensity and temperature of the fire. The two major fire characteristics that have an effect on abiotic controls in ecosystems are combustion temperature duration and rate of spread of the burn. Since temperature variations are fundamental elements of ecosystem recovery and development, fire detection sensors should be capable of recording high temperature sources. Issac and Hopkins (1937) and Smith (1970), have shown that the highest temperatures recorded for burning biomass are in heavy slash fires in coniferous forests where temperatures above 1000°C were noted above the forest floor. In chaparral ecosystems, Bentley and Fenner (1958) and DeBano and Conrad (1978) found temperatures in the 350°C range at the surface. The authors have recorded temperatures above 1200°C in a controlled burn in dead, cut and dried chaparral of Southern California (Cofer *et al*., 1988) The high temperature fires that sustain extended burn durations are the most lethal to the biota redevelopment. It therefore becomes imperative that high temperature variations can be accurately differentiated and recorded.

Active fires present a complex target for any remote sensing system with the resolution to discriminate individual frame fronts. This burning can be characterized as a gradient of temperatures and radiances from spreading fire front to residual flaming, smoldering combustion and cooling ash (Riggan *et al*., 1993). Peak temperatures measured one meter above the ground can range from 1300°K in chaparral (F. Weinrick, 1993) to 900°K in

Brazilian savanna (A. Miranda, 1994). Ash surfaces in chaparral can remain above 700 K for one to two minutes following the flame front and under solar heating reach temperatures as high as 345°K (Riggan et al., 1993). Flames and ash at peak temperatures are very bright targets in the intermediate infrared wavelengths. For example, blackbody radiation at 1173 K is maximized at 2.5 micrometers, and 38 times higher than the same temperature energy emittance at 11.0 microns.

Satellite Analysis Of Fires

The local area coverage (1.0 kilometer spatial resolution) and global area coverage (4 km) of the Advanced Very High Resolution Radiometer (AVHRR) satellite scanning system have been used successfully for detecting fires. However, the detectors have been found to have signal saturation in band 3 (3.55 - 3.93 µm) over actively burning areas because of the high radiance produced by the flame and hot ash. This saturation problem prevents any calculation of fire intensity or accurate description of flame front size and smoldering area. Specular reflectance from bare ground and dry foliage produces bright targets in AVHRR band 3, resulting in false positives for fire detection. Although the AVHRR satellite system was not specifically designed for fire detection, it has proved useful in detecting and assessing large inaccessible fires in remote areas. Also, the large scale of the data makes accurate fire locational analysis difficult.

The higher resolution, lower orbit Landsat series of satellites offers the ability to monitor fire activity in short- and long-wavelength infrared channels. Using the TM sensor on-board both Landsat 4 and 5, 30-meter data (120-meter thermal infrared) are collected over a repeated site every 16 days. Channels 5 (1.55 - 1.75 µm), 6 (10.4 - 12.5 µm) and 7 (2.08 - 2.35 µm) provide most of the information for active fire characterization. Channels 5 and 6 are particularly effective for mapping burn scars by thresholding the spectral response. However, using only a thresholding technique for scar detection requires local knowledge of scar characteristics and some spectral training. Also, one of the main drawbacks to using the Landsat series satellites is the 16 day repeat orbit; much greater return time is needed to accurately monitor fires. The thermal channel included in the Thematic Mapper series scanners is calibrated for low earth ambient temperatures (132°C) and therefore does not have the radiometric sensitivity to differentiate fire fronts from background ash layers at high temperatures. Also, data cannot be directly transmitted to a common receiving station, and is therefore received up to a month after the fires have occurred.

Landsat TM data continue to be used for describing and mapping actively burning fires. The area and temperature of the flame front within a pixel can be estimated using two wavelengths, and if the target behaves as a blackbody, the radiance of the background can be estimated from nearby pixels and the signal does not saturate the detector (Matson and Dozier, 1981). However, upwelling radiation, common to large fires, will saturate TM channels 6 and 7 (Brustet et al., 1991). Therefore, a statistical distribution of fire radiances derived from TM data would be biased (Riggan et al., 1993). Any attempt to use Matson and Dozier's algorithm fire and background algorithm, when a pixel encompassed the active fire front, ash layer, and background, would result in the Planck Function (evaluated at the composite temperature), being a poor estimation of fire size and temperature.

Airborne Analysis of Fires

Brustet et al., (1991) recommends employing airborne remote sensing systems to develop an improved understanding of fires. A number of airborne systems have been used to detect and map fires. A majority of these systems are prototypes of satellite systems currently in operation for general land use / land cover mapping. Generally, these line-scanning systems collect data in broad spectral wavelength bands using low temperature blackbody reference sources onboard for system calibration. Historically, these systems had multiple problems when used to measure radiances from active fires. Following Wein's Displacement Law (Wolfe, 1989), the peak radiant emittance for temperatures of 900K occurs between 3.0 and 4.0 micrometers, and for ambient earth temperatures, between 10.0 and 11.0 micrometers. Airborne systems are usually not configured for characterization of fires because they lack channels in this part of the electromagnetic spectrum.

High temperature calibration, as is required for active fire characterization, is a second problem encountered when using airborne scanning systems. Ambrosia and Brass (1988) documented the problems with onboard calibration and the relationship of signal brightness to temperature based on two low temperature blackbody sources. As evidenced by Planck's Function, the linear relationship defined by the blackbody sources does not hold for temperatures produced by active fire fronts and super-heated soils.

Signal saturation is the third problem area found in many airborne systems. As discussed earlier, radiance from active fires can be 50 times that found at ambient temperatures. For many airborne scanner system designs, the dynamic range of the signal digitizer is too small to process both an ambient signal and a high radiance signal emitting from a fire front. Airborne systems, designed for earth science research and applications, are generally not modified to detect high temperature emittance, and therefore the result is signal saturation in the range of 100°C or lower (Ambrosia and Brass, 1988). Fire detection and burn area estimations are possible with these existing systems, however, fire front dimensions, temperature and intensity (duration) cannot be derived from the signal saturated data set (Brass et al., 1987).

AIRDAS DESIGN SPECIFICATIONS

The AIRDAS scanner was designed to be capable of quantitative measurements of very high emitted radiances (temperatures) associated with wildfires. The scanner incorporates the optical system of a Texas Instruments RS-25 thermal line scanner and was reconfigured at NASA-Ames to acquire 4-channel, signed 16-bit data. The AIRDAS incorporates four detector elements for data collection: channel 1: silicon (Si), channel 2: indium-gallium-arsenide (InGaAs), channel 3: indium antimonide (InSb), and channel 4: mercury-cadmium telluride (HgCdTe). Channels 3 and 4 are sandwiched detectors with detector 3 (InSb) on top of detector 4 (HgCdTe). The four detectors are used for the AIRDAS band wavelength regions found in Table 1. Each of the four bands can be filtered to narrower bandpass regions.

Table 1. AIRDAS Channels, Detectors, And Wavelength Regions.

AIRDAS CHANNEL	DETECTOR TYPE	WAVELENGTH (µm)
1	Si	0.61 - 0.68
2	InGaAs	1.57 - 1.70
3	InSb	3.60 - 5.50
4	HgCdTe	5.50 - 13.0

The AIRDAS channel 1 acquires data from the visible red portion of the EM spectrum, channel 2 from the shorter-wavelength near-infrared portion, channel 3 from the mid-infrared portion, and channel 4 from the long-wave thermal-infrared portion of the EM spectrum. For fire analysis, the visible band (channel 1) allows for determination, extent and movement of the smoke plume, as well as distinguishing surface cultural and vegetative features. Channel 2 allows for the analysis of vegetative composition, as well as very hot fire fronts, while still penetrating most associated smoke plumes. Channel 2 becomes sensitive to fires and hot spots at temperatures above 573K (300°C), and the response shifts away from reflective to emitted energy detection. Channel 3 was specifically designed for analysis of distinct fire temperatures while penetrating the associated smoke column. Channel 4 is designed to collect thermal data on earth ambient temperatures and on the lower temperature soil heating conditions behind fire fronts

AIRDAS High Temperature Calibration
The three AIRDAS infrared bands (band 2, 3, and 4) are calibrated for high temperature recording. The mid-infrared band 3 (3.6 - 5.5 μm) of the AIRDAS has been laboratory calibrated to temperatures above 873K (600°C). An extended blackbody thermal target was designed for instrument laboratory calibration consisting of a heating element and a metal plate with embedded thermocouples to record the temperature gradient across the surface of the plate. The thermocouple signals are processed by an Omega data-logger and transmitted to the AIRDAS embedded computer over a serial interface during the calibration run. The target element is composed of a 23cm -by- 48cm -by- 0.6cm stainless steel plate, to which was applied a 3M solar absorber coating to simulate a blackbody source. The solar coating has a normal emissivity of 0.95. Nine 3M solar absorber painted thermocouples are attached to the front of the plate in a regular array in order to record temperature variations as it is heated from below. Temperatures of the plate are recorded every second and stored to a data-logger in order to reference the time variation in the heating. The target is then heated from below by an array of quartz heaters until maximum temperatures are reached. The scanner is placed over the thermal plate, so that the calibration plate is arranged within the field-of-view of the scanning mirror. The temperature variations recorded by the thermocouples and stored to the Omega data-logger, are then uploaded to Lab-Tech Notebook software housed on a 486 personal computer. The data, recorded by the thermocouples is then time-matched to the spectral data collected by the AIRDAS scanner.

Two-Step Pre-Amplifier
High temperatures are recorded by the AIRDAS through the use of a two-step, linear curve pre-amplifier. The two-step linear pre-amplifier divides the complete dynamic range into two linear segments with an adjustable break-point. The two-step pre-amps are used on channels 2, 3 and 4. This configuration allows for the recording of a greater range of temperature profiles than is possible with standard signal pre-amplifiers. The pre-amp is included to provide the transducer interface with an electrical signal boost. The selection of low-gain/high-gain break-points allow for a fine temperature resolution (high gain), at ambient temperatures below the break-point, while still providing a more coarse temperature resolution (low gain) above the break point. The high gain temperature resolutions are generally configured for 1.0°C/count (or less), while the high temperature, low gain values are set for ~5°C/count. This dynamic range, coupled with 16-bit digitization provides an adequate overall system dynamic range for accommodating low-end ambient temperature resolutions as well as high-end temperature resolutions.

The two system blackbody reference sources are used to calibrate the four AIRDAS channels. For channels 1 and 2, the blackbody reference sources represent the zero (0) radiance for the tare value. For channels 3 and 4, the blackbody reference sources represent a known temperature. All signals are referenced to the BB1 source to restore the reference of the pre-amps on each scan, and to minimize pre-amp drift. The two blackbodies fill the aperture on each scan and are located opposite each other just before the video gate. Blackbody 1 and 2 are imbedded in the AIRDAS system, but the external plate reference is used as the upper calibration source for high temperatures. This configuration allows for the fulfillment of all mission requirements without compromising data collection over non-fire areas.

Spectral Calibration

The AIRDAS scanner is routinely calibrated both radiometrically and spectrally to ensure accurate data gathering. The radiometric calibration incorporates a Tungsten-Halogen lamp (used for channels 1 and 2) and a Nernst glow bar light source (channels 3 and 4), a grating monochrometer, an optical chopper, and collimating optics. The collimator output light beam was directed into the scanner with a large turning mirror. After passing through the scanner optics, the beam falls on each of the four detectors. The detector output signal is then sent to the two-step, linear pre-amplifier where it is conditioned to a higher gain stage and then sent to the electronic lock-in amplifier. The lock-in amplifier measures the signal amplitude at the optical chopper frequency and improves the signal-to-noise ratio of the detector. The electronic response is then converted to a digital count or number (DN) and stored. The spectral calibration is accomplished with a pair of reference detectors (Silicon (Si) and Pyro-electric). The light from the collimated beam which is detected by the AIRDAS is also detected by these reference detectors at the fringe of the beam path. The reference detector signal and the scanner output signals are combined on a personal computer spreadsheet program to obtain the relative spectral response of the scanner. The AIRDAS channels 1 and 2 have very good calibration with less than 5% noise. Channels 3 and 4 had a higher noise factor. This noise is largely due to the inherent noise of the pyro-electric reference detector, which is roughly 100 times less sensitive than the InSb or HgCdTe detectors. A calibrated integrating sphere was also used to derive the absolute spectral response. Information on the most current calibration can be obtained from the authors.

Airframe Integration and System Composition

The AIRDAS system is designed to be easily incorporated into numerous, compatible aircraft by one person in under one hour. The system is self contained, requiring a minimum of integration tools and power requirements. The system requires a 7-inch x 14-inch open scanner port in the belly of the aircraft, with enough room for control rack and operator. At present the scanner has been successfully integrated on a number of airframes including the Los Angeles County Fire Department Navajo, the National Center for Atmospheric Research (NCAR) King-Air, the NASA Lear jet (model 23), a Brazilian Lear Jet (Model 35), a Navy P-3, a Cessna 310 and 206, and the NASA C-130B. The system composition, including the weight of the head and rack and power requirements, are found in Table 2.

Table 2. AIRDAS System Composition.

Power Requirements:	28V DC @ 20 amps
Weight, Head:	80 pounds
Rack:	190 pounds
Total:	270 pounds
Scanner Port Size:	7 inches x 14 inches

The AIRDAS system is self-contained and portable and includes the Texas Instruments RS-25 optical head, the two-step linear pre-amplifiers, a sixteen bit digitizer, dichroic filters for the band passes, an Ampro 386 system control computer, an Exabyte 8500, 5Gb tape output device, an integrated Trimble TN2000 Global Positioning System (GPS) unit, and a two-axis gyro. The filters for narrowing the bandpass width of the individual channels are changeable and easily incorporated into the system. The GPS is integrated into the scanner output and delivers encoded location information on relative position to the header file for each flight segment (scanline). The two-axis gyro sends encoded information on pitch and roll to the control system in order to allow for post-flight correction. A magnetic compass assists in determining heading and allowing for geometric correction. The barometric altimeter data is also incorporated in the header. The system accommodates additional serial interfaces to integrate additional avionics navigation systems on airframes that acquire such information.

The Field-of-View (FOV) of the scanning optics is 108 cross-track, with an Instantaneous Field Of View (IFOV) of 2.62 milliradians. The system, at a designed scan rate of 5-20 scans/second, can operate in a flight envelope of 3000 to over 34,000 foot AGL, at aircraft ground speeds of 100 to 260 knots. The AIRDAS has a digitized swath width of 720 pixels in the cross-track direction, with continuous data flow acquired in the along-track direction. These parameters provide a ground spatial resolution of 26 feet at an aircraft altitude of 10,000 feet AGL.

Data Archiving and Storage

The AIRDAS data is collected and stored on Exabyte 8mm data storage tapes from an 8500 device. In-flight tape changes are possible, allowing for continuous, uninterrupted data collection. The data down-loading and integration software, developed by the U.S. Forest Service, Riverside Fire Research Lab, California and further enhanced at Ames Research Center, operates on a SUN SPARC workstation. The software is composed of two programs which extract the flight data segments from tape, formats the data , lists header information, and allows for roll correction. The input data are organized as housekeeping data (header information) followed by image data in Band Interleaved by Pixel (BIP) format. Incorporated in the extraction and image creation software are image statistics generation which relate to the minimum, maximum and mean aircraft (scanner) roll, as well as the aircraft (scanner) pitch which are derived from the two-axis gyro. Roll correction of the data is allowed and controlled by the user as a pre-processing function. The system digital output is a signed 16-bit integer data stream (SPARC short and DOS integer data). These values are derived by dividing the original tape data by eight in order to exclude the three low-order bits which are associated with system noise. Another program allows for the scan angle correction of the data set to compensate for off-angle viewing. Image data can then be analyzed on standard image processing systems.

The original system design incorporates a telemetry interface, which would include a high speed digital down-link. That down-link would deliver scanner data from the AIRDAS platform to a receiving station on the ground, such as a Fire Incident Command Center (FICC). Development of an Airphone digital data relay from the aircraft to a ground phone data link is currently under consideration. That system will be integrated and tested in mid-1996. The housekeeping information, including the aircraft attitude parameters and GPS locations would be sent serially through the link also. Data could then be received and processed in real or near-real time, allowing for rapid fire decision management strategy implementation. The telemetry link is under planning and development for the AIRDAS system. Further refinements of the prototype AIRDAS will also follow.

REMARKS

The Airborne Infrared Disaster Assessment System (AIRDAS) is designed as a prototype remote sensing scanner capable of filling a critical gap in the scientific instrumentation of fires. The scanner incorporates new technologies, such as the two-step linear pre-amplifiers, in order to accurately resolve fire intensities up to 873K (600°C). Besides the thermal infrared bands, the scanner incorporates other channels in the visible and near-infrared portion of the electro-magnetic spectrum. These channels allow for the study of vegetative conditions and other factors not associated with fires. The AIRDAS system can be rapidly integrated in a number of airframes, allowing for rapid deployment and utilization by a large user community. The system can be requested for flight support on a user's compatible airframe, while avoiding the high costs and time delays of larger aircraft, which might be committed to other scientific missions during critical data collection periods or unavailable on extremely short notice in an emergency response framework. The AIRDAS has, and will continue to provide the scientific support demanded for fire research, thermal analysis, and combustion related research topics and applications.

REFERENCES

Ambrosia, V.G. and J.A. Brass, "Thermal Analysis of Wildfires and Effects on Global Ecosystem Cycling," *Geocarto International*, Vol. 3, No. 1, pp. 29-39, 1988.

Bentley, J.R. and R.L. Fenner, "Soil Temperatures During Burning Related To Post-Fire Seed Beds on Woodland Range," *Journal of Forestry*, Vol. 56, pp. 737-774, 1958.

Brass, J.A., V.G. Ambrosia, P.J. Riggan, J.S. Myers, and J.C. Arvesen, "Aircraft and Satellite Thermographic Systems For Wildfire Mapping and Assessment," *American Institute of Aeronautics and Astronautics*, AIAA-87-0187, pp. 1-7, January 1987.

Brustet, J.M., J.B. Vickos, J. Fontan, A. Podaire, F. Lavenu, "Characterization of Active Fires in West Africa Savannas By Analysis of Satellite Data: Landsat Thematic Mapper," In J.S. Levine (ed.) *Proceedings of the Chapman Conference on Global Biomass Burning: Atmospheric, Climate, and Biospheric Implications*, MIT Press, Cambridge, Mass., pp. 53-60, Williamsburg, Virginia, 19-23 March 1990.

DeBano, L.F. and C.E. Conrad, "The Effect of Fire on Nutrients in a Chaparral Ecosystem," *Ecology*, Vol. 59, pp. 489-497, 1978.

Cofer, W.R., III, J.S. Levine, P.J. Riggan, D.I. Sebacher, E.L. Winstead, E.F. Shaw, Jr., J.A. Brass, and V.G. Ambrosia, "Trace Gas Emissions From Mid-Latitude Prescribed Chaparral Fire, *Journal of Geophysical Research*, Vol. 93, No. D2, pp. 1653-1658, February 1988.

Issac, L.A. and H.G. Hopkins, "The Forest Soil of the Douglas Fir Region, and Changes Brought Upon It By Logging and Slash Burning," *Ecology*, Vol. 18, pp. 264-279, 1937.

Matson, M. and J. Dozier, "Identification of Sub-resolution High Temperature Sources Using A Thermal IR Sensor," *Photogrammetric Engineering and Remote Sensing*, Vol. 47, pp. 1311-1318, 1981.

Meyers, N., "The Greenhouse Effect: A Tropical Forestry Response," *Biomass*, Vol. 18, pp. 73-78, 1989.

Miranda, A., Personal Communication, University of Brasilia, Brazil, 1994.

Riggan, P.J., J.A. Brass, and R.N. Lockwood, "Assessing Fire Emissions From Tropical Savanna and Forests of Central Brazil," *Photogrammetric Engineering and Remote Sensing*, Vol. 59, No. 6, pp. 1009-1015, June 1993.

Smith, D.W., "Concentration of Soil Nutrients Before and After Fire," *Canadian Journal of Soil Science*, Vol. 50, pp. 17-29, 1970.

Weinrick, F., "Unpublished Report," USDA Forest Service, PSW Forest and Range Experiment Station, Riverside, CA. 1993.

Wolfe, W.L., "Radiation Theory." In *The Infrared Handbook, Revised Edition*, eds. Wolfe, W.L. and G.J. Zissis, The Infrared Information Analysis (IRIA) Center, Environmental Research Institute of Michigan, Chapter 1, p. 1.2-1.30, 1989.

FOREST SERVICE REMOTE SENSING CONFERENCE
POSTER ABSTRACTS

HYPERSPECTRAL IMAGING - A TOOL FOR RESOURCE MANAGEMENT

Debbie Beiso
TRW
One Space Park
MS E2/4051
Redondo Beach, CA 90278

TRW has built a data exploitation system for analyzing and extracting information from hyperspectral data collected using the family of TRW Imaging Spectrometers (TRWIS). The system includes interactive tools for examining images and spectra and for classifying images. It is based on COTS software and is compatible with LANDSAT, Airborne Visible/Infrared Imaging Spectrometer (AVRIS) and other multiband data as well as the TRWIS data.

TRW offers an end-to-end systems approach with space and airborne remote sensing for quick response at high resolution and space-based systems for overview coverage.

RESEARCH AND TECHNOLOGY DEVELOPMENT FOR ECOSYSTEM MANAGEMENT

Doug Bond
Del Terra, Inc.
2042 Market Street
Redding, CA 96001
and
Kathleen Harcksen
Pacific Southwest Research Station
2400 Washington Street
Redding, CA 96001

The Pacific Southwest Research Station and Del Terra, Inc. are working together to develop new information and technology needed for ecosystem management.

Currently, PSW has initiated two large-scale, interdisciplinary research projects in Northern California. These research projects have been developed at scales adequate to answer questions ranging from soil processes through biodiversity and sustainable productivity. Digital Orthophotography, GPS, and GIS technologies are being used to integrate information through time and at a variety of geographic scales. Del Terra, Inc. has established very accurate permanently monumented geographic reference systems for the research projects. These reference systems are used to geographically control remotely sensed data and to provide the link for analysis of spatial and temporal information.

A new remote sensing technology (Low Altitude Mapping Process) is being evaluated for applicability and use in ecological research and ecosystem management.

THE MONITE SITE: A HAZMAT APPLICATION OF REMOTE SENSING TECHNOLOGY

Allen Cook
TRW
BLM National Applied Resource Science Center
Denver Federal Center, Bldg 50
Denver, CO 80226

The poster will reference the Monite Site, a hazmat site on BLM land near Sparks, NV. The poster will demonstrate how aerial photography, GPS and historical archive searches were used to build a site GIS. The GIS was used to reconstruct the periods and type of land use, extent, and spatial placement of structures on an industrial site cleared and abandoned in 1966. In addition, the GIS enables the integration of site surveys, such as chemical analyses and magnetometer surveys, into a cohesive dataset that is being used for public meetings and planning of site cleanup. This is perhaps novel use of existing technology, that should generate some interest from the participants. By the way, the hazardous material of interest is explosives: dynamite, TNT and DNT.

SIGNAL PROCESSING REQUIREMENTS FOR THE USFS FIREFLY SYSTEM: PRESENT AND FUTURE

Roger Crump, Krishna Myneni & Mark Hose
Science Applications International Corp.
6725 Odyssey Drive
Huntsville, AL 35806

High resolution dual-waveband digital imagery is obtained and processed by the Forest Service's airborne fire detection and mapping system, known as Firefly. Raw data, acquired at a rate of 100 kHz in the present system, is processed in the digital domain using dedicated computer hardware. Due to limitations of the processing hardware when the original system was designed, angular resolution and signal dynamic range were reduced to maintain manageable data rates. Modern digital processing systems such as digital signal processors (DSPs) and high speed multi-processor PCS will enable higher data rates to be processed, thereby allowing higher resolution and wider dynamic range measurements to be performed in future upgrades to the system. These enhancements will improve the fire detection and mapping capabilities of the Firefly system.

DEVELOPMENT OF A HIGH-RESOLUTION PHOTOGRAMMETRIC BASE MAPPING PROGRAM FOR FOREST MULTI-SCALE ECOSYSTEM MANAGEMENT RESEARCH

Larry J. Edwards
Aero-Metric, Inc.
4708 South College Avenue
Fort Collins, CO 80525

AERO-METRIC, INC. of Fort Collins, Colorado, under contract to USFS Region 2 Geometronics Section, developed a GIS base mapping plan for the Fraser Experimental Forest, Rocky Mountain Forest and Range Experimentation Station. Large scale topographic mapping and high resolution digital orthophotography were needed to serve as geographic base to supplement existing satellite imagery and geo-referenced research plot locations in support of the Experiment Station's multi-scale ecosystem management research at the 23,000-acre Fraser Experimental Forest, Fraser Colorado. The comprehensive project approach provided the research scientists at the experiment Forest with

- a 3-dimensional survey control network

- simultaneous 1:27,500 scale panchromatic and color aerial photography

- co-centric color infrared aerial photography

- 1:5000 scale topographic mapping

- 2 meter contours

- digital terrain model

- digital orthophotography with ½ meter pixel resolution

- full Arc/Info data set of all topographic coverages and orthophotography

A permanently monumented Second Order geodetic control network consisting of 16 three-dimensional control points was established around and throughout the Forest to provide the necessary control for photogrammetric mapping, and to serve as a basis for various surveys which will be performed in the course of ongoing research activities to insure that the surveyed research data will be compatible with and conform to the GIS mapping datum. The success of this phase of the project was due in large measure to the cooperative effort between the contractor, USFS Fraser Experimental Forest staff and Denver Water Board staff who all worked together to emplace and target the sixteen control points, several of which had to be reached by pack trails. The geodetic survey was accomplished by the contractor, AERO-METRIC, INC., using fast-static GPS methods.

Three types of aerial photography were obtained for the project. Panchromatic and normal color were obtained simultaneously using a dual camera system. False color infrared was obtained in a second pass of the aircraft over the forest. Exposure centers were maintained at coincident locations for both passes through the use of a GPS navigation and camera control system. The variation of the photo centers from the planned spatial locations was within 5 meters in the direction of flight as the camera was tripped by the GPS system, and within 100 meters transverse to the direction of flight, a factor of piloting of the aircraft to the indicated GPS line of flight.

Digital mapping was completed at a scale of 1:5000 (1"=417') with contours at an interval of 2 meters through analytical photogrammetric methods using the color aerial photography. The resultant 19 map sheets contain the traditional detail expected at the final mapping scale of 1:5000 shown with traditional map symbology (i.e. roads, trails, buildings, hydrographic features, wooded areas). The hypsography (contours) was complete through the application of digital terrain modeling (DTM). This method is particularly well suited for forest mapping where continuous photographic view of the ground surface is restricted by the forest cover. The DTM was also used for the digital orthophotography processing, and was furnished as a byproduct data layer as a triangulated irregular network (TIN) for use in GIS modeling.

The panchromatic photography was scanned at a resolution of approximately 1500 dpi creating raster image files. These digital images were then processed analytically along with aerial camera spatial orientation data, lens distortion data and the DTM data to remove all photographic distortions, creating a digitally differentially rectified image file, each pixel in its correct geographic location within National Map Accuracy Standards. The digital orthophoto files were resampled and formatted to a consistent ½-meter pixel size and output as geo-referenced ERDAS format files for use with the Experiment Station's Arc/Info GIS system. The digital image files were also output to hard copy images using a continuous tone laser film writer. The images corresponded to the coverage of each of the 19 topographic map sheets. Finished orthophoto map sheets were printed with and without the 2-meter contours superimposed.

All digital map data and digital map graphics were converted to Arc/Info file format for use in the Experiment Station's GIS system.

These digital imagery and map graphics provide an unusually high-resolution GIS data set for multi-scale ecosystem management research. Over 50 years of existing plot or site-level research measurements will be incorporated. This will enable the integration of new and pre-existing data into multi-scale analyses of patterns and processes across the inter-nested mosaic of watersheds and forested areas across the 23,000-acre landscape. Integration with satellite imagery places this landscape in the context of adjacent wilderness areas, but juxtaposed to the encroaching urban-wildland interface throughout the upper Fraser River Valley.

OVERVUE™ FOR U.S. FOREST SERVICE APPLICATIONS
Marshall B. Faintich
TRIFID Corporation
680 Craig Road, Suite 308
St. Louis, MO 63141

OverVue™ is the 25 meter resolution, natural color layer of the TruVue™ digital geographic information product line. OverVue™ is produced to 1:100,000 Class A map accuracy, and includes Digital Line Graph, Digital Elevation Models, feature names, and image database management software. The OverVue™ product is a geometrically and radiometrically seamless orthorectified mosaic and is available over the entire contiguous United States. U.S. Forest Service applications include, but are not limited to:

- 1:100,000 scale, natural color hardcopy maps
- automatic control for new Landsat imagery
- large area forest remote sensing analyses

The sample plots shown are a full resolution, 1:100,000 scale image map over part of the Mark Twain National Forest, and a reduced resolution image map of the entire United States. This OverVue™ layer was produced from 434 Landsat TM images acquired between 1991 - 1994.

ASPECTS OF AIRBORNE VIDEO
Ronald Hall
USDA Forest Service
Tongass-Stikine Area
Box 309
Petersburg, AK 99833

I would like to show airborne video products of color and black-and-white prints we currently can supply to aid other Forest Service departments in their work. Some examples would be:

1. A scanned black-and-white orthophoto with a color airborne video capture georeferenced and overlaid on it to show the current condition of that area.
2. An example of a color print from the Kodak DCS420 digital camera.
3. A color print of the airborne video mount showing the Panasonic CCD video camera, the Kodak DCS420 digital camera and the Panasonic micro navigation camera.

SEASTAR, REAL-TIME GLOBAL COLOR IMAGES FROM SPACE: A STATUS REPORT FROM ORBITAL IMAGING CORPORATION

Chris Hill and Mark Pastrone
Orbital Imaging Corporation (ORBIMAGE)
21700 Atlantic Boulevard
Dulles, VA 20166

Orbital Imaging Corporation (ORBIMAGE), a wholly owned subsidiary of Orbital Sciences Corporation (OSC), is the exclusive provider of SeaStar data to commercial and operational users. This paper will focus on the remote sensing specifications of the SeaStar satellite and the means of distributing the data to commercial and non-commercial users. The SeaStar spacecraft will be launched in 1996 by Orbital Sciences Corporations' Pegasus XL launch vehicle.

SeaStar's Sea-viewing Wide Field-of-view Sensor (SeaWiFS) instrument is a 1.1 km multispectral sensor designed to distinguish between subtle color variations on the Earth's surface. The SeaWiFS sensor uses six visible and two near infrared channels for broad area applications such as fishing operations, harmful algae bloom monitoring, sediment monitoring, offshore oil and gas operations and agriculture and forestry management. Originally designed to image only open ocean regions, SeaWiFS' imaging capability has evolved to also image land and costal zone regions.

SeaStar image products will be available in near real-time through ORBIMAGE's active archive service, OrbNet. The OrbNet Active Archive will network SeaStar ground receiving stations worldwide to provide fast reliable local, regional and global image-sets via an Internet-based on-line catalog (http://www.orbimage.com).

EVALUATION OF A COLOR INFRARED DIGITAL CAMERA SYSTEM FOR NATURAL RESOURCE APPLICATIONS

Paul Ishikawa Jr., Mike Hoppus and Jenny Alban
USDA Forest Service
Remote Sensing Applications Center
2222 West 2300 South
Salt Lake City, UT 84119

The Remote Sensing Applications Center (RSAC) played a key role in the development of the color infrared digital still camera currently being manufactured and marketed by the Eastman Kodak Company. The USDA Forest Service, WO Engineering, RSAC was responsible for defining the technical specifications for the first color infrared digital camera that Kodak produced under a contract with the Forest Service. The color infrared digital camera is designed to capture spectral information similar to color infrared film, but in a digital format. The camera is currently being evaluated for a variety of forest resource and ecosystem management applications.

INTERIOR COLUMBIA RIVER BASIN SPATIAL DATA LAYERS: PAST, PRESENT AND FUTURE VEGETATION CONDITIONS

Robert E. Keane, James Menakis, Donald Long
USDA Forest Service
Intermountain Research Station
Intermountain Fire Sciences Laboratory
P.O. Box 8089
Missoula, MT 59807
and
Wendel J. Hann
USDA Forest Service
Northern Region
P.O. Box 7669
Missoula, MT 59807

A scientifically-based appraisal, called the Interior Columbia River Basin (ICRB) Scientific Assessment project, was completed in the winter of 1995 for all lands in and adjacent to this 80 million acre watershed. This immense project resulted in the creation, compilation and modification of a multitude of continuous, coarse scale, spatial data layers to assess ICRB natural resources and ecosystem processes. In addition, many computer models were developed or modified specifically for the ICRB Scientific Assessment. This poster presents a small but important fraction of the coarse scale data layers developed specifically for this effort. These layers describe historical, current, and future vegetation conditions in the ICRB. These layers are available from the Interior Columbia Basin Ecosystem Management Project (ICBEMP) in the Walla Walla, Washington, USA. Historical conditions were assessed from archived data and maps, current conditions were quantified from satellite imagery and future conditions were predicted using the simulation model CRBSUM. CRBSUM is a deterministic succession model with a stochastic treatment of disturbance regimes.

THERMAL INFRARED FIRE DETECTION AND MAPPING

Robert Kennedy
USDA Forest Service
National Interagency Fire Center
3833 S. Development Avenue
Boise, ID 83704-5354

The USDA Forest Service Infrared Operations at Boise, Idaho provides infrared imagery of wildfires to the fire incident. The IR interpreter assigned to the fire uses the imagery to make a fire map for the Incident Commander. Two Forest Service-owned aircraft, a KingAir 200 and a Saberliner, are equipped with state of the art infrared line scanners that produce the IR imagery. The Boise IR Operations services all 50 states and when requested, Canada.
Contact personnel:
Bob Kennedy, USFS, (208)387-5648

Tom Gough, USFS, (208)387-5647
Woody Smith, BLM (208)387-5647

GPS: MAPPING A COMPLEX GROVE EVANS COMPLEX
Cherie Klein
USDA Forest Service
Sequoia National Forest
Hume Lake Ranger District
35860 E. Kings Cyn
Dunlap, CA 93621

The project was part of a forest wide task to map all giant sequoia groves on the Sequoia National Forest. A "status" report was presented at the fifth F.S. Remote Sensing Applications conference in Portland, Oregon. This poster illustrates the significant changes in accurately mapping the location of grove boundaries in the area now referred to as the Evans Complex. Fig. 1 shows the earliest known data to 1988. Fig. 2 shows NFAP data from aerial photo interpretation 1991. Fig. 3 combines the data from 1 and 2 with an overlay of the GPS data. Fig. 4 is a 3-D terrain model showing the Evans Complex, derived from the GPS data. Six separate groves were once shown here, but due to their close proximity it was impossible to tell where one ended and the other began. Concern over grove locations due to different mapping techniques from varying sources led to the need for accurate maps captured through a universally acceptable means. This kind of verifiable data was only attainable through the use of GPS.

And

GPS: REVEALING "HIDDEN" TRUTHS

The Hume Lake District was using GPS to traverse treatment units on an upcoming timber sale (Fig. 1 & 2). At the same time a crew was performing GPS traverses on Giant Sequoia groves on the district (Fig. 3 & 4). During post-processing it became evident that a previously unknown grove boundary was on the same ridge as a timber unit boundary. GPS traverses were imported to GIS immediately and overlaid (Fig.5). As a result, the timber unit boundary was moved back to account for the grove and its influence zone buffer (Fig. 6). In addition, the timber unit was using a road on the quadrangle map for its west boundary. After the GPS traverse it became clear that the road was not accurately represented on the quad map and that it was actually about 500 feet east of the map location. Accordingly, original estimated unit acres proved to be significantly reduced. As a result of using GPS we now have more accurate data on our groves and are alerted to incorrect data such as road locations that are otherwise not readily apparent

EVALUATION OF A COLOR INFRARED DIGITAL CAMERA SYSTEM FOR FOREST HEALTH PROTECTION APPLICATIONS

Andrew Knapp
USDA Forest Service
Forest Health Protection
1750 Front Street
Boise, ID 83702
and
Mike Hoppus
USDA Forest Service
Remote Sensing Applications Center
2222 West 2300 South
Salt Lake City, UT 84119

In the Intermountain West fire exclusion, grazing, and past forest management activities have created extensive areas of overly dense stands of trees which are highly susceptible to attack by forest pests and infection by forest diseases. A major component of forest health management is the detection, monitoring, and quantification of forest insect and disease activity.

Color infrared digital camera systems are a new remote sensing tool that have the potential to improve temporal, spatial, and quantification limitations of existing remote sensing technologies used for forest pest detection and monitoring. A color infrared digital camera system, the Kodak DCS-420, was evaluated in Idaho and Utah during 1995. Images were collected over areas affected by various forest insect and disease pests and over areas of recent wildfire activity. Images were compared to visual observations and to small and medium format photographs. Results indicate that the camera system can be successfully used to supplement existing operational remote sensing techniques currently used for monitoring and quantifying forest pest activity.

INTERREGIONAL ECOSYSTEM MANAGEMENT COORDINATION GROUP (IREMCG) PROCUREMENT OF REMOTE SENSING DATA

Henry Lachowski, Tim Wirth, Vicky Varner and Paul Maus
USDA Forest Service
Remote Sensing Applications Center
2222 West 2300 South
Salt Lake City, UT 84119

The InterRegional Ecosystem Management Coordination Group (IREMCG) has approved the procurement and use of remotely sensed information to support field implementation of ecosystem management. A team chartered by the IREMCG developed a description of the products, data, and time frames to make this information available to Regions and Experiment Stations over the next two years. The primary purpose of this procurement is to assist in broad area assessments, and is not intended to replace ongoing Regional imagery utilization programs.

Products to be purchased are currently being used by other federal agencies, and are available through USGS EROS Data Center in Sioux Falls, SD. The image data sets are as follows:

IMAGE TYPE	SCALE	RESOLUTION
Weather Satellite Imagery (AVHRR)	1:1,000,000	1 km
Landsat MSS Triplicates (1973, 1986 & 1991)	1:250,000	80 m (1 acre)
Landsat TM (1990's)	1:100,000	30 m (0.25 acres)

This procurement will be phased over two years to meet the anticipated needs for large area assessments, and other spatial analysis requirements. The data and products will be delivered on tapes and/or on CD-ROM's. All the products will be georeferenced to facilitate efficient use and to allow input into geographic information systems.

This poster display describes the image procurement and provides examples of data types.

GEOMORPHOLOGY, HYDROLOGY, AND LAND USE IN MOUNTAIN MEADOWS: REMOTE SENSING APPLICATIONS

Jim McKean
USDA Forest Service
Region 5 Engineering
2245 Morello Avenue
Pleasant Hill, CA 94523
and
Laurel Collins
Consulting Geologist
2338 Valley Street
Berkeley, CA 94702

Large high-elevation meadows occur in the upper South Fork Kern River drainage on the Kern Plateau in the southern Sierra Nevada, California. A project has been started to understand the response of these meadows to changes in outside factors and then make recommendations for land management. In the past 150 years, conditions in the meadow appear to have changed from broad, wet floodplains containing narrow, meandering channels with relatively stable banks, to the present state of shallow, wide channels that are actively migrating laterally in a narrower floodplain that is inset within at least six terraces. The current channels are poor habitat for the California state fish, the golden trout, which is endemic to this area. Under investigation as possible causes of these changes are climate fluctuations, major floods, seismicity, and grazing. Airborne videography, digital camera imagery, and Hasselblad photography are being used for detailed topographic and vegetation mapping, mapping abandoned channels, mapping spatial patterns of channel degradation and aggradation, selecting sampling sites for age-dating of geomorphic features, and monitoring channel and vegetation changes over time. In lower Monache Meadow, the main channel has migrated laterally up to 250 meters between 1976 and 1994. However, fossil pollen analysis and ^{210}Pb dating of another abandoned channel in lower Monache Meadow indicates that channel was abandoned at least 3350 years ago. ^{14}C dating of a high terrace in Mulkey Meadow shows that the surface of that terrace is at least 6780 years old. These results illustrate the complex nature of the probably multiple causes of meadow evolution.

MAPPING ASPEN CHANGES USING REMOTE SENSING

Jim McNamara, Kevin Suzuki, Patrick Riordan and Ron Brohman
USDA Forest Service
Beaverhead-Deerlodge National Forests
420 Barrett Street
Dillon, MT 59725
and
Tim Wirth and Jay Powell
USDA Forest Service
2222 West 2300 South
Salt Lake City, UT 84119

The Beaverhead and Deerlodge National Forests are conducting a landscape level inventory of aspen coverage using remote sensing. Changes in aspen coverage were assessed and mapped by comparing the maps created from a 1992 TM satellite image to a map created from historical black and white photographs from 1947. Overall patterns in the location and distribution of aspen and are evident and mappable from this type of data.

ACCURACY ASSESSMENT FOR
THE GAP ANALYSIS COVER-MAP OF UTAH

Gretchen G. Moisen
USDA Forest Service
Intermountain Research Station
507 25th Street
Ogden, UT 84401

Thomas C. Edwards, Jr.
USDA National Biological Service
Utah Cooperative Fish and Wildlife Research Unit
Utah State University
Logan, UT 84322

and
D. Richard Cutler
Department of Mathematics and Statistics
Utah State University
Logan, UT 84322

With the increasing demand for broad-scale vegetation maps for ecosystem management and conservation planning comes the need for cost-effective means to assess the thematic accuracy of these maps. Here, we illustrate the approaches taken to assess the accuracy of the 21.9 million ha cover-map developed for the Utah Gap Analysis. As part of our design process, we first reviewed the effect of intracluster correlation and a simple cost function on the relative efficiency of cluster sample designs to simple random designs. Our design ultimately combined clustered and subsampled field data stratified be ecological modeling unit and accessibility. We next outline estimation formulae for simple map accuracy measures and report results for 8 major cover-types and the 3 ecoregions mapped as part of the Utah Gap Analysis. We examine gains in efficiency of our design over a simple random sample approach for reporting traditional map accuracy measures, and discuss logistical constraints facing attempts to assess accuracy of large area cover-maps. Finally, we extend the analysis of map error beyond traditional measures by exploring the relationship between thematic accuracy and various topographical and heterogeneity components of that map using generalized linear mixed models. Specifically, map error is modeled as a function of elevation, aspect, slope, local heterogeneity, and distance to vegetation boundaries.

APPLICATION OF REMOTE SENSING AND GIS ON THE GRASSLANDS
OF THE SAMUEL R. MCKELVIE NATIONAL FOREST

Mike Morrison
Pueblo Integrated Resources Inventory
1920 Valley Drive
Pueblo, CO 81008
and
Vicky Varner
USDA Forest Service
Remote Sensing Applications Center
2222 West 2300 South
Salt Lake City, UT 84119
and
Paul Maus
USDA Forest Service
Remote Sensing Applications Center
2222 West 2300 South
Salt Lake City, UT 84119

The Rocky Mountain Region of the Forest Service and the Remote Sensing

Applications Center (RSAC) have been doing a cooperative study on the Application of Remote Sensing and GIS on the Grasslands of the Samuel R. McKelvie National Forest. The Pueblo Integrated Resource Inventory (IRI) Center has utilized various aspects of that study to develop a supervised classification of Landsat (TM) imagery separating out the unique occurrences of species and/or cover types for integration into the IRI Common Vegetation Unit.

The Pueblo IRI Center and RSAC are presenting a joint effort for a poster display. This will cover the following four themes:

1) The various remote sensing technologies utilized by RSAC, how the data was collected, and what technologies show potential for future use on the grassland ecosystems.

2) The supervised classification and how some of the remote sensing technologies utilized by RSAC were then used by the Pueblo IRI Center toward the development and refinement of the classification process. In particular the use of airborne video and Xybion multispectral video.

3) How the supervised classification of unique types is integrated into the final IRI Common Vegetation Unit (CVU) and how the tabular data is tied to the polygons for GIS application.

4) A GIS application displaying a photo realistic (trees, shrubs, grass, rock, etc.) 3D representations of various parts of the study area. This application utilizes the spatial and tabular data of the supervised classification or the IRI CVU polygons to drape the ecosystem over a DEM. This includes the ability to display in a 3D environment various species, sizes, densities, and layers within a single CVU polygon or classified cover type.

NASA'S AIRBORNE IMAGE TELEMETRY SYSTEMS

Jeffrey Myers
Aircraft Data Facility
High Altitude Missions Branch
NASA-Ames Research Center
Mail Stop 240-6
Moffett Field, CA 94035

Over a period of 15 years, the NASA airborne science program has implemented a variety of systems for relaying digital imagery (and other data) from aircraft to ground stations, in real time. This poster will detail several currently operational systems, including K- and S- Band aircraft-to-satellite uplinks, and a UHF direct downlink. Included will be a description of a low cost auto-tracking ground station for the direct broadcast system. Potential areas of technology transfer and interagency cooperation will be discussed as well.

REMOTE SENSING EVALUATION AND DEVELOPMENT FOR FOREST HEALTH MONITORING

Dick Myhre and Ross Pywell
USDA Forest Service
Forest Health Technology Enterprise Team - Ft. Collins
3825 E. Mulberry Street, Room 228
Fort Collins, CO 80524
and
Paul Ishikawa Jr., and Jenny Alban
USDA Forest Service
Remote Sensing Applications Center
2222 West 2300 South
Salt Lake City, UT 84119

The Forest Health Technology Enterprise Team (FHTET) - Fort Collins and the Remote Sensing Applications Center (RSAC) have a long history of cooperation in the field of remote sensing within the Forest Service. In recent years numerous remote sensing systems and image processing techniques have been developed and evaluated jointly to support forest health monitoring applications. Some of the imagery collected with these systems include: 1) The Panasonic S-VHS natural color video with GPS, 2) Xybion multispectral video, and 3) Kodak color infrared digital still camera imagery.

POINT AND CLICK GIS USING AVENUE, ARCVIEW'S OBJECT ORIENTED PROGRAMMING LANGUAGE

David Neufeld
Department of Forest Sciences
Colorado State University
Fort Collins, CO 80521

This application written in Avenue, ArcView's object oriented programming language, allows users with little or no previous GIS experience to quickly display and query forestry data from the Arapahoe-Roosevelt National Forest. A series of dialog boxes guide the user through three different views of the forest at gradually larger scales. Forest stands from the Boulder Ranger District can be selected based on percent slope, distance from streams, and average tree size. The selected stands are displayed and the total area in acres is reported back to the user. All data for the application were collected from publicly accessible Internet sites.

MONITORING BLACK-TAILED PRAIRIE DOG POPULATIONS WITH REMOTE SENSING

Frederic E. Nichols
Remote Sensing Research, a non-profit 501 © 3 corporation
P.O. Box 1949
Ft. Collins, CO 80522

Large-scale aerial photography provides a method for rapidly estimating the prairie dog density in large colonies and can also be used to accurately determine prairie dog populations in smaller colonies. Remote Sensing Research Inc. employs electric-powered remote piloted vehicles to obtain high resolution imagery over sample plots in large colonies and complete coverage over smaller colonies. The City of Fort Collins Natural Resources Department utilizes data obtained from these aerial surveys to manage the prairie dog colonies in and around Fort Collins.

THE AIRBORNE VIDEO TOOLKIT -
A SYSTEM FOR THE AUTO-MOSAICKING OF AIRBORNE VIDEO
IMAGERY AND ITS APPLICATION IN FOREST PEST MANAGEMENT

Jeanine L. Paschke
Management Assistance Corporation of America
In residence at:
USDA Forest Service
Forest Health Technology Enterprise Team
3825 E. Mulberry Street
Fort Collins, CO 80524
and
David S. Linden
Colorado State University
Fort Collins, CO 80521

Airborne video systems have been in use throughout the U.S. Forest Service for several years. Among other applications, these low-cost remote sensing systems, combined with video image processing and geographic information systems (GIS), have proven to be powerful tools for forest pest detection and monitoring. However, the operator time required to digitally mosaic and register video flightline imagery can be quite extensive, especially for imagery collected over a dense forest canopy which reveals few landmark features. An innovative software package has now made it possible to automatically mosaic video imagery using some additional hardware components and data from global positional systems (GPS) and aircraft attitude sensors. This poster discusses these hardware add-ons and the Airborne Video Toolkit (AVT) software.

The AVT software allows for the managing, analyzing, displaying and digital processing of video flight mission imagery and data. It integrates aircraft locational data and calculates specific ground coordinates for each video frame, allowing for a quick and accurate mosaic of the flightline imagery. The employment of these hardware and software enhancements to the basic airborne video system has been shown to dramatically reduce mosaicking and registration time, especially over heavily forested terrain. This decrease in processing time can be a critical factor in the timely detection and remediation of rapidly moving disturbance agents.

COMPACT AIRBORNE MULTISPECTRAL IMAGER

Alex Pertica
Lawrence Livermore National Laboratory
PO Box 808, L-183
Livermore, CA 94550

The Compact Airborne Multispectral Imager (CAMI) is a framing imager designed for airborne spot survey applications based on collection of ground reflectivity data. The system has applications in environmental monitoring and pollution assessment. CAMI employs two electronically-tunable liquid-crystal-based narrow bandpass filters in conjunction with intensified CCD focal planes to generate images in 62 distinct bands in the range 400-1000 nm. Images are obtained successively through the filter as the filter passband is tuned to selected wavelengths. Images are acquired at a 10 Hz rate and each filter can be tuned to any wavelength in random order across its respective tuning range of either 400 - 700 nm or 700 - 1000 nm. The system is extremely compact; the camera payload is housed in a four-axis gyro-stabilized turret that is aircraft ready. The image handling system incorporates a frame grabber that digitizes the analog output of the two cameras. The framing (as opposed to pushbroom or whiskbroom) architecture of this imager supports data collection modes which are consistent with real time hyperspectral image processing. Thus, this system can be used as a finder to cue other instruments.

CAMI has been built and flight tested on an Air Force RC-135 aircraft. CAMI is housed in a 14" gimbal that is compatible with many small fixed wing aircraft. Our goal is to demonstrate capability for a variety of applications.

MAPPING APHID DAMAGE IN KENYA CYPRESS

Jay Powell
USDA Forest Service
Remote Sensing Applications Center
2222 West 2300 South
Salt Lake City, UT 84119
and
Denny Ward
USDA Forest Service
Forest Health - Region 8
1720 Peachtree Road NW
Atlanta, GA 30367
and
Chuck Dull
USDA Forest Service
Washington Office Engineering
14th and Independence Ave., SW
Washington D.C. 20250

Landsat Thematic Mapper was used to map aphid pest damage on cypress forest plantations in Kenya as part of a cooperative project between the US Forest Service and the Food and Agriculture Organization of the United Nations (FAO). Maps produced from this project were used by the Kenya Department of Resources to guide ground sampling in damaged areas and to devise management prescriptions.

USING LANDSAT TM DATA FOR ECOSYSTEM MANAGEMENT

Edward Reilly
USDA Forest Service
Rogue River National Forest
Applegate Ranger District
6941 Upper Applegate Road
Jacksonville, OR 97530

The Applegate watershed encompasses 500,000 acres and spans three Ranger Districts on two National Forests, two Resource areas on the Medford BLM District and almost 1/3 private land. The Applegate Adaptive Management area was designated in the Clinton Northwest Forest Plan as an area to do continuing scientific and social research in ecosystem management with strong local community involvement in the decision making process.

Landsat TM data was used to create a vegetation data set detailing size, density and species composition across all ownerships. ARC/INFO and ARC GRID were used to perform the subsequent analysis. The classified image data was used to produce fire hazard maps using combined plant communities, aspect, slope and elevation factors. County tax lot data was used to determine land use and development values along with added areas of biological sensitivity. These maps when combined show the area most at risk from large fire events. The federal agencies are taking preventative measures to reduce the fire risk on these lands.

Other projects include the use of polygon data derived from Landsat TM grids to produce maps of northern spotted owl nesting, roosting, foraging and dispersal habitat. Comparisons of the TM data set with a biologist's aerial photo mapping will be shown. Maps were also created to predict deer winter range including thermal cover and hiding cover.

EOS SENSORS: APPLICABILITY TO FORESTRY, ECOSYSTEM MANAGEMENT

Carl Wheatley
Hughes Information Technology Systems
1616 McCormick Drive
Upper Marlboro, MD 20774-5372

HITS is the prime contractor for EOSDIS, the Data and Information System for

EOS, the NASA-sponsored Mission to Planet Earth (MTPE) Earth Observing System. This poster highlights the NASA MTPE EOS sensors and their potential to Forestry, Forest Management, and Ecosystem Management.

EOS is part of MTPE and is composed of a satellite constellation, a scientific research program, and EOSDIS. Scientific data from eight instruments are organized within EOS into over 150 standard data products. Our poster focuses on the categories of information from each instrument and the utility of this information for Forest Service remote sensing applications. The most applicable instruments are the VNIR imagers that include MODIS, ASTER, LANDSAT 7, and MISR. Additional instruments of applicability include CERES, TMI*, PR*, and EOS Model Output. Full technical descriptions of these instruments and their platforms/missions will be provided. Specific Forest Service application potential is presented through a matrix that maps applications to EOS instruments. Interactive analysis of this matrix will enable 1) Foresters to better understand the value of EOS instrument data to their domain; 2) EODIS representatives to better understand the needs of Foresters; and 3) discovery of previously unrecognized instrument/application synergies.

and

CONCEPT FOR REAL-TIME FOREST FIRE MANAGEMENT TOOL: MOBILE FIREFIGHTER DATA SYSTEM

This poster highlights a concept to integrate fire monitoring systems, Hughes DirecPC technology, and real-time remote sensing data integration service for real-time remote sensing data collection and delivery to the field. DirecPC can deliver prepackaged data, images and information at 3 Mbps rates to remote location fire monitoring and analysis equipment. Online Internet service delivery can be provided at 400 kbps rates, and a continuous video/audio stream can accommodate service delivery of real-time news and information services. The technology uses the Galaxy IV satellite and a 24-inch remote site satellite dish. Conventional http/WWW protocols or secure special-purpose protocols can be easily implemented within this information infrastructure.

The Mobile Firefighter Data System provides instantaneous instrument management and real-time data collection services for firefighters through a Real-time Data Integration Center. Using orbital tracking software, key remote sensing instruments are tracked over the fire management area to ensure timely data ordering and delivery of high-resolution fire data. Government and commercial remote sensing data is collected and uplinked through a Network Operation Center for 3Mbps delivery to multiple remote field and mobile offices. Internet WWW and ftp data to supplement fire management can be requested via cellular or land-based modem and delivered through a return channel at up to 400Kbps rate. The system is further enhanced through the integration of a continuous video/audio stream to the field, supporting firefighter access to information services such as CNN and the Weather Channel and the delivery of other high-volume data. The system can be rapidly deployed and relocated as changing conditions require.

The Mobile Firefighter Data System concept will encourage dialog to 1) enhance Forester awareness and understanding of advanced wireless communication systems; 2) assess the needs and benefits of the system concept; and 3) provide an open forum platform to apply the concept technologies to other Forest Service application areas.

DINOSAUR TRACKSITE PHOTOGRAPHY AND MAPPING
David Wolf
USDA Forest Service
R2 Geometronics
740 Simms Street
Golden, CO 80401

Illustration of use of very large scale aerial photography for recording and topographic mapping of a dinosaur tracksite on the Comanche National Grassland, Colorado. Project involved taking 1:1300 - scale, panchromatic aerial photos with a forward image motion compensation camera, targeting and GPS survey of ground control points, and topographic mapping of the site by photogrammetric methods. The poster display would include a 4 X enlargement of the central photograph overlaid with the topographic map, and a map showing the flight line layout and ground control configuration. A video flight has also been planned for the site, and a hard copy of a video frame may also be included.

SIXTH BIENNIAL USDA FOREST SERVICE
REMOTE SENSING APPLICATIONS CONFERENCE
Holiday Inn Denver Southeast
3200 S. Parker Road
Aurora, Colorado 80014
(303) 695-1700

<u>Monday, April 29</u>

Conference Registration 9:00 AM - 1:00 PM

"People in Partnership With Technology"

Opening Session

Moderator: Dave Wolf, Region 2 Geometronics

1:00 PM	Tom Bobbe, USDA Forest Service, Remote Sensing Applications Center, Salt Lake City, Utah, Conference Chair: **Opening Remarks.**
1:15 PM	Elizabeth Estill, USDA Forest Service, Region 2 Regional Forester, Denver, Colorado: **Welcome and Regional Perspective.**
1:45 PM	Janice McDougle, USDA Forest Service, Associate Deputy Chief, National Forest System, Washington, DC: **Washington Office Perspective.**
2:15 PM	Chuck Dull, USDA Forest Service, WO Engineering Remote Sensing Coordinator, Washington, DC: **Current Forest Service Remote Sensing Activities.**
2:45 - 3:00 PM	Break
3:00 PM	Tom Loveland, US Geological Survey, EROS Data Center, Sioux Falls, South Dakota: **The U.S. Geological Survey Land Cover Characterization Program.**
3:30 PM	Maury Nyquist, USDI National Biological Service, Lakewood, Colorado: **The National Biological Service and National Park Service Vegetation Mapping Program - Review and Update.**
4:00 PM	Jim Turner, USDI Bureau of Land Management, National Applied Resource Sciences Center, Lakewood, Colorado: **The Bureau of Land Managment Modernization Resources Option.**

Tuesday, April 30

Remote Sensing and GIS for Ecosystem Management

Moderator: Mike Hamby, Region 8 Geometronics

8:00 AM   Stan Bain, USDA Forest Service, Remote Sensing Applications Center, Salt Lake City, Utah: **Project 615 - Implementation and Add On Technologies.**

8:30 AM   Roger Hoffer, Colorado State University, Department of Forest Sciences, Fort Collins, Colorado: **Geomatics at Colorado State University.**

9:00 AM   Henry Lachowski and Paul Maus, USDA Forest Service, Remote Sensing Applications Center, Salt Lake City, Utah: **Remote Sensing Data Purchases for Ecosystem Management.**

9:30 AM   Bill Wilen, U.S. Fish and Wildlife Service, National Wetlands Inventory, Washington DC: **Assessment of Remote Sensing/GIS Technologies to Improve on National Wetlands Inventory Maps.**

10:00 - 10:30 AM   Break and Poster Exhibit

10:30 AM   Cindy Williams, USDA Forest Service, Pacific Northwest Research Station, Institute of Northern Forestry, Fairbanks, Alaska: **Exploratory Use of Synthetic Aperture Radar (SAR) - Conclusions From Forest, Wetland, and Geologic Mapping.**

11:00 AM   Melinda Walker, USDI Bureau of Land Management, National Applied Resource Sciences Center, Lakewood, Colorado: **The Southwest Colorado Vegetation Classification Project - An Example of Interagency Cooperation.**

11:30 AM   Virginia Emly, USDA Forest Service, Region 2, Nebraska National Forest, Bessey Ranger District, Halsey, Nebraska and Vicky Varner, USDA Forest Service, Remote Sensing Applications Center, Salt Lake City, Utah: **Integrating Remote Sensing into Classification of Seral Stages with Structure.**

12:00 - 1:00 PM   Lunch

Remote Sensing and GIS for Ecosystem Management

Moderator: Ebeth McMullen, Region 1 Geometronics

1:00 PM   Kim Mayeski, USDA Forest Service, Region 4, Caribou National Forest, Pocatello, Idaho: **Technological Building Blocks for Deriving Timberland Suitability Answers.**

1:30 PM   Jeff Milliken, USDA Forest Service, Region 5 Remote Sensing Laboratory, Sacramento, California: **Integration of Remote Sensing with Vegetation Mapping, Integrated Resource Inventories, and GIS Database Projects.**

2:00 PM   Tim Hill, Geographic Resource Solutions, Arcata California: **Forest Biometrics From Space.**

2:30 - 3:00 PM   Break and Poster Exhibit

3:30 PM   Lowell Lewis, USDA Forest Service, Forest Health Technology Enterprise Team, Fort Collins, Colorado: **The Implementation of GIS in a USDA Forest Service Project to Assess the Role of Insects and Pathogens in Forest Succession.**

3:30 PM   Robert Keane, USDA Forest Service, Intermountain Research Station, Intermountain Fire Sciences Laboratory, Missoula, Montana: **Spatially Explicit Ecological Inventories for Ecosystem Management Planning Using Remote Sensing and Gradient Modeling.**

4:00 PM   John Steffenson, Environmental Systems Research Institute, Inc., Boulder, Colorado: **Trends in Spatial Information Technology - What's Hot, What's Not.**

Wednesday, May 1

Creating GIS Data Bases, Performing Change Detection, and Advanced Image Processing Applications.

Moderator: Roger Crystal, Region 6 Geometronics

8:00 AM   Jim Schramek, USDA Forest Service, Region 10 Tongass National Forest, Stikine Area, Petersburg, Alaska: **Imagery in Support of GIS on the Stikine Area of the Tongass National Forest.**

8:30 AM   Carl Markin, Hughes STX Corporation, U.S. Geological Survey/EROS Alaska Field Office, Anchorage, Alaska: **Development of a GIS for the Chugach National Forest.**

9:00 AM   Lisa Levien, USDA Forest Service, Region 5 Forest Pest Management, Scaramento, California: **Large Area Change Detection in California Using LANDSAT Satellite Imagery.**

9:30 AM   Ron Brohman, USDA Forest Service, Eastside R1 Zone Ecosystem Planning, Helena, Montana, and Tim Wirth, USDA Forest Service, Remote Sensing Applications Center, Salt Lake City, Utah: **Monitoring Aspen Decline Using Remote Sensing and GIS.**

10:00 - 10:30 AM   Break and Poster Exhibit

10:30 AM   Kass Green, Pacific Meridian Resources, Emeryville, California: **Change Detection Using Remotely Sensed Data.**

11:00 AM   Bill Cooke, USDA Forest Service, Southern Research Station, Forestry Sciences Laboratory, Starkville, Mississippi: **Evaluating the Utility of Partitioning Remotely Sensed Data.**

11:30 AM   Joseph Spruce, Lockheed Martin Stennis Operations, Stennis Space Center, Mississippi: **Improving Vegetation Maps From LANDSAT TM Imagery Through Subpixel Analysis.**

12:00 - 1:00 PM   Lunch

Moderator: Paul Greenfield, Remote Sensing Applications Center

Emerging Remote Sensing Technologies

1:00 PM   Michael Bullock, Booz, Allen & Hamilton, Inc., Colorado Springs, Colorado: **High Resolution Commercial Satellite Imagery and Data Exploitation for Forestry Applications.**

1:30 PM   Dale Johnson, Positive Systems, Whitefish, Montana: **Digital Aerial Photography Used in River and Riparian Corridor Analysis - Blackfoot River Study.**

2:00 PM   Andrew Knapp, USDA Forest Service, Forest Health Protection, Boise, Idaho: **Evaluation of a Color Infrared Digital Camera System for Forest Pest Management Applications.**

2:30 - 3:00 PM   Break and Poster Exhibit

3:00 - 4:30 PM   Panel Discussion: **Accuracy Assessment for Remotely Sensed Data.**

Panel Members:

Henry Lachowski, USDA Forest Service, Remote Sensing Applications Center, Salt Lake City, Utah.

Ray Czaplewski, USDA Forest Service, Rocky Mountain Range and Experiment Station, Fort Collins, Colorado.

Bill Clerke, USDA Forest Service, Region 8 Remote Sensing, Atlanta, Georgia.

John Teply, USDA Forest Service, Region 6 Timber Management, Portland, Oregon.

Steve Stehman, Forestry Department, SUNY, Syracuse, New York.

Russell Congalton, Department of Natural Resources, University of New Hampshire, Durham, New Hampshire.

Thursday, May 2

GIS and GPS Data Collection

Moderator: Dave Schultz, Region 3 Geometronics

8:00 AM   Gyde Lund, USDA Forest Service, Forest Inventory, Economics, and Recreation Research Staff, Washington, DC: **Bread Making and Designing Resource Inventories: The GIS Connection (The Whole Loaf).**

8:30 AM   Carlos Rodriguez, USDA Forest Service, International Institute of Tropical Forestry, Rio Piedras, Puerto Rico: **Technology Transfer, An Important Role - GPS in Uruara, Brazil.**

9:00 AM   Jeff Moll, USDA Forest Service, San Dimas Technology and Development Center, San Dimas, California: **Laser System Remote Sensing Applications.**

9:30 AM   Carl Sumpter, USDA Forest Service, Region 2 Medicine Bow-Routt National Forest, Laramie Wyoming: **Sub-meter GPS Positioning Over Long Baselines.**

10:00 - 10:30 AM   Break and Poster Exhibit

10:30 AM   Dick Mangan, USDA Forest Service, Missoula Technology and Development Center, Missoula, Montana: **Using Hand-held GPS in Wildland Fire Management.**

11:00 AM   Carol Brady, USDA Forest Service, Geometronics Service Center, Salt Integrating GPS Data Into C Lake City, Utah: **Cartographic Feature Files.**

11:30 AM   William Michalson and Joshua Single, Worcester Polytechnic Institute, Worcester, Massachusetts: **A Real-time GPS System for Monitoring Forestry Operations.**

12:00 - 1:00 PM   Lunch

Aerial Photography and Digital Imagery Applications

Moderator: Ron Skillings, Region 10 Geometronics

1:00 PM   Richard Grotefendt, Forest Consultant and Photogrammetrist, North Bend, Washington: **Fixed-base Large Scale Aerial Photography Applied to Individual Tree Dimensions, Forest Plot Volumes, Riparian Buffer Strips, and Marine Mammals.**

1:30 PM   Jeffrey Barry, University of Washington, College of Forest Resources, Seattle, Washington: **Use of Historical Aerial Photos to Evaluate Stream Channel Migration.**

2:00 PM   Calvin O'Niel, U.S. Geological Survey, Southern Science Center, Lafayette, Louisiana: **1995 Scanned Aerial Photography of the Kisatchie National Forest in Louisiana.**

2:30 - 3:00 PM   Break and Poster Exhibit

3:00 PM   Ken Winterberger, USDA Forest Service, Pacific Northwest Research Station, Forestry Sciences Laboratory, Anchorage, Alaska: **Using "Pseudo" Digital Orthophotography or the Good, the Bad and the Ugly.**

3:30 PM   Ward Carson, Oregon State University, College of Forestry, Corvallis, Oregon: **Accuracy of USGS Digital Elevation Models in Forested Areas of Oregon and Washington.**

4:00 PM   Robert McGaughey, USDA Forest Service, Pacific Northwest Research Station, Forestry Sciences Laboratory, Seattle, Washington: **Applying UTOOLS/UVIEW to Natural Resource Analysis Problems.**

Fire Ecology and Thermal Infrared Remote Sensing

Moderator: Chuck Dull, WO Engineering Remote Sensing Coordinator

8:00 AM   Susan Boudreau, USDA Forest Service, Region 4 Payette National
            Forest, McCall, Idaho: **An Ecological Approach to Assess
            Vegetation Change After Large Scale Fires on the Payette
            National Forest.**

8:30 AM   Bill Rush, USDA Forest Service, Region 4 Boise National Forest,
            Boise, Idaho: **Resources at Risk - Fire Based Hazard/Risk
            Assessment for the Boise National Forest.**

9:00 AM   Ron Wicks, U.S. Army Missile Command, Redstone Arsenal, Alabama:
            **Technical Upgrades to the Forest Service Fire Fly Infrared
            Mapping System.**

9:30 AM   Bill Krausmann, USDA Forest Service, Region 3 Southwestern Region
            Remote Sensing, Albuquerque, New Mexico: **The Application of
            Forward Looking Infrared Imaging Systems in Wildland Fire
            Suppression.**

10:00 - 10:15 AM   Break

10:15 AM   Nathan Poage, Oregon State University, Department of Forest
            Sciences, Corvallis, Oregon: **Application of Thermal Infrared FLIR
            and Visible Videograpghy to the Monitoring and Restoration of
            Salmonid Habitat in the Pacific Northwest.**

10:45 AM   Jim Hoffman, Space Instruments, Inc., Encinitas, California: **New
            Technologies for Infrared Remote Sensing and Fire Mapping.**

11:15 AM   Vince Ambrosia, NASA Ames Research Center, Moffett Field,
            California: **Development of a Four Channel Scanner for Wildland
            Fire Research and Applications.**

11:45 AM   Tom Bobbe, Conference Chair: **Closing Remarks**

Scenes from the 1996 Forest Service Remote Sensing Applications Conference

The poster sessions stimulated lively conversations between presentations

Numerous posters provided information on the latest use of current technologies

Information and ideas were shared throughout the week

Vendors provided lively on-site demonstrations

The conference included forums for special interest groups

Presentations were well attended

Interesting, informed and well prepared speakers made this conference worthwhile

Questions, suggestions, and idea exchanges often lead to remote sensing breakthroughs

Jeff Campbell demonstrates new remote sensing software

A forum on accuracy assessment was hosted by some of the foremost experts in the field

Andy Knapp and Yeda Maria Malheiros de Oliveria, from Brazil, discussing recent project work in South America

Passion for our natural heritage drives eager discussions of aerial imagery

Chuck Dull, US Forest Service Remote Sensing Coordinator and Tom Bobbe, Program Manager of the US Forest Service Remote Sensing Applications Center confer about another successful conference

The conference provides fertile ground for exchanging ideas

COMMERCIAL VENDOR EXHIBITS

ESRI
380 New York St.
Redlands, CA 92373
(909)793-2853
Fax (909)307-3039

Eastman Kodak (Aerial Systems)
1447 St. Paul Street
Rochester, NY 14653-7128
(716)253-1813
Fax (716)253-0705

PCI Remote Sensing Corp.
1925 N. Lynn Street Suite 803
Arlington, VA 22209
(703)243-3700
Fax (703)243-3705

Geographic Resource Solutions
1125 16th Street
Suite 213
Arcata, CA 95521
(707)822-8005
Fax (707)822-2864

SPOT Image Corporation
1897 Preston White Drive
Reston, VA 22091
(703)715-3100
Fax (703)648-1813

NASA Ames Research Center
Aircraft Data Facility
Mail Stop 240-6
Moffett Field, CA 94035
(415)604-6252
Fax (415)604-4987

Pacific Meridian Resources
421 SW 6th, Suite 850
Portland, OR 97204
503)228-8708
Fax (503)228-8751

Trimble Navigation
3630 Sinton Blvd, Suite 200
Colorado Springs, CO 80907
(719)471-1474
Fax (719)475-1916

Positive Systems, Inc.
250 2nd Street East
Whitefish, MT 59937
(406)862-7745
Fax (406)862-7759

ERDAS
2801 Buford Highway NE
Atlanta, GA 3032939
(404)248-9000
Fax (404)248-9909

Image Scans Inc
7651 W. 1st Ave, Ste 200
Wheat Ridge, CO 80033
(303) 422-2227
Fax (303) 422-2958

Core Software
675 S. Arroyo Parkway, 2nd floor
Pasadena, CA 91105
(818) 796-6155
Fax (818) 796-8574

REGISTER OF PEOPLE ATTENDING THE SIXTH
FOREST SERVICE REMOTE SENSING CONFERENCE

Cele Aguirre
USDA FS Rocky Mt. Forest
& Range Experiment Sta.
240 West Prospect Road
Fort Collins, CO 80526-2098

Vince Ambrosia
JCWS/NASA-AMES Research Ctr.
MS 242-4
Moffett Field, CA 94035

Delmar Anderson
749 Blue Mountain Dr
Fort Collins, CO 80526

Richard Arnold
BLM CO State Offices
2850 Youngfield St.
Lakewood, CO 80215

Stan Bain
USDA Forest Service
Remote Sensing Applications Center
2222 West 2300 South
Salt Lake City, UT 84119

Yolanda Barnett
USDA Forest Service
Klamath Ranger District
1936 California Ave
Klamath Falls, OR 97601

Jeff Barry
Univ of Washington
College of Forest Resources
PO Box 352100
Seattle, WA 98195-2100

Larry Batten
ESRI
4875 Pearl East Circle, Ste 200
Boulder, CO 80301-6103

Jenny Alban
USDA Forest Service
Remote Sensing Applications Center
2222 West 2300 South
Salt Lake City, UT 84119

Karen Andersen
DSL Consulting
4799 White Rock Cir.
Boulder, CO 80301

Peter Archibald
PCI Remote Sensing
2138 Toronto Street
Falls Church, VA 22043

Phil Austin
Positive Systems
250 Second Street East
Whitefish, MT 59937

Janet Bare-Broen
Lockheed Martin
PO Box 179 - MS DC 4001
Denver, CO 80201

Rebekah Barrish
NPIC/IAG/EED/EIB
214 Adams Avenue
Alexandria, VA 22301

Scott Bartling
NCWCD
1250 North Wilson Avenue
Loveland, CO 80538

Paul Beaty
ERDAS
2801 Buford Highway NE
Atlanta, GA 30329

Ron Behrendt
Positive Systems
250 Second Street East
Whitefish, MT 59937

Bill Belton
USDA Forest Service
WO Engineering
201 - 14th Street SW
Washington, DC 20250

Todd Birdsall
Eastman Kodak Company
1447 St Paul Street
Rochester, NY 14653-7128

Tom Bobbe
USDA Forest Service
Remote Sensing Applications Center
2222 West 2300 South
Salt Lake City, UT 84119

Douglas Bond
Del Terra, Inc.
2042 Market St.
Redding, CA 96001

Susan Boudreau
USDA Forest Service
Payette National Forest
PO Box 1026
McCall, ID 83638

Carol Brady
USDA-Forest Servuce
Geometronics Service Center
2222 W. 2300 South
Salt Lake City, UT 84119

Jim Brass
NASA Ames Research Center
MS 242-4
Moffett Field, CA 94035

Ron Brohman
USDA Forest Service
East Zone, Region 1
2880 Skyway Dr.
Helena, MT 59601

David A. Brown
ADR Associates Inc.
1155 Kelly Johnson Blvd., Ste 110
Colorado Springs, CO 80920

Paige Brown
CEPAT/USGS Advanced System Ctr.
3201 Silverstone Ct.
Oakton, VA 22124

William Buchholz
Lockheed Martin Corp
751 Vandenburg Rd.
Bldg A - Room 10A37
King of Prussia, PA 19406

Terry Busch
Spot Image Corp
1897 Preston White Dr.
Reston, VA 22091-4368

Craig Busskohl
USDA Forest Service
Umatilla National Forest
2517 SW Hailey
Pendleton, OR 97801

Amy Cade
Colorado Division of Wildlife
Habitat Resources Section
6060 N. Broadway
Denver, CO 80216

Joe Campbell
USDA Forest Service
Ecosystem Management
635 S. Eldridge St.
Lakewood, CO 80228

Jeff Campbell
Pacific Meridian Resources
421 SW 6th, Suite 850
Portland, OR 97204

Ron Cannarella
Hawaii Forestry & Wildlife
1151 Punchbowl St., Room 325
Honolulu, HI 96813

Scott Carlson
Science Applications Inc
4501 Daly Drive, Suite 400
Chantilly, VA 22021-3707

Bill Carr
Laser Technology, Inc.
6195 Gharrett
Missoula, MT 59803

Robin Carroll
Geometronics Service Center
2222 W. 2300 S.
Salt Lake City, UT 84119

Ward Carson
Oregon State Univ.
Corvallis, OR 97331-5703

Jule Caylor
USDA Forest Service
Remote Sensing Applications Center
2222 West 2300 South
Salt Lake City, UT 84119

Patrick Chase
British Columbia Institute of
Technology
33408 Berg Ln.
Pike, CO 80470

Char-Fong Chen
Conrad Aviation Technologies Inc.
4177 Melissa Dr
Lake Oswego, OR 97034-7203

Jian Chen
USDA Forest Service
Remote Sensing Applications Center
2222 West 2300 South
Salt Lake City, UT 84119

Jan Cipra
Colorado State University
Soil & Crop Sciences
Plant Science Bldg.
Ft. Collins, CO 80523

Bill Clerke
USDA Forest Service
Region 8
1720 Peachtree Rd. NW
Atlanta, GA 30367

Ted Coffelt
TRIFID Corp.
2315 Juniper
Boulder, CO 80304

Bruce Coffland
ATAC/NASA Ames Research Center
Aircraft Data Facility
Mail Stop 240-6
Moffett Field, CA 94035-1000

Andre' J. Coisman
USDA Forest Service
Engineering 3 S.E.
201 - 14th Street SW.
Washington, DC 20250

Allen Cook
TRW/BLM
200 Union Blvd., Suite 100
Lakewood, CO 80226

William H. Cooke
USDA Forest Service
Southern Research Station
201 Lincoln Green
Starkvelle, MS 39760-0928

Melanie Coppen
ERDAS Inc.
2801 Bufford Highway NE
Atlanta, GA 30329

Roger Crump
Science Applications Inc
6725 Odyssey
Huntsville, AL 35806

Roger Crystal
USDA Forest Service
Region 6
333 SW First Ave.
Portland, OR 97204

Larry Cunningham
NARSC RS-120 BLM
BLDG. 50 Denver Federal Center
Denver, CO 80225

Raymond Czaplewski
USDA Forest Service
Rocky Mt. Forest & Range Exp. Sta.
240 W. Prospect Road
Fort Collins, CO 80526

John Daley
Remote Sensing Research
1705 Heatheridge #G302
Fort Collins, CO 80526

Yeda Maria Malheiros de Oliveria
EMBRAPA
University of Oxford
10 Bradmore Road - Room 6
OX2 6QN Oxford United Kingdom

Denis Dean
Colorado State University
113 Forestry
Fort Collins, CO 80525

Gene DeGayner
USDA Forest Service
Tongass-Stikine Area
Box 309
Petersburg, AK 99833

Robert DeSawal
USGS, NMD, RMMC
11256 W. Hawaii Pl.
Lakewood, CO 80232

Paul DeWolf
Booz-Allen & Hamilton Inc.
7851 Teller St.
Arcada, CO 80003

Chuck Dull
USDA Forest Service
WO Engineering
14th and Independence Ave, SW
Washington, DC 20250

Glenn Dunno
ERDAS Inc.
2801 Buford Highway NE
Atlanta, GA 30329

Larry Edwards
Aero-Metric, Inc.
4708 South College Avenue
Fort Collins, CO 80525

Tom Edwards
National Biological Service
Utah Cooperative Research Unit
Utah State University
Logan, UT 84322-5210

Robert Ekstrand
Remote Sensing Image
Interpretation Services
5567 Copeland Place
San Jose, CA 95124-6102

Virginia Emly
USDA Forest Service
Nebraska NF, Bessey RD
Spur 86B off Hwy. 2
Box 38
Halsey, NE 69142

Don Evans
USDA Forest Service
Remote Sensing Applications Center
2222 West 2300 South
Salt Lake City, UT 84119

Chris Fayad
USDA Forest Service
Region 6
333 SW First Ave.
Portland, OR 97204

Bryon Foss
USDA Forest Service
WO Engineering
201 - 14th Street SW.
Washington, DC 20250

William Frament
USDA Forest Service
Northeastern Area FHP
PO Box 640
Durham, NH 03824

Tom Furst
Walsh Environmental
4888 Pearl East Circle
Bolder, CO 80301

Pat Gardiner
USDA FS GIS CoE
Intermountain Region
324 25th Street
Ogden, UT 84401

Mick Garrett
Trimble Navigation
3630 Sinton Rd.
Colorado Springs, CO 80521

Dave George
USDA FS GIS CoE
Intermountain Region
324 25th Street
Ogden, UT 84401

Diann Gese
10200 W. Warren Drive
Lakewood, CO 80227

Bruce Goldetsky
Trimble Navigation
645 North Mary Ave.
Sunnyvale, CA 94041

Jessi Gonzales
USDA FS San Juan-Rio Grande NF
Mancos-Dolores RD, IRI Center
PO Box 210
Dolores, CO 81323

Thomas E. Gough
USDA FS Infrared
3833 S. Development Ave.
Boise, ID 83705

Kass Green
Pacific Meridian Resources
5915 Holis, Bldg. B
Emeryville, CA 94608

Paul H. Greenfield
USDA Forest Service
Remote Sensing Applications Center
2222 West 2300 South
Salt Lake City, UT 84119

Jerry Greer
USDA Forest Service
Payette NF
P.O. Box 1026
McCall, Idaho 83638

Richard A. Grotefendt
Forest Consultant & Photogrammetrist
13504 432nd Ave. SE
North Bend, WA 98045

Louise Gunderson
Gunderson & Gunderson
212 E. 3rd Ave.
Denver, CO 80203

Jim Gunderson
Gunderson & Gunderson
212 E. 3rd Ave
Denver, CO 80203

Mike Hamby
USDA Forest Service
Southern Region
1720 Peachtree Street, NW
Atlanta, GA 30367

Chris Hanson
Questar Productions
1058 Weld County Road, 23 1/2
Brighton, CO 80601

Kathleen Harcksen
USDA Forest Service
Pacific Southwest Station
2400 Washington Ave.
Redding, CA 96001

Tim Hill
Geographic Resource Solutions
1125 16th Street, Suite 213
Arcata, CA 95521

Chris Hill
ORBIMAGE/Orbital Sciences Corp
21700 Atlantic Boulevard
Dulles, VA 20166

Everett Hinkley
USDA Forest Service, Region 10
709 West 9th St.
P.O. Box 21628
Juneau, Alaska 99802

Roger Hoffer
Colorado State University
College of Natural Resources
Dept. of Forest Sciences
Fort Collins, CO 80523-1470

James W. Hoffman
Space Instruments, Inc
4403 Manchester Ave., Suite 203
Encinitas, CA 92024

Carolyn Holland
USDA Forest Service
WO Engineering
201 - 14th Street SW
Washington, DC 20250

Mike Hoppus
USDA Forest Service
Remote Sensing Applications Center
2222 West 2300 South
Salt Lake City, UT 84119

Gary Huber
Questar Productions
1058 Weld County Rd. 23 1/2
Brighton CO, 80601

Mike Hutt
USGS
3812 Simms St.
Wheat Ridge, CO 80033-3800

Paul Ishikawa
USDA Forest Service
Remote Sensing Applications Center
2222 West 2300 South
Salt Lake City, UT 84119

Jan Johnson
USDA Forest Service
Remote Sensing Applications Center
2222 West 2300 South
Salt Lake City, UT 84119

Dale Johnson
Positive Systems
250 Second Street East
Whitefish, MT 59937

Mohammed A. Kalkhan
Natural Resource Ecology Lab
NREL-A219
Colorado State Univ.
Ft. Collins, CO 80523

Bob Kennedy
USDA Forest Service
National Interagency Fire Center
3833 South Development Ave.
Boise, ID 83705

Bill Krausmann
USDA Forest Service
Southwestern Region
517 Gold Ave, SW
Albuquerque, NM 87102

Gary Kuchel
SPOT Image Corp.
1897 Preston White Dr.
Reston, VA 22091-4368

Henry Lachowski
USDA Forest Service
Remote Sensing Applications Center
2222 West 2300 South
Salt Lake City, UT 84119

Eldon Jessen
US Geological Survey
PO Box 25046, MS-516
DFC, Building 810
Denver, CO 80225

Susan J. Johnson
USDA Forest Service
Rocky Mountain Region
740 Simms Street
Golden, CO 80104

Paul Jurasin
Earth Watch Inc.
1900 Pike Road
Longmont, CO 80501

Bob Keane
USDA Forest Service
Intermountain Fire Sciences Lab
PO Box 8089
Missoula, MT 59807

Andy Knapp
USDA Forest Service
Forest Health Protection
1750 Front Street Rm 202
Boise, ID 83704

Karen Kristin
CEPAT/USGS
Advanced Systems Ctr.
44022 Gala Circle
Ashburn, VA 22011

Nils Kunces
Trimble Navigation
645 N Mary Ave.
Sunnyvale, CA 94086

Lisa M. Levien
USDA Forest Service
Region 5 S&PF, FPM
1920 20th Street
Sacramento, CA 95814

Lowell Lewis
USDA FS / MACA
c/o FHTET
3825 E. Mulberry St.
Ft. Collins, CO 80524

David Linden
DSL Consulting Inc.
1520 River Oak Dr.
Ft. Collins, CO 80525

Thomas Loveland
USGS
EROS Data Center
Mundt Federal Building
Sioux Falls, SD 57198

Ben J. Lowman
USDA Forest Service, MTDC
Ft. Missoula, Building 1
Missoula, MT 59801

Douglas E. Luepke
USDA Forest Service
Region 8
1720 Peachtree Road, Rm. 716
Atlanta, GA 30367

Gyde Lund
USDA Forest Service, FIERR
Auditors Building
14th & Independence Ave. SW
Washington, DC 20250

Mike Lunt
USDA FS GIS CoE
Intermountain Region
324 25th Street
Ogden, UT 84401

Douglas MacCleery
USDA Forest Service, WO
14th & Independence, Ave SW
201 14th Street, SW
Washington, DC 20250

Richard MacDonald
Earth Watch Inc.
1900 Pike Road
Longmont, CO 80501

Dick Mangan
USDA Forest Service, MTDC
Ft. Missoula, Bldg 1
Missoula, MT 59801

Michael Marchase
BLM-NARSC
PO Box 25047
Lakewood, CO 80225-0047

Bill Marinelli
Physical Sciences Inc.
20 New England Business Center
Andover, MA 01810

Carl J. Markon
EROS Alaska Field Office
4230 University Drive, Suite 320
Anchorage, AK 99508-4994

Paul Maus
USDA Forest Service
Remote Sensing Applications Center
2222 West 2300 South
Salt Lake City, UT 84119

Kim Mayeski
USDA Forest Service
Caribou National Forest
250 South 4th Avenue
Federal Bldg. Ste. 172
Pocatello, ID 83201

Tim McConnell
USDA Forest Service
Forest Health Protection
Box 7669
Missoula, MT 59807

Janice McDougle
USDA Forest Service. WO
401 Independence Ave SW
Washington, DC 20250

Robert McGaughey
USDA FS PNW Research Station
University of Washingtron
PO Box 352100
Seattle, WA 98195-2100

Jim McKean
USDA Forest Service, ENG
2245 Morello Ave.
Pleasant Hill, CA 94523

Ebeth McMullen
USDA Forest Service
Region 1
200 E. Broadway
Missoula, MT 59802

William Michalson
Worchester Polythecenic Institute
100 Institute Road, ECE Dept.
Warchester, MA 01609

Jeff Milliken
USDA Forest Service
Region 5 Remote Sensing Lab
1920 20th Street
Sacramento, CA 95814

Carol Mladnich
US Geological Survey
PO Box 25046, MS 516
Lakeview, CO 80225

Gretchen Moisen
USDA Forest Service
Intermountain Research Station
507 25th Street
Ogden, UT 84401

Jeff Moll
USDA Forest Service
SDTDC
444 East Bonita Ave
San Kimas, CA 91773

Mike Morrison
USDA FS R2-Pueblo IRI Center
Pike & San Isabel National Forest
1920 Valley Drive
Pueblo, CO 81008

Todd Mowrer
USDA Forest Service
Rocky Mountain Station
240 W. Prospect
Ft. Collins, CO 80526-2098

Jeff Myers
ATAC/NASA Ames Research Center
Aircraft Data Facility
Mail Stop 240-6
Moffett Field, CA 94035-1000

Dick Myhre
FS - Forest Health Protection/FHTET
3825 East Mulberry
Fort Collins, CO 80524

Krishna Myneni
Science Applications Inc
6725 Odyssey
Huntsville, AL 35806

Karen Nabity
Geometronics Service Center
2222 West 2300 South
Salt Lake City, UT 84119

Lisa Nelson
Johnson Controls, Int'l.
NBS-MESC
4512 McMurry Avenue
Ft. Collins, CO 80525-3400

443

David Neufeld
Colorado State University
924 Gay Street
Longmont, CO 80501

Don Norris
USDA Forest Service
Francis Marion/Sumter NF
4931 Broad River Rd.
Columbia, SC 29210

Calvin P. O'Neil
National Biological Service
Southern Science Center
700 Cajundome Blvd.
Lafayette, LS 70506

Alex Pertica
Lawrence Livermore Nat'l Lab
7000 East Avenue, L-183
Livermore, CA 94550

Nathan Poage
OSU/PNW
3200 SW Jefferson Way
Corvallis, OR 97331

Ross Pywell
USDA Forest Service
FHTET/ Ft. Collins
3825 E. Mulberry Street
Fort Collins, CO   80524

Julie Ranie
Positive Systems
250 Second Street East
Whitefish, MT 59937

Edward Reilly
USDA Forest Service
Rogue River NF, Applegate RD
6941 Upper Applegate Road
Jacksonvelle, OR 97530

Fred Nichols
Remote Sensing Research
1050 South Summit View
Fort Collins, CO 80524

Maurice Nyquist
NBS Tech Trans Center
PO Box 25387, DFC
Denver, CO 80225-0387

Jeanine L Paschke
USDA FS / MACA
3825 Mulberry St.
Ft. Collins, CO 80525

Lawrence R. Pettinger
USGS Office of External Affairs
590 National Center
Reston, VA 22092-0001

Ronold Podmilsak
Autometric Inc.
5301 Shawnee Rd.
Alexandria, VA 22312

Roberta Quigley
USDA Forest Service
WO ENG
201 - 14th Street SW
Washington, DC 20250

Mary Rasmussen
USDA Forest Service
Willamette Naional Forest
211 E. 7th Ave
Eugene, OR 97401

Steve Reutebuch
USFS PNW Research Station
University of Washington
PO Box 352100
Seattle, WA 98195-2100

Kim Rivard
USDA Forest Service
Region 6
333 SW First Ave
Portland, OR 97204

Kathy Roche
USDA Forest Service
Greenhorn Ranger District
PO Box 6
Kernville, CA 93238

Mark E. Romano
Eagle Scan Incorporated
1770-B Range Street
Boulder, CO 80301

John Runyon
Dynamac, Inc.
200 SW 35th Street
Corvallis, OR 97333

Walter R. Salazar
USDA Forest Service
Forest Health Protection
1720 Peachtree Rd. NW Ste 925 N
Atlanta, GA 30367

James R. Schramek
USDA Forest Service
Tongass-Stikine Area
PO Box 309
Petersburg, AK 99833

Carol A. Scott
USDA Forest Service
Forest Health Protection
1720 Peachtree Rd. Room 925 N
Atlanta, GA 30367

Laura L. Rivers
USDA Forest Service
Lowman Ranger Station
H.C. 77 Box 3020
Lowman, ID 83637

Carlos D. Rodriguez
USDA Forest Service
Intn'l Institute of Tropical Forestry
Univ of Puerto Rico Botanical Garden
Rio Piedras to Caguas Road
Rio Piedras, PR 00927

Robert Rosenthal
USDA Forest Service
Region 9
310 W. Wisconsin Ave
Milwaukee, WI 53203

Bill Rush
USDA Forest Service
Boise National Forest
1750 Front St.
Boise, ID 83702

Allen Salzmann
SHC
PO Box 958
Wheat Ridge, CO 80033

Don Schrupp
Colorado Division of Wildlife
Habitat Resources Section
6060 North Broadway
Denver, CO 80216

Drew Selig
National Forests in North Carolina
160a Zillicoa St.
Asheville, NC 28802

Gail Shaw
USDA Forest Service
Remote Sensing Applications Center
2222 West 2300 South
Salt Lake City, UT 84119

Joe Shemon
Eastman Kodak
1447 St. Paul Street
Rochester, NY 14653-7128

Paul Simmons
USDA FS
WO Engineering
201 - 14th Street SW
Washington, DC 20250

Joshua Single
Worchester Polytechnic Institute
100 Institute Road, ECE Dept
Worchester, MA 01609

Gary Sjolander
811 West Kettle Ave.
Littleton, CO 80120

Ron Skillings
USDA Forest Service, Region 10
709 West 9th St.
P.O. Box 21628
Juneau, Alaska 99802

Chan C. Smith
Eagle Scan Incorporated
1770-B Range Street
Boulder, CO 80301

Jim Sorenson
USDA FS - Fire & Aviation Mgt.
1720 Peachtree St.
Atlanta, GA 30367

George Southard
Leica Inc.
2 Inverness Dr East #106
Englewood, CO 80112

Michael Spadazzi
Worchester Polytechnic Institute
100 Institute Road, ECE Dept
Worchester, MA 01609

Rich Spradling
USDA Forest Service
Region 5 Geometronics
630 Sansome St.
San Francisco, CA 94111

Joseph P. Spruce
Lockheed Martin
Building 1210
Stennis Space Center, MS 39529

John Steffenson
ESRI
4875 Pearl East Circle. Ste. 200
Boulder, CO 80301-6103

Steve Stehman
SUNY - ESF
320 Bray Hall
Syracuse, NY 13210

Carl Sumpter
USDA Forest Service
Medicine Bow-Routt NF
2468 Jackson St.
Laramie, WY 82070-6535

Harold W. Thistle
USDA Forest Service, MTDC
Building 1, Fort Missoula
Missoula, MT 59801

Jeffrey Tobolski
Conrad Aviation Technologies Inc.
4177 Melissa Dr.
Lake Oswego, OR 97034-7203

Nancy Tubbs
BLM National Applied Resource
Sciences Ctr
3387 S Danube St
Aurora, CO 80013

Matt Turner
USDA Forest Service
White River NF - GIS/Planning
900 Grand Avenue, PO Box 948
Glenwood Springs, CO 81601

Vicky Varner
USDA Forest Service
Remote Sensing Applications Center
2222 West 2300 South
Salt Lake City, UT 84119

Melinda Walker
Bureau of Land Management
NARSC - RS 140
Denver Federal Center, Bldg 50
Denver, CO 80225-0047

Charles L. Walthall
USDI-ARS RSML
008 Bldg. 007, BARC-West
10300 Baltimore Ave
Beltsville, MD 20705-2350

Liz Wegenka
USDA Forest Service
Geometronics Service Center
2222 W. 2300 South
Salt Lake City, UT 84119

Tom Weithman
Hughes Info Tech
1616 McCormic Drive
Upper Marlboro, MD 20774

Frank Toomer
SAF/ST (CEPAT)
4501 Daly Drive, Suite 400
Chantilly, VA 22021-3707

James Turner
BLM-NARSC
PO Box 25047 (RS-100A)
Bldg 50 DFC
Denver, Co 80225

Dave Van Mouwerik
ERDAS Inc.
2801 Buford Highway NE
Atlanta, GA 30329

Steve Wagner
Booz-Allen & Hamilton Inc.
1050 S. Academy Blvd, Ste 148
Colorado Springs, CO 80910

Wanda Wallace
USDA Forest Service
WO, IS&T, Rm 808, RPE
PO Box 96090
Washington, DC 20090-6090

David Watkins
USDA Forest Service
Pacific Southwest Research Station
2400 Washington Ave.
Redding, CA 96001

Michael A. Wehr
USDA Forest Service
Forestry Sciences Lab
410 MacInnes Dr.
Houghton, MI 49931-1199

Sara Wesser
National Park Service
2525 Gambell St.
Anchorage, AK 99503

Carl Wheatly
Hughes Info Tech
1616 McCormic Drive
Upper Marlboro, MD 20774

Bill Wilen
US Fish & Wildlife
2202 Hollow Lane
Bowie, MD 20716

Bill Wilson
USDA Forest Service
Timber Management
709 W. 9th Street
Juneau, AK 99801

Ken Winterberger
USDA Forest Service
Forestry Sciences Lab
3301 C Street, Suite 200
Anchorage, AK 99503

Tim Wirth
USDA Forest Service
Remote Sensing Applications Center
2222 West 2300 South
Salt Lake City, UT 84119

Patrick Woodruff
Airborne Data Systems
Rural Route 2, Box 38
Wabasso, MN 56293

Robert Wright
Atterbury Consultants, Inc
3800 SW Cedar Hills BLVD, Ste #280
Beaverton, OR 97005

Ronald C. Wicks
US Army Missile Command
ATTN: AMSMI-RD-MG-TD
Redstone Arsenal, AL 35898-5234

Cindy Williams
PNW, USDA Forest Service
Institure of Northern Forestry
308 Tanana Drive
Fairbanks, AK 99775

Brad Wind
NCWCD
1250 North Wilson Ave.
Loveland, CO 80539

Steven Winward
USDA Forest Service
Region 4
324 25th Street
Ogden, UT 84401

Dave Wolf
USDA Forest Service
Region 2
740 Simms St.
Golden, CO 80401

Dave Woods
Eastman Kodak
1447 St. Paul Street
Rochester, NY 14653-7128